Lecture Notes Editorial Policies

Lecture Notes in Statistics provides a format for the informal and quick publication of monographs, case studies, and workshops of theoretical or applied importance. Thus, in some instances, proofs may be merely outlined and results presented which will later be published in a different form.

Publication of the Lecture Notes is intended as a service to the international statistical community, in that a commercial publisher, Springer-Verlag, can provide efficient distribution of documents that would otherwise have a restricted readership. Once published and copyrighted, they can be documented and discussed in the scientific literature.

Lecture Notes are reprinted photographically from the copy delivered in camera-ready form by the author or editor. Springer-Verlag provides technical instructions for the preparation of manuscripts. Volumes should be no less than 100 pages and preferably no more than 400 pages. A subject index is expected for authored but not edited volumes. Proposals for volumes should be sent to one of the series editors or addressed to "Statistics Editor" at Springer-Verlag in New York.

Authors of monographs receive 50 free copies of their book. Editors receive 50 free copies and are responsible for distributing them to contributors. Authors, editors, and contributors may purchase additional copies at the publisher's discount. No reprints of individual contributions will be supplied and no royalties are paid on Lecture Notes volumes. Springer-Verlag secures the copyright for each volume.

Lecture Notes in Statistics **142**

Edited by P. Bickel, P. Diggle, S. Fienberg, K. Krickeberg,
I. Olkin, N. Wermuth, S. Zeger

Springer Science+Business Media, LLC

György Terdik

Bilinear Stochastic Models and Related Problems of Nonlinear Time Series Analysis

A Frequency Domain Approach

Springer

György Terdik
Center for Informatics and Computing
Kossuth University of Debrecen
Debrecen 4010, PF 58
Hungary

Library of Congress Cataloging-in-Publication Data

Terdik, György.
Bilinear stochastic models and related problems of nonlinear time series analysis : a frequency domain approach / György Terdik.
p. cm. -- (Lecture notes in statistics ; 142)
Includes bibliographical references and index.
ISBN 978-0-387-98872-6 ISBN 978-1-4612-1552-3 (eBook)
DOI 10.1007/978-1-4612-1552-3
1. Time-series analysis. 2. Nonlinear theories. I. Title.
II. Series: Lecture notes in statistics (Springer-Verlag) ; v. 142.
QA280.T44 1999
519.5'5--dc21 99-23873

Printed on acid-free paper.

Originally published by Springer-Verlag New York, Inc. in 1999

Camera-ready copy provided by the author.

9 8 7 6 5 4 3 2 1

ISBN 978-0-387-98872-6

Contents

List of Figures

Introduction

"Ninety percent of inspiration is perspiration."

[31]

The Wiener approach to nonlinear stochastic systems [146] permits the representation of single-valued systems with memory for which a small perturbation of the input produces a small perturbation of the output. The Wiener functional series representation contains many transfer functions to describe entirely the input-output connections. Although, theoretically, these representations are elegant, in practice it is not feasible to estimate all the finite-order transfer functions (or the kernels) from a finite sample. One of the most important classes of stochastic systems, especially from a statistical point of view, is the case when all the transfer functions are determined by finitely many parameters. Therefore, one has to seek a finite-parameter nonlinear model which can adequately represent nonlinearity in a series. Among the special classes of nonlinear models that have been studied are the bilinear processes, which have found applications both in econometrics and control theory; see, for example, Granger and Andersen [43] and Ruberti, et al. [4]. These bilinear processes are defined to be linear in both input and output only, when either the input or output are fixed. The bilinear model was introduced by Granger and Andersen [43] and Subba Rao [118], [119]. Terdik [126] gave the solution of a lower triangular bilinear model in terms of multiple Wiener–Itô integrals and gave a sufficient condition for the second order stationarity. An important representation in terms of the generalized transfer functions was given by Priestley [98] and in terms of higher order spectra by Brillinger [17].

The present work is intended to be a systematic statistical analysis of bilinear processes in the frequency domain. The first two chapters are devoted to the basic theory of nonlinear functions of stationary Gaussian processes; Hermite polynomials; cumulants; higher order spectra; multiple Wiener–Itô integrals; and finally, chaotic Wiener–Itô spectral representation of subordinated processes. Chapter 3 contains the results concerning bilinear processes. For an easier understanding of the technique of chaotic representation, three levels of bilinear processes are considered: the simple bilinear model, the general bilinear model with scalar value, and the multiple bilinear model. In each case explicit assumptions of second order stationarity and expression for the second order spectrum are given. The assumptions of the existence of the $2n$th order moments are proved for the general bilinear model with scalar value, and an expression for the bispectrum is obtained. The Generalized Autoregressive Conditionally Heteroscedastic (GARCH(1,1)) model is investigated by the same methods as the bilinear one, and its basic spectral properties are shown. The bilinear realization for Hermite degree-N homogeneous polynomial model and its minimal realization are also considered. There are two chapters for general nonlinear time series problems. Chapter 4 covers the non-Gaussian estimation. It was Brillinger [21] who suggested using for parameter estimation not only the spectrum but the bispectrum as well. We give explicit expression for the asymptotic variance of this estimator and prove the asymptotic normality and consistency. The asymptotic variance in the case of linear non-Gaussian processes is expressed in terms of skewness and kurtosis. This method is used for the parameter estimation of bilinear processes. The other general problem is in Chapter 5, where we consider the linearity of a time series. We use a weak notion of linearity of a time series and give a bispectrum-based test for checking it.

Further references and historical comments on the frequency domain approach to the time series analysis and nonlinear models are provided in the works of D. R. Brillinger and T. Subba Rao.

Data under consideration

There are several fields of data where the linear model does not provide a satisfactory result. Our aim is to check the linearity of the data, and in case of nonlinearity, to show the higher order spectral properties and the use of bilinear fitting. Each of the data sets below are found on the Internet.

The S&P 500 Index

The primary objective of the Standard & Poor's 500 Composite Stock Price Index, known as the S&P 500, is to be the performance benchmark for the U.S. stock market. The Index is a market value-weighted index (shares outstanding times stock price) in which each company's influence in Index performance is directly proportional to its market value. The origins of the S&P 500 Index go back to 1923 when Standard & Poor introduced a series of indices that included 233 companies and covered 26 industries. The Index, as it is now known, was introduced in 1957. Today, the S&P 500 encompasses 500 companies, representing 90 specific industry groups. The Index is widely regarded as the standard for broad stock market performance The data of S&P 500 Index was found at the Web site of the Chadwick Investment Group by the address

http://chdwk.com/stock.html

among the Historical Stock Price Data. More information about the Index is listed at

http://www.cftech.com/BrainBank/FINANCE/SandPIndexCalc.html.

Recently, it has been shown [80] the non-Gaussianity of this index, and it was also pointed out that the probability density functions of GARCH(1,1) models are quite different from that observed data.

IBM stock prices

The data of the IBM stock prices come also from the Historical Stock Price Data library of the above named web site. It has been mentioned by the classic book of time series by Box and Jenkins [14] and also by Tong [140].

Geomagnetic indices

K indices isolate solar particle effects on the earth's magnetic field. Over a 3-hour period, they classify into disturbance levels the range of variation of the more unsettled horizontal field component. Each activity level relates almost logarithmically to its corresponding disturbance amplitude. Three-hour indices discriminate conservatively between true magnetic field perturbations and the quiet-day variations produced by ionospheric currents. The A-index ranges from 0 to 400 and represents a K-value converted to a linear scale in gammas (nanoteslas)—a scale that measures equivalent disturbance amplitude of a station at which K=9 has a lower limit of 400 gammas. The subscript p means planetary and designates a global magnetic activity index. The following 13 observatories, which lie between 46 and 63 degrees north and south geomagnetic latitude, now contribute to the planetary indices: Lerwick (UK), Eskdalemuir (UK), Hartland (UK), Ottawa (Canada), Fredericksburg, Virginia (USA), Meanook (Canada), Sitka, Alaska (USA), Eyrewell (New Zealand), Canberra (Australia), Lovo (Sweden), Brorfelde (Denmark), Wingst and Niemegk (Germany). For details see at National Geophysical Data Center

http://www.ngdc.noaa.gov/wdc/.

The aa-index is a simple global index of magnetic activity it is produced in France from the K indices of two nearly antipodal magnetic observatories in England and Australia. This index aa, is the 3-hourly equivalent amplitude antipodal index. Daily average aa may be derived similarly to ap. A historical advantage to using aa is that these indices have been extended back in time through scaling of magnetic activity from magnetograms of earlier observations. The aa indices are derived from 1868 to the present.

Magnetic field data

An example of bilinear systems comes from the nuclear magnetic resonance (NMR) spectroscopy studying the response of a changing magnetic field. The phenomenon is quantum mechanical and the concerning equation is bilinear it is called as Bloch equation. Brillinger, [22], [26], considered the analysis procedures of NMR spectroscopy when both the input and the output of system are observed. He estimated the unknown parameters of the bilinear equation. The data we consider here is a component of the multi-resolution magnetic field of the sun, measured by a spacecraft called Ulysses. The COHOWeb (http://nssdc.gsfc.nasa.gov/) provides access to hourly resolution magnetic field and plasma data from each of several heliospheric spacecraft. The hourly averages of parameters for the interplanetary magnetic field between October 25, 1990 and June 30, 1997 were chosen among the data available at the COHOWeb site. The principal investigator of magnetic field data was Dr. A. Balogh, Imperial College, London, UK. Three components of the magnetic field hour average are the following:

Magnetic field hour average of R component (nT).
Magnetic field hour average of T component (nT).
Magnetic field hour average of N component (nT).

The RTN system is fixed at a spacecraft (or the planet). The R axis is directed radially away from the sun; the T axis is the cross product of the solar rotation axis and the R axis, and the N axis is the cross product of R and T.

Data sets for the paper "A Single-Blind Controlled Competition Among Tests for Nonlinearity and Chaos" [9]

The data is simulated data, produced from five different generating models. One model, and hence two of the data sets, is purely deterministic (and chaotic). The other four models, and hence eight of the data sets, are stochastic processes, in which the randomness was produced by Monte Carlo methods. One of the stochastic processes was linear, while the other three were nonlinear, but not chaotic. The data were generated at Washington University in St. Louis, see http://wuecon.wustl.edu/~barnett/.

Acknowledgments

- The author wishes to thank David R. Brillinger and T. Subba Rao for several helpful discussions.
- Special thanks go to my colleagues Endre Iglói, Márton Ispány, and János Máth for joint work.
- The author highly appreciates the work of several research centers not only for measuring and analyzing the data but for providing them on-line for further investigations for all the Internet community, in particular the National Space Science Data Center NASA, USA; Chadwick Investment Group, USA; International Service of Geomagnetic Indices, CETP, France; University of Goettingen, Germany and professor William Barnett, Washington University in St. Louis, USA.
- This research is partially supported by the Hungarian National Science Foundation OTKA No. T 019501.
- The author wishes to thank the referees and Gy. Pap and L. Szeidl for their comments and suggestions.

Notations

The following notations are used

1. $\vec{1}$ denotes a row vector having all ones in its coordinates, i.e., $\vec{1} = (1, 1, ..., 1)$ with appropriate dimension.

2. Capitals X, Y, $Z \ldots$ stand for random variables.

3. Subscripting. Put $(1 : n) = (1, 2, ..., n)$. The vector $(X_1, X_2, ..., X_n)$ will be denoted by $X_{(1:n)}$. In general if the subscript is a set (ordered) of natural numbers, say $a_{(1:n)} = (a_1, a_2, ..., a_n)$, then $X_{a_{(1:n)}}$ denotes a vector with components indexed by the elements of that set, i.e. $(X_{a_1}, X_{a_2}, ..., X_{a_n})$. In general if the subscript is a set (ordered) of natural numbers, say $K = \{a_1, a_2, ..., a_n\}$, then X_K denotes a vector with components indexed by the elements of that set, i.e. $(X_{a_1}, X_{a_2}, ..., X_{a_n})$. It sholuld cause no confusion to denote X_K also by $X_{a_{(1:n)}}$.

4. Let $\mathcal{K}$ denote some set of index sets, i.e. $\mathcal{K} = \{K, M, N\}$ where say $K = a_{(1:k_1)}$, $M = b_{(1:k_2)}$, $N = c_{(1:k_3)}$. The $X_{\mathcal{K}}$ denotes the vector of products with respect to the subsets of $\mathcal{K}$, e.g.,

$$X_{\mathcal{K}} = (\prod_{j=1}^{k_1} X_{a_j}, \prod_{j=1}^{k_2} X_{b_j}, \prod_{j=1}^{k_3} X_{c_j}).$$

5. The product and the sum associated with an index set have the following shorter notation

$$\Pi X_{a_{(1:n)}} = \prod_{j=1}^{n} X_{a_j}, \quad \Sigma X_{a_{(1:n)}} = \sum_{j=1}^{n} X_{a_j}.$$

6. $(a_{(1:n)}, x_{(1:n)})$ denotes the usual inner product of vectors, i.e.,

$$(a_{(1:n)}, x_{(1:n)}) = \sum_{k=1}^{n} a_k x_k.$$

7. Repetition. A vector having the same components, $(X_1, X_1, X_2, X_2, X_2) = X_{(1,1,2,2,2)}$, say, will be denoted by $X_{(1:2)[2,3]}$, as well. In general, the set with natural numbers in brackets [ ] denotes the number of the same components of the previous ordered set. In that sense

$$(X_1, X_1, X_2, X_2, X_2, X_3, X_3) = X_{\{1,1,2,2,2,3,3\}} = X_{(1:3)[2,3,2]}.$$

8. Exponent.

$$(X_1, X_2, ..., X_n)^{(t_1, t_2, ..., t_n)} = X_{(1:n)}^{t_{(1:n)}} = \prod_{k=1}^{n} X_k^{t_k},$$

i.e., the exponent of a vector by a vector with the same dimension is the product of the exponents, following this role

$$X_{(1:n)}^{r\vec{1}} = X_{(1:n)}^{(r,r,...,r)} = \prod_{k=1}^{n} X_k^r.$$

9. Permutations. $\mathfrak{P}_n$ denotes the set of all permutations of the numbers $(1:n) = (1, 2, \ldots, n)$, if $\mathfrak{p} \in \mathfrak{P}_n$ then $\mathfrak{p}(1:n) = (\mathfrak{p}(1), \mathfrak{p}(2), \ldots, \mathfrak{p}(n))$.

10. Partitions. The set of all partitions of the numbers $(1:n) = (1, 2, \ldots, n)$, is denoted by $\mathcal{P}_{(1:n)}$, if $\mathcal{L} \in \mathcal{P}_{(1:n)}$ then $\mathcal{L} = \{K_1, K_2, \ldots, K_k\}$ such that $K_1, K_2, \ldots, K_k$ are disjoint and $\cup K_j = \{1, 2, \ldots, n\}$. In particular the set of all partitions into pairs of the set $(1, 2, ...n)$ is denoted by $\mathcal{P}_{(1:n)}^{\mathrm{II}}$ and the set of all partitions having one or two elements is $\mathcal{P}_{(1:n)}^{\mathrm{I,II}}$. $\mathcal{K}^{\mathrm{II}}$ and $\mathcal{K}^{\mathrm{I,II}}$ are the elements of $\mathcal{P}_{(1:n)}^{\mathrm{II}}$ and $\mathcal{P}_{(1:n)}^{\mathrm{I,II}}$, respectively.

11. Gaussian characteristic function is denoted by Ψ.

12. Gaussian distribution function is denoted by G.

13. Complex Gaussian stochastic spectral measure is denoted by W.
14. Spectral measure is Φ.
15. Cumulant spectral density is denoted by S.
16. $\overline{L_\Phi^n}$ and $\widetilde{L_\Phi^n}$ are Hilbert spaces.
17. Transfer functions are denoted by h, $g \ldots$.
18. $\mathcal{B}$ is the Borel σ-algebra with elements A, $B, \ldots$.
19. H_n is Hermite polynomial of order n.
20. $\mathcal{X}$ is Gaussian system if any subset of the elements of $\mathcal{X}$ contains jointly Gaussian random variables.
21. $\mathfrak{L}^2(\mathcal{X})$ is the Hilbert space of all random variables depending on the Gaussian system $\mathcal{X}$, see page 7.
22. $L_k^2(\mathcal{X})$ is the Hilbert space generated linearly by all possible Hermite polynomials with order k of the Gaussian system $\mathcal{X}$.
23. The time shift operator U_s for a stationary X_t, is defined by the equality $\mathsf{U}_s X_t = X_{t+s}$ for every s, t and is extended to a group of unitary transformations over $L_1^2(\mathcal{X})$.
24. The covariance and the second order cumulant of random variables Z_1, Z_2 with complex values are different

$$\begin{aligned}\operatorname{Cov}(Z_1, Z_2) &= \mathrm{E}(Z_1 - \mathrm{E}\, Z_1)\,\overline{(Z_2 - \mathrm{E}\, Z_2)},\\ \operatorname{Cum}(Z_1, Z_2) &= \mathrm{E}(Z_1 - \mathrm{E}\, Z_1)(Z_2 - \mathrm{E}\, Z_2)\,.\end{aligned}$$

25. $\bmod (1)$ denotes a relation on real numbers such that $x = y \bmod (1)$ if the fracional parts of x and y are equal.
26. $\delta_1(\cdot)$ is Dirac delta with periodic extension of $\bmod (1)$, called also Dirac comb.
27. $\delta_{\{\}}$ is Kronecker delta.
28. $\chi_A(\omega) = 1$ if $\omega \in A \bmod (1)$ and zero otherwise.
29. The Fock space is denoted by $\mathsf{Exp}\left(\widetilde{L_\Phi^2}\right)$.
30. $z_k = e^{i2\pi\omega_k}$, and by #8 above $z_{(1:n)}^{t\vec{1}} = e^{i2\pi t \sum_{k=1}^n \omega_k}$.
31. $\mathrm{w}_t, t \in \mathbb{Z}$, is standard Gaussian white noise, e.g., w_t is an independent Gaussian series with mean 0 and variance 1.

32. $\mathcal{N}(\mu, \sigma^2)$ denotes the family of Gaussian random variables with mean μ and variance σ^2.

33. The semifactorial is $(2k-1)!! = 1 \cdot 3 \cdot 5 \cdots (2k-1)$.

34. Symmetrized version of f_n is defined by

$$\operatorname{sym} f_n \left(\omega_{(1:n)}\right) = f_n^{\mathfrak{P}} \left(\omega_1, \omega_2, \dots, \omega_n\right) = \widetilde{f_n} \left(\omega_1, \omega_2, \dots, \omega_n\right)$$
$$= \frac{1}{n!} \sum_{\mathfrak{p} \in \mathfrak{P}_n} f_n \left(\omega_{\mathfrak{p}(1)}, \omega_{\mathfrak{p}(2)}, \dots, \omega_{\mathfrak{p}(n)}\right).$$

35. Symmetrized version by the variables $\omega_{(1:m)}$, where $n \le m$, of a function f_n of variables $\omega_{(1:n)}$ is denoted by $\underset{\omega_{(1:m)}}{\operatorname{sym}} f_n$ and defined by

$$f_m^{\mathfrak{P}} \left(\omega_1, \omega_2, \dots, \omega_n\right) = \underset{\omega_{(1:m)}}{\operatorname{sym}} f_n \left(\omega_1, \omega_2, \dots, \omega_n\right)$$
$$= \frac{1}{m!} \sum_{\mathfrak{p} \in \mathfrak{P}_m} f_n \left(\omega_{\mathfrak{p}(1)}, \omega_{\mathfrak{p}(2)}, \dots, \omega_{\mathfrak{p}(n)}\right).$$

36. $\hat{f}(u_1, u_2)$ is the Fourier transform of $f(x_1, x_2)$, i.e.,

$$\hat{f}(u_1, u_2) = \frac{1}{(2\pi)^2} \int_{-\infty}^{\infty} \int_{-\infty}^{\infty} f(x_1, x_2) exp\{-i(x_1 u_1 + x_2 u_2)\}\, dx_1\, dx_2.$$

37. Restricted Fourier transforms of cumulants of jointly stationary time series X_t, Y_t, and V_t

$$S_{X,Y}^{+}(z) = \sum_{s=1}^{\infty} \operatorname{Cum}\left(X_{t+s}, Y_t\right) z^{-s},$$
$$S_{X,Y,V}^{+}(z_1, z_2) = \sum_{k=1}^{\infty} \sum_{l=1}^{\infty} \operatorname{Cum}(X_{t+k+l}, Y_{t+l}, V_t) z_1^{-k-l} z_2^{-l}.$$

38. $S_{2,X}^{*n}$ denotes the nth order convolution of $S_{2,X}$, i.e.,

$$S_{2,X}^{*n}(\omega) = \int_{[0,1]^{n-1}}^{2} S_{2,X}\left(\omega - \omega_2\right) \prod_{k=2}^{n-1} S_{2,X}\left(\omega_k - \omega_{k+1}\right) \cdot \times S_{2,X}\left(\omega_n\right) d\omega_{(2:n)}. \tag{1}$$

1

Foundations

There are several ways to approach Hermite polynomial systems, cumulants and their relationship. Our treatment starts with a general method of calculating the expectation of nonlinear function of Gaussian random variables, see [13] and [127]. Then we define the classical Hermite polynomials and their generalization with several variables. A rather simple introduction to cumulants is given. The diagram formulae are used to show the basic connections between cumulants and Hermite polynomials. These facts will be important for the multiple Wiener–Itô stochastic integrals in the next Chapter. Some general well known theory of stationary processes as spectral representation and higher order spectra are also considered. An approximation of the spectrum and the bispectrum of some nonlinear function of a Gaussian stationary process closes the Chapter.

1.1 Expectation of nonlinear functions of Gaussian variables

Consider a Gaussian vector $X_{(1,2)} = (X_1, X_2)$. It might have no density function, for example when $X_1 = X_2$. Therefore it can be more convenient to handle a Gaussian random vector by its characteristic function, i.e.,

$$\Psi_{X_{(1,2)}}(u_1, u_2) = exp\left\{i(\mu_1 u_1 + \mu_2 u_2) - 1/2(\sigma_{11}u_1^2 + 2\sigma_{12}u_1u_2 + \sigma_{22}u_2^2)\right\},$$

where

$$\mu_j = \mathsf{E}X_j \text{ and } \sigma_{jk} = \mathrm{Cov}(X_j, X_k) = \mathsf{E}(X_j - \mu_j)(X_k - \mu_k), \quad j,k = 1,2.$$

Let $G_{X_1}(x_1)$ and $G_{X_2}(x_2)$ denote the marginal distribution functions and $G_{X_{(1,2)}}(x_1, x_2)$ the joint one of (X_1, X_2) respectively. If $f(x_1, x_2)$ is an appropriate real valued function of its variables, then the standard method to calculate the expectation of the random variable $f(X_1, X_2)$ is to evaluate the integral

$$\mathsf{E}f(X_1, X_2) = \int_{-\infty}^{\infty} \int_{-\infty}^{\infty} f(x_1, x_2)\, dG_{X_{(1,2)}}(x_1, x_2).$$

Now, if $f(x_1, x_2)$ does not increase faster than a polynomial and differentiable, see [108] then the following calculation can be performed, we rewrite the integral by means of the Plancherel formula and apply the relation

$$\widehat{\left[\frac{\partial^k}{\partial x_1^k} f\right]}(u_1, u_2) = (iu_1)^k \hat{f}(u_1, u_2),$$

where $\hat{f}(u_1, u_2)$ is the Fourier transform of $f(x_1, x_2)$, i.e.,

$$\hat{f}(u_1, u_2) = \frac{1}{(2\pi)^2} \int_{-\infty}^{\infty} \int_{-\infty}^{\infty} f(x_1, x_2) exp\{-i(x_1 u_1 + x_2 u_2)\}\, dx_1\, dx_2,$$

and get

$$\begin{aligned}
\mathsf{E}f(X_1, X_2) &= \int_{-\infty}^{\infty} \int_{-\infty}^{\infty} f(x_1, x_2)\, dG_{X_{(1,2)}}(x_1, x_2) \\
&= \int_{-\infty}^{\infty} \int_{-\infty}^{\infty} \hat{f}(u_1, u_2)\, \Psi_{X_{(1,2)}}(-u_1, -u_2)\, du_1\, du_2 \\
&= \sum_{k=0}^{\infty} \frac{\sigma_{12}^k}{k!} \int_{-\infty}^{\infty} \int_{-\infty}^{\infty} (iu_1)^k (iu_2)^k \hat{f}(u_1, u_2)\, \Psi_{X_1}(-u_1) \Psi_{X_2}(-u_2)\, du_1\, du_2 \\
&= \sum_{k=0}^{\infty} \frac{\sigma_{12}^k}{k!} \int_{-\infty}^{\infty} \int_{-\infty}^{\infty} \frac{\partial^k}{\partial x_1^k} \frac{\partial^k}{\partial x_2^k} f(x_1, x_2)\, dG_{X_1}(x_1)\, G_{X_2}(x_2).
\end{aligned}$$

The conclusion of this calculation is that the expectation $\mathsf{E}f(X_1, X_2)$ can be given by the help of the partial derivatives of f and by two independent Gaussian random variables, say, (Y_1, Y_2) with the same mean and variance as (X_1, X_2), respectively.

Theorem 1 *Suppose that f is an appropriate function and let (Y_1, Y_2) be orthogonalized copies of Gaussian random variables (X_1, X_2), i.e., (Y_1, Y_2) is Gaussian with $\mathsf{E}Y_k = \mathsf{E}X_k$, $\mathrm{Var}(Y_k) = \mathrm{Var}(X_k)$, $k = 1,2$ and $\mathrm{Cov}(Y_1, Y_2) = 0$, then*

$$\mathsf{E}f(X_1, X_2) = \sum_{k=0}^{\infty} \frac{\sigma_{12}^k}{k!} \mathsf{E}f_{kk}(Y_1, Y_2), \tag{1.1}$$

where

$$f_{kk}(x_1, x_2) = \frac{\partial^k}{\partial x_1^k}\frac{\partial^k}{\partial x_2^k} f(x_1, x_2).$$

The orthogonality, i.e., the independence, of Y_1 and Y_2 is particularly important when f is some linear combination of the products of the functions of its variables.

Example 2 *If* $\mathsf{E}X = 0$, *then for every* $n = 1, 2, \ldots$

$$\begin{aligned} \mathsf{E}X^{2n-1} &= 0, \\ \mathsf{E}X^{2n} &= (2n-1)!!\sigma^{2n}, \end{aligned}$$

where $(2n-1)!! = 1 \cdot 3 \cdot 5 \cdots (2n-1)$.

Proof. Put $(X_1, X_2) = (X, X)$. Therefore $\sigma_{jk} = \sigma^2$, $j, k = 1, 2$. The expectation $\mathsf{E}X^4$ is calculated by (1.1) as follows. Start first with

$$\mathsf{E}X^3 = \mathsf{E}X_1^2 X_2 = \mathsf{E}Y_1^2 \mathsf{E}Y_2 + 2\sigma^2 \mathsf{E}Y_1 = 0,$$

so

$$\mathsf{E}X^4 = \mathsf{E}X_1^3 X_2 = \mathsf{E}Y_1^3 \mathsf{E}Y_2 + 3\sigma^2 \mathsf{E}Y_1^2 = 3\sigma^4.$$

Assume as an induction step that

$$\begin{aligned} \mathsf{E}X^{2k-1} &= 0, \\ \mathsf{E}X^{2k} &= (2k-1)!!\sigma^{2k}, \end{aligned}$$

is fulfilled for $k = 1, 2, \ldots n$ where $(2k-1)!! = 1 \cdot 3 \cdot 5 \cdots (2k-1)$. Now we use (1.1) to show this statement for $k = n + 1$. First of all

$$\mathsf{E}X^{2n+1} = \mathsf{E}X_1^{2n} X_2 = \mathsf{E}Y_1^{2n} \mathsf{E}Y_2 + 2n\sigma^2 \mathsf{E}Y_1^{2n-1} = 0,$$

moreover

$$\mathsf{E}X^{2n+2} = \mathsf{E}X_1^{2n+1} X_2 = \mathsf{E}Y_1^{2n+1} \mathsf{E}Y_2 + (2n+1)\sigma^2 \mathsf{E}Y_1^{2n} = (2n+1)!!\sigma^{2n+2}.$$

We have therefore proved the well known fact that for a Gaussian random variable X with $\mathsf{E}X = 0$ all the odd moments are zero and the even ones are given by

$$\mathsf{E}X^{2n} = (2n-1)!!\sigma^{2n}.$$

■

Example 3 *If* $f(X_1, X_2) = f_1(X_1) f_2(X_2)$ *then*

$$\mathsf{E} f_1(X_1) f_2(X_2) = \sum_{k=0}^{\infty} \frac{\sigma_{12}^k}{k!} \mathsf{E} f_1^{(k)}(X_1) \mathsf{E} f_2^{(k)}(X_2),$$

where

$$f_j^{(k)}(x) = \frac{d^k}{dx^k} f_j(x).$$

It is quite straightforward to generalize the formula (1.1) to the d dimensional case. Let $(X_1, X_2, \ldots, X_d)$ be an arbitrary Gaussian random vector with $\mu_j = \mathsf{E} X_j$ and

$$\sigma_{jk} = \mathrm{Cov}(X_j, X_k) = \mathsf{E}(X_j - \mu_j)(X_k - \mu_k), \quad j, k = 1, 2, \ldots, d.$$

Let $(Y_1, Y_2, \ldots, Y_d)$ denote the orthogonal copy of $(X_1, X_2, \ldots, X_d)$, i.e., $(Y_1, Y_2, \ldots, Y_d)$ is Gaussian with $\mathsf{E} Y_j = \mu_j$ and

$$\mathrm{Cov}(Y_j, Y_k) = \delta_{\{j=k\}} \sigma_{jk}, \quad j, k = 1, 2, \ldots, d,$$

where $\delta_{\{j=k\}}$ is the Kronecker symbol.

Theorem 4 *If* f *is an appropriate function with* d *variables then the expectation*

$$\mathsf{E} f(X_1, X_2, \ldots, X_d)$$

$$= \mathsf{E} \left[\prod_{\substack{j<k \\ k=2,3\ldots d}} \left(\sum_{l=0}^{\infty} \frac{\sigma_{jk}^l}{l!} D_{y_j}^l D_{y_k}^l \right) \right] f(Y_1, Y_2, \ldots, Y_d), \tag{1.2}$$

where D_{y_j} *is the partial differential operator* $\frac{\partial}{\partial y_j}$.

Example 5 *Take the monomial* $X_1 \cdot X_2 \cdots X_d$. *Then the formula (1.2) gives, for the expectation*

$$\mathsf{E} X_1 \cdot X_2 \cdots X_d = \prod_{j=1}^{d} \mathsf{E} X_j + \sum_{j<k} \sigma_{jk} \prod_{r \neq j,k} \mathsf{E} X_r$$

$$+ \sum_{\substack{j<k,\ j \neq l,m; \\ l<m,\ k \neq l,m}} \sigma_{jk} \sigma_{lm} \prod_{r \neq j,k,l,m} \mathsf{E} X_r + \ldots .$$

The above formula gives, for $d = 3$,

$$\mathsf{E}X_1X_2X_3 = \prod_{j=1}^{3} \mathsf{E}X_j + \sigma_{12}\mathsf{E}X_3 + \sigma_{13}\mathsf{E}X_2 + \sigma_{23}\mathsf{E}X_1.$$

If $\mathsf{E}X_j = 0$ and d is an odd number then

$$\mathsf{E}X_1X_2 \cdots X_d = 0,$$

moreover if d is even, then

$$\mathsf{E}X_1X_2 \cdots X_d = \sum_{(d-1)!!} \prod_{l=1}^{d/2} \sigma_{j_l k_l},$$

where the summation is taken over all the possible choices of covariances σ_{jk} of pairs X_j and X_k such that every variable is taken exactly once. Note that each of the products above consists of covariances of such pairs that gives a partition of $\{1, 2, \ldots d\}$ into pairs. Further, the summation taken over all such partitions. The number of them is

$$\frac{1}{(d/2)!}\binom{d}{2}\binom{d-2}{2}\cdots\binom{4}{2}\binom{2}{2} = (d-1)(d-3)\cdots 3\cdot 1$$
$$= (d-1)!!. \tag{1.3}$$

Particularly for $d = 4$ the partitions of $\{1, 2, 3, 4\}$ to the pairs are

$$\mathcal{L}_1 = \{(1,2),(3,4)\},\ \mathcal{L}_2 = \{(1,3),(2,4)\},\ \text{and}\ \mathcal{L}_3 = \{(1,4),(2,3)\},$$

therefore

$$\mathsf{E}X_1X_2X_3X_4 = \sigma_{12}\sigma_{34} + \sigma_{13}\sigma_{24} + \sigma_{14}\sigma_{23},$$

taking all the variables to be identical here the statement of Example 2 follows.

1.2 Hermite polynomials

1.2.1 *Hermite polynomials of one variable*

Hermite polynomials $H_n(x)$ of degree n and with a positive leading coefficient were originally defined by the following conditions

$$\int_{-\infty}^{+\infty} H_n(x)H_m(x)\frac{1}{\sqrt{2\pi}}exp(-x^2/2)\,dx = n!\delta_{\{n=m\}}, \quad n, m = 0, 1, 2, \ldots,$$

see [97]. These polynomials form a closed orthogonal system in the Hilbert space $\mathfrak{L}^2\left(\mathbf{R}, \mathcal{B}, \frac{1}{\sqrt{2\pi}} exp(-x^2/2)dx\right)$, see [125], i.e., every Borel measurable function g for which

$$\int_{-\infty}^{+\infty} g^2(x) \frac{1}{\sqrt{2\pi}} exp(-x^2/2)\, dx < \infty,$$

can be put into the Hermite series expansion

$$g(x) = \sum_{k=0}^{\infty} \frac{g_k}{k!} H_k(x), \tag{1.4}$$

by the Hermite polynomials H_k, where

$$g_k = \int_{-\infty}^{+\infty} g(x) H_k(x) \frac{1}{\sqrt{2\pi}} exp(-x^2/2)\, dx.$$

The convergence of (1.4) is taken in the sense of
$\mathfrak{L}^2\left(\mathbf{R}, \mathcal{B}, \frac{1}{\sqrt{2\pi}} exp(-x^2/2)dx\right)$, i.e.,

$$\lim_{n\to\infty} \int_{-\infty}^{+\infty} \left(g(x) - \sum_{k=0}^{n} \frac{g_k}{k!} H_k(x) \right)^2 \frac{1}{\sqrt{2\pi}} exp(-x^2/2)\, dx = 0.$$

One can check the following properties of the Hermite polynomials, see [125] p. 102.

- *Property 1.*

$$e^{-x^2/2} H_n(x) = (-1)^n \left(\frac{d}{dx}\right)^n e^{-x^2/2}.$$

- *Property 2.*

$$H_0 = 1, \quad \frac{d}{dx} H_n(x) = n H_{n-1}(x).$$

- *Property 3.*

$$\sum_{n=0}^{\infty} \frac{H_n(x)}{n!} a^n = exp(xa - a^2/2).$$

Note that *Property 3* is understood as Taylor series by a for fixed x and an $\mathfrak{L}^2$ Hermite series, see (1.4), when a is fixed.

The translation of these facts into the terminology of probability leads to the following idea. Let X be a Gaussian random variable with mean

0 and variance 1. The set of all random variables Y which are measurable with respect to X with finite variance, sometimes called variables *depending on* X, forms a Hilbert space. Denote it by $\mathfrak{L}^2(X)$. The inner product in $\mathfrak{L}^2(X)$ is the covariance as usual, i.e., if $Y_1, Y_2 \in \mathfrak{L}^2(X)$ then $\langle Y_1, Y_2 \rangle = \operatorname{Cov}(Y_1, Y_2)$. Now it is easy to see that the random variables $H_n = H_n(X)$, $n = 0, 1, 2, \ldots$ form a closed orthogonal system in $\mathfrak{L}^2(X)$, see [78]. Therefore if $Y \in \mathfrak{L}^2(X)$, then there exists a Borel measurable function g such that $Y = g(X)$ and also

$$Y = \sum_{k=0}^{\infty} \frac{g_k}{k!} H_k,$$

where $g_k = \operatorname{Cov}(Y, H_k)$ and

$$\lim_{n \to \infty} \operatorname{Var}\left(Y - \sum_{k=0}^{n} \frac{g_k}{k!} H_k\right) = 0.$$

An interesting question to address is the following. Take a Gaussian system $\mathcal{X} = \{X_t,\ t \in T\}$, which means that every finite linear combination

$$\sum_{k=0}^{n} a_k X_{t_k}, \quad a_k \in \mathbf{R}, \quad t_k \in T, \quad k = 1, 2, \ldots n, \quad n \in \mathbb{Z},$$

of X_t is Gaussian. Put $\mathfrak{L}^2(\mathcal{X})$ as the Hilbert space of all random variables depending on $\mathcal{X}$, i.e., measurable with respect to the system $\mathcal{X}$ and having a finite variance. Note that the Hilbert space $\mathfrak{L}^2(\mathcal{X})$ is sometimes called nonlinear to distinguish between $\mathfrak{L}^2(\mathcal{X})$ and the linear Hilbert space $L_1^2(\mathcal{X})$ generated by the system $\mathcal{X}$ which contains only the finite linear combinations and their limits in the mean square sense. So every element in the linear space $L_1^2(\mathcal{X})$ is Gaussian but this is not the case in $\mathfrak{L}^2(\mathcal{X})$. Now the question is whether there exist a closed orthogonal system, the elements of which are polynomials in X_t, in $\mathfrak{L}^2(\mathcal{X})$, and what it looks like. This problem is investigated in the next sections. See also Ibragimov and Rosanov [61], p. 27.

1.2.2 Hermite polynomials of several variables

Consider a Gaussian system $\mathcal{X} = \{X_t,\ t = 1, 2, \ldots\}$ with $\mathsf{E}X_t = 0,\ t = 1, 2, \ldots$ and put $\sigma_{j,k} = \operatorname{Cov}(X_j, X_k)$. Each of the following rules defines the same polynomial system, see [59] for futher generalization.

1. Recursion I.

- $H_0 = 1$,
- $\frac{\partial}{\partial x_k} H_n(x_{(1:n)}) = H_{n-1}(x_{(1:k-1,k+1:n)})$,

- $H_n(x_{(1:n)})$ is a symmetric function and $\mathsf{E}H_n(X_{(1:n)}) = 0$.

2. Recursion II.

- $H_0 = 1, \quad H_1(x) = x,$
- If $n > 1$

$$H_n(x_{(1:n)}) = x_n H_{n-1}(x_{(1:n-1)}) - \sum_{j=1}^{n-1} \sigma_{j,n} H_{n-2}(x_{(1:j-1,j+1:n-1)}).$$

where if $j = 1$ then the variable x_1 is missing from H_{n-2}, i.e., $H_{n-2}(x_{(2:n-1)})$.

3. Exponent.

$$H_n(x_{(1:n)}) = \frac{\partial^n}{\partial a_{(1:n)}^{\bar{1}}} \exp\left((a_{(1:n)}, x_{(1:n)}) - \frac{1}{2} \sum_{j,k=1}^{n} \sigma_{j,k} a_j a_k \right)\Bigg|_{a_{(1:n)}=0},$$

where $(a_{(1:n)}, x_{(1:n)})$ denotes the usual inner product of vectors, i.e.,

$$(a_{(1:n)}, x_{(1:n)}) = \sum_{k=1}^{n} a_k x_k.$$

Note that in this case the Hermite polynomials depend on the covariance structure of the underlying Gaussian system. Hermite polynomials of a complex valued Gaussian system are defined in [55]. The variables of Hermite polynomials are taken from a Gaussian random system and their covariance structure also belongs to the definition of Hermite polynomials. The list of the first 5 polynomials follows.

1. $H_0 = 1,$
2. $H_1(X_1) = X_1,$
3. $H_2(X_{(1:2)}) = X_1 X_2 - \sigma_{1,2},$
4. $H_3(X_{(1:3)}) = X_1 X_2 X_3 - \sigma_{1,2} X_3 - \sigma_{1,3} X_2 - \sigma_{2,3} X_1,$
5. $H_4(X_{(1:4)}) = X_1 X_2 X_3 X_4 - \sigma_{1,2} X_3 X_4 - \sigma_{1,3} X_2 X_4 - \sigma_{1,4} X_2 X_3,$ $-\sigma_{2,3} X_1 X_4 - \sigma_{2,4} X_1 X_3 - \sigma_{3,4} X_1 X_2 + \sigma_{1,2}\sigma_{2,3} + \sigma_{1,3}\sigma_{2,4} + \sigma_{1,4}\sigma_{2,3}.$

It is easy to see that if $X_{(1:k)}$ and $X_{(k+1:n)}$ are statistically independent then

$$H_n(X_{(1:n)}) = H_k(X_{(1:k)}) H_{n-k}(X_{(k+1:n)}).$$

It is symmetric and multilinear, i.e.,

$$H_n(X_{(1:n)}) = H_n(X_{\mathfrak{p}(1:n)}),$$

for any permutation $\mathfrak{p} \in \mathfrak{P}_n$ and

$$H_{n+1}(aY + bZ, X_{(1:n)}) = aH_{n+1}(Y, X_{(1:n)}) + bH_{n+1}(Z, X_{(1:n)}),$$

if $\{Y, Z, X_{(1:n)}\}$ is a Gaussian system and a, b are real numbers. As $X_t = X, \quad t = 1, 2, ...$ is a Gaussian system therefore the above definitions are generalizations of those given in the previous section, i.e., put $\operatorname{Var}(X) = 1$ then

$$H_n(\underbrace{X_1, X_1, ..., X_1}_{n}) = H_n(X_1),$$

for example $H_2(X, X) = X^2 - 1$. Repeat the notation for the vector of such entries that some of them are the same

$$x_{(1:n)[m_{(1:n)}]} = (\underbrace{x_1, x_1, ..., x_1}_{m_1}, \underbrace{x_2, x_2, ..., x_2}_{m_2}, ... \underbrace{x_n, x_n, ..., x_n}_{m_n}).$$

It is easy to see that the following series expansion of the exponent is valid in L_2 for every fixed a and it is the Taylor series expansion with respect to a when X is fixed.

$$\exp\left((a_{(1:n)}, X_{(1:n)}) - \frac{1}{2}\sum \sigma_{j,k} a_j a_k\right) = \sum_{m_{(1:n)} \geq 0} \frac{a_{(1:n)}^{m_{(1:n)}}}{m_{(1:n)}!} H_{\Sigma m_{(1:n)}}\left(X_{(1:n)[m_{(1:n)}]}\right).$$

The Hermite polynomials depend strongly on the covariance structure of the Gaussian variables under consideration therefore we shall often put Gaussian variables as the variables of the polynomials to make this fact clear. We highlight this by the following examples. By the definition in Section 1.2.1

$$H_2(x) = x^2 - 1,$$

and by the Section 1.2.2

$$H_2(X) = X^2 - \operatorname{Var}(X),$$

also by Section 1.2.1

$$2^{n/2} H_n(x + y) = \sum_{k=0}^{n} \binom{n}{k} H_k(\sqrt{2}x) H_{n-k}(\sqrt{2}y),$$

while by Section 1.2.2 if X and Y are independent then

$$H_n(aX + bY) = \sum_{k=0}^{n} \binom{n}{k} a^k b^{n-k} H_k(X) H_{n-k}(Y).$$

1.3 Cumulants

Moments and cumulants are basic tools of the theory of random series, see Brillinger [20], Leonov and Shiryaev [73], Shiryaev [114], [23] and [25]. A similar treatment to the definition of moments, see Appendix A, can be used for the definition of cumulants.

1.3.1 *Definition of cumulants*

The use of the different definitions of moments, see page 211, will help us towards an easier understanding of cumulants. Let us keep the assumptions of the previous section and now consider the Taylor series expansion around zero of the logarithm of the characteristic function of $X_{(1:n)}$, i.e.,

$$\ln \phi_X(\lambda_{(1:n)}) = \sum_{\Sigma k_{(1:n)} \le m} \frac{i^{\Sigma k_{(1:n)}}}{k_{(1:n)}!} \operatorname{Cum}\left(X_{(1:n)[k_{(1:n)}]}\right) \lambda_{(1:n)}^{k_{(1:n)}} + o(|\lambda_{(1:n)}|^m). \tag{1.5}$$

The cumulant $\operatorname{Cum}\left(X_{(1:n)[k_{(1:n)}]}\right)$ is defined as the coefficient of $\lambda_{(1:n)}^{k_{(1:n)}}$ via the equation (1.5). Because of the equality of the derivatives

$$\left.\frac{\partial^{\Sigma k_{(1:n)}}}{\partial \lambda_{(1:n)}^{k_{(1:n)}}} \ln \phi_{X_{(1:n)}}(\lambda_{(1:n)})\right|_{\lambda_{(1:n)}=0} = \left.\frac{\partial^{\Sigma k_{(1:n)}}}{\partial \lambda_{(1:\Sigma k_{(1:n)})}^{\mathbf{1}}} \ln \phi_1\left(\lambda_{(1:\Sigma k_{(1:n)})}\right)\right|_{\lambda_{(1:\Sigma k_{(1:n)})}=0}.$$

The Taylor series expansion of the logarithm of the characteristic function $\phi_1\left(\lambda_{(1:\Sigma k_{(1:n)})}\right)$ of $X_{(1:n)[k_{(1:n)}]}$ corresponds to the Taylor series expansion of the logarithm of the characteristic function $\phi_{X_{(1:n)}}(\lambda_{(1:n)})$ of $X_{(1:n)}$. Therefore the definition is correct and that formula allows us to act as if all the components of the vector X were different so it is satisfactory to define the cumulant for different variables only as

$$\operatorname{Cum}(X_{(1:n)}) = \left.\frac{(-i\partial)^n}{\partial \lambda_{(1:n)}^{\mathbf{1}}} \ln \phi_X(\lambda_{(1:n)})\right|_{\lambda_{(1:n)}=0}.$$

Example 6 *The characteristic function of variable X with gamma distribution $\Gamma(\alpha, \mu)$ is*

$$\phi_X(\lambda) = \left(1 - \frac{i\lambda}{\alpha}\right)^{-\mu}.$$

Therefore the nth *order cumulant* $\operatorname{Cum}\Big(\underbrace{X, X, \ldots, X, X}_{n}\Big)$ *of* X *can be obtained either as the* n^{th} *derivative of* $\ln \phi_X(\lambda)$ *at zero or as the* $\partial^n / \partial \lambda_{(1:n)}^{\vec{1}}$ *partial derivative at zero of the characteristic function of*

$$\ln \phi_{X_{(1:n)}}(\lambda_{(1:n)}) = -\mu \ln \left(1 - \frac{i \sum \lambda_{(1:n)}}{\alpha}\right),$$

both giving

$$\operatorname{Cum}\Big(\underbrace{X, X, \ldots, X, X}_{n}\Big) = \frac{(n-1)!\mu}{\alpha^n},$$

where $0! = 1$ *as usual.*

Example 7 *The characteristic function of a Gaussian random vector* $X_{(1,2)}$ *is*

$$\Psi_{X_{(1,2)}}(u_1, u_2) = exp\left\{i(\mu_1 u_1 + \mu_2 u_2) - 1/2(\sigma_{11} u_1^2 + 2\sigma_{12} u_1 u_2 + \sigma_{22} u_2^2)\right\}.$$

The first order cumulant is

$$\operatorname{Cum}(X_j) = \mu_j,$$

the second order is

$$\operatorname{Cum}(X_j, X_k) = \sigma_{jk},$$

and any higher order one is zero.

1.3.2 Basic properties

Symmetry

$$\operatorname{Cum}(X_{(1:n)}) = \operatorname{Cum}(X_{\mathfrak{p}(1:n)}),$$

where $\mathfrak{p}(1:n) = (\mathfrak{p}(1), \mathfrak{p}(2), \ldots, \mathfrak{p}(n))$, $\mathfrak{p} \in \mathfrak{P}_n$ and $\mathfrak{P}_n$ denotes the set of all permutations of the integers $(1:n)$.

Multilinearity

For any constants $c_{(1:2)} = (c_1, c_2)$

$$\operatorname{Cum}(c_1 Y_1 + c_2 Y_2, X_{(1:n)}) = c_1 \operatorname{Cum}(Y_1, X_{(1:n)}) + c_2 \operatorname{Cum}(Y_2, X_{(1:n)}).$$

Independence

If $X_{(1:n)}$ is independent of $Y_{(1:m)}$ where $n, m > 0$ then

$$\mathrm{Cum}(X_{(1:n)}, Y_{(1:m)}) = 0.$$

In particular if $n = m$ then

$$\mathrm{Cum}(X_{(1:n)} + Y_{(1:n)}) = \mathrm{Cum}(X_{(1:n)}) + \mathrm{Cum}(Y_{(1:n)}).$$

This formula is the additive version of the formula (A.2) of the expectation of the product of independent variables.

Gaussianity

The random vector $X_{(1:n)}$ is Gaussian if and only if for all subset $k_{(1:m)}$ of $(1:n)$

$$\mathrm{Cum}(X_{k_{(1:m)}}) = 0, \ m > 2.$$

Expression of the cumulant via moments

Now if we apply formula (A.1) and the Taylor series expansion of the function $\ln(1+y)$ for $n = 2$ and $m = 2$ we have

$$\begin{aligned}\ln \phi(\lambda_{(1:2)}) &= \ln [1 + i\mathsf{E}X_1\lambda_1 + i\mathsf{E}X_2\lambda_2 - \mathsf{E}X_1X_2\lambda_1\lambda_2 \\ &\quad - 1/2\mathsf{E}X_1^2\lambda_1^2 - 1/2\mathsf{E}X_2^2\lambda_2^2 + o\left(|\lambda_{(1:2)}|^2\right)] \\ &= i\mathsf{E}X_1\lambda_1 + i\mathsf{E}X_2\lambda_2 - (\mathsf{E}X_1X_2 - \mathsf{E}X_1\mathsf{E}X_2)\lambda_1\lambda_2 \\ &\quad - 1/2\left(\mathsf{E}X_1^2 - (\mathsf{E}X_1)^2\right)\lambda_1^2 - 1/2(\mathsf{E}X_2^2 - (\mathsf{E}X_2)^2)\lambda_2^2 \\ &\quad + o\left(|\lambda_{(1:2)}|^2\right).\end{aligned}$$

Therefore, by formula (1.5)

$$\mathrm{Cum}(X_t) = \mathsf{E}X_t, \ t = 1, 2;$$

$$\mathrm{Cum}(X_{t_1}, X_{t_2}) = \mathrm{Cov}(X_{t_1}, X_{t_2}), \quad t_1, t_2 = 1, 2. \tag{1.6}$$

The same reasoning leads to the general formula

$$\mathrm{Cum}(X_{(1:n)}) = \sum_{k=1}^{n} (-1)^{k-1}(k-1)! \sum_{\Sigma_{j=1}^{k} \vec{a}_j = \vec{1}} \prod_{j=1}^{k} \mathsf{E}X_{(1:n)}^{\vec{a}_j}, \tag{1.7}$$

where the second summation is taken over all possible sets of n dimensional vectors $\{\vec{a}_j, \ j = 1, 2, \ldots, k\}$ with entries 0, 1 such that their sum is $\vec{1}$. Such a vector system $\vec{a}_j, \ j = 1, 2, \ldots k$ corresponds to a partition $\mathcal{L}$ of the set $(1:n) = (1, 2, \ldots, n)$, therefore the double sum in (1.7) is over all partitions of $\mathcal{L} \in \mathcal{P}_{(1:n)}$, where $\mathcal{P}_{(1:n)}$ is the set of all partitions of the numbers $(1:n)$.

Example 8 *First note that if $n = 3$, then the kth order partitions of $(1:3)$ are*

for $k = 1$; $\mathcal{L}_1 = \left\{\vec{a}_1 = \vec{1}\right\}$,

for $k = 2$; $\mathcal{L}_2 = \{\vec{a}_1 = (1,0,0), \vec{a}_2 = (0,1,1)\}$,

$\mathcal{L}_3 = \{\vec{a}_1 = (0,1,0), \vec{a}_2 = (1,0,1)\}$, $\mathcal{L}_4 = \{\vec{a}_1 = (0,0,1), \vec{a}_2 = (1,1,0)\}$,

for $k = 3$; $\mathcal{L}_5 = \{\vec{a}_1 = (1,0,0), \vec{a}_2 = (0,1,0), \vec{a}_3 = (0,0,1)\}$.

Now we apply (1.7):

$$\begin{aligned}\mathrm{Cum}(X_{(1:3)}) &= \mathsf{E}X_1X_2X_3 - \mathsf{E}X_1\mathsf{E}X_2X_3 - \mathsf{E}X_2\mathsf{E}X_1X_3 \\ &\quad - \mathsf{E}X_3\mathsf{E}X_1X_2 + 2\mathsf{E}X_1\mathsf{E}X_2\mathsf{E}X_3 \\ &= \mathsf{E}\prod_{i=1}^{3}(X_i - \mathsf{E}X_i),\end{aligned}$$

In particular,

$$\begin{aligned}\mathrm{Cum}(X,X,X) &= \mathsf{E}X^3 - 3\mathsf{E}X\mathsf{E}X^2 + 2(\mathsf{E}X)^3 \\ &= \mathsf{E}(X - \mathsf{E}X)^3,\end{aligned}$$

The first three cumulants, see also (1.6), equal the central moments but this is not true for higher order cumulants. One might easily check this for the case of cumulants of order four. Indeed,

$$\begin{aligned}\mathrm{Cum}\left(X_{(1:4)}\right) &= \mathrm{Cum}\left(X_{(1:4)} - \mathsf{E}X_{(1:4)}\right) \\ &= \mathsf{E}\prod_{i=1}^{4}(X_i - \mathsf{E}X_i) - \mathrm{Cov}(X_1,X_2)\,\mathrm{Cov}(X_3,X_4) \\ &\quad - \mathrm{Cov}(X_1,X_3)\,\mathrm{Cov}(X_2,X_4) - \mathrm{Cov}(X_1,X_4)\,\mathrm{Cov}(X_2,X_3) \\ &\neq \mathsf{E}\prod_{i=1}^{4}(X_i - \mathsf{E}X_i),\end{aligned} \tag{1.8}$$

unless all covariances are zero.

Expression of the moment via cumulants

We consider the moment $\mathsf{E}X_{(1:n)}^{\vec{1}}$ as the general case because the moment $\mathsf{E}Y_{(1:m)}^{k_{(1:m)}}$ can be put into the form $\mathsf{E}X_{(1:n)}^{\vec{1}}$, where $X_{(1:n)} = Y_{(1:m)[k_{(1:m)}]}$.

$$\mathsf{E}X_{(1:n)}^{\vec{1}} = \sum_{\mathcal{L}\in\mathcal{P}_{(1:n)}} \prod_{K_j\in\mathcal{L}} \mathrm{Cum}(X_{K_j}), \tag{1.9}$$

where the summation is over all partitions $\mathcal{L} = \{K_1, K_2, \ldots, K_k\}$ of $(1:n)$.

An example of the use of the formula (1.9) can be seen when $n = 3$.

Example 9 *Each K_j corresponds to a vector $\vec{a}_j$ with entries 0 and 1, as in the previous Example.*

$$\begin{aligned}\mathsf{E}X_{(1:3)}^{\mathbf{1}} = &\operatorname{Cum}(X_{(1:3)}) + \operatorname{Cum}(X_1)\operatorname{Cum}(X_2, X_3)\\ &+ \operatorname{Cum}(X_2)\operatorname{Cum}(X_1, X_3) + \operatorname{Cum}(X_3)\operatorname{Cum}(X_1, X_2)\\ &+ \operatorname{Cum}(X_1)\operatorname{Cum}(X_2)\operatorname{Cum}(X_3),\end{aligned}$$

Now in particular

$$\mathsf{E}X^3 = \operatorname{Cum}(X, X, X) + 3\operatorname{Cum}(X)\operatorname{Cum}(X, X) + \operatorname{Cum}(X)^3.$$

Expression of the cumulant of products via products of cumulants

Let us start solving this problem by finding $\operatorname{Cum}(X_1, X_2X_3)$. The expressions (1.7) and (1.9) are applied, so

$$\begin{aligned}\operatorname{Cum}(X_1, X_2X_3) &= \mathsf{E}X_1X_2X_3 - \mathsf{E}X_1\mathsf{E}X_2X_3\\ &= \operatorname{Cum}(X_{(1:3)}) + \operatorname{Cum}(X_1)\operatorname{Cum}(X_2, X_3)\\ &\quad + \operatorname{Cum}(X_2)\operatorname{Cum}(X_1, X_3) + \operatorname{Cum}(X_3)\operatorname{Cum}(X_1, X_2)\\ &\quad + \operatorname{Cum}(X_1)\operatorname{Cum}(X_2)\operatorname{Cum}(X_3)\\ &\quad - \operatorname{Cum}(X_1)\operatorname{Cum}(X_2, X_3)\\ &\quad - \operatorname{Cum}(X_1)\operatorname{Cum}(X_2)\operatorname{Cum}(X_3)\\ &= \operatorname{Cum}(X_{(1:3)}) + \operatorname{Cum}(X_2)\operatorname{Cum}(X_1, X_3)\\ &\quad + \operatorname{Cum}(X_3)\operatorname{Cum}(X_1, X_2).\end{aligned} \tag{1.10}$$

The point actually is that the moments are taken by the products according to the variables given by the cumulant. Therefore the cumulant of the products can be expressed by the products of cumulants of the variables themselves. Moreover the cumulant of X_1 and X_2X_3 can be considered as cumulant of products according to a partition of all the variables considered. To the partition $\mathcal{K} = \{(1),\ (2,3)\}$ there corresponds the random vector $X_{\mathcal{K}} = (X_1, X_2X_3)$. The result we got above is a sum of the products of cumulants each term of which also corresponds to a partition of the set $(1,2,3)$. They are $\mathcal{K}_1 = \{(1,2,3)\}$, $\mathcal{K}_2 = \{(1,3),(2)\}$ and $\mathcal{K}_3 = \{(1,2),(3)\}$. The main feature of these partitions is that there is no other partition but the largest one $(1,2,3)$ which can be combined by the help of the elements, i.e., blocks, of both $\mathcal{K}$ and $\mathcal{K}_j$ at the same time.

To be more precise consider the set $\mathcal{P}_{(1:n)}$ of all the partitions of $(1:n)$. One says, the partition $\mathcal{K}$ is **contained** in partition $\mathcal{L}$, $\mathcal{K} \subseteq \mathcal{L}$ if the elements of $\mathcal{L}$ are the unions of the elements of $\mathcal{K}$. The largest, i.e., coarsest (crudest) partition in that sense is $\mathcal{I} = \{(1:n)\}$ and the smallest, i.e., finest one is $\mathcal{O} = \{(1),(2),...(n)\}$. The **union** $\mathcal{K} \cup \mathcal{L}$ of the partitions $\mathcal{K}$ and $\mathcal{L}$ is defined as the smallest one which contains both of them. We shall call the partition $\mathcal{K}$ **indecomposable** with respect to the partition $\mathcal{L}$ if $\mathcal{K} \cup \mathcal{L} = \mathcal{I}$. It is easy to see that $\mathcal{K}$ and $\mathcal{L}$ are indecomposable if and only if **there are no two nonempty subsets of** $(1:n)$ **such that each of them can be composed as the union of the blocks of both partitions** $\mathcal{K}$ **and** $\mathcal{L}$. Note that the partitions indecomposable according to $\mathcal{I}$ and $\mathcal{O}$ are the set of all partitions and $\mathcal{I}$, respectively.

All the indecomposable partitions of $(1,2,3)$ according to the partition $\mathcal{K} = \{(1),(2,3)\}$ are $\mathcal{L}_1 = \{(1,2,3)\}$, $\mathcal{L}_2 = \{(1,3),(2)\}$ and $\mathcal{L}_3 = \{(1,2),(3)\}$.

Let $X_{\mathcal{K}}$ denote the vector of entries taken by the partition $\mathcal{K}$, i.e., if $\mathcal{K} = \{K_1, K_2, \ldots, K_k\}$ then $X_{\mathcal{K}} = (\prod X_{K_1}, \prod X_{K_2}, \ldots, \prod X_{K_k})$. Now the cumulant of the products
$(\prod X_{K_1}, \prod X_{K_2}, \ldots, \prod X_{K_k})$ can be expressed by the cumulants of the individual variables X_M, $M \in \mathcal{L}$, such that $\mathcal{K} \cup \mathcal{L} = \mathcal{I}$, i.e.,

$$\operatorname{Cum}(X_{\mathcal{K}}) = \sum_{\mathcal{K}\cup\mathcal{L}=\mathcal{I}} \prod_{M\in\mathcal{L}} \operatorname{Cum}(X_M), \tag{1.11}$$

where X_M denotes the vector containing the items X_s, $s \in M$, see (1.10).

1.4 Diagrams, and moments and cumulants for Gaussian systems.

1.4.1 Diagrams

The characteristic function of Gaussian random variables depends only on the means and covariances, therefore all the higher order moments and cumulants of Gaussian random variables can be expressed by the first and second order ones. So the partitions into pairs of the set $(1:n) = (1,2,\ldots n)$ have a particular importance. The set of all such partitions will be denoted by $\mathcal{P}^{\mathrm{II}}_{(1:n)}$ and the elements of $\mathcal{P}^{\mathrm{II}}_{(1:n)}$ are $\mathcal{K}^{\mathrm{II}}$. The set $\mathcal{P}^{\mathrm{II}}_{(1:n)}$ is empty if n is odd and it has $(n-1)!!$ elements if n is even, i.e., all the partitions by pairs can be obtained if one fix an element of numbers $(1:n)$, say k and choose a pair for it from the remaining $n-1$ numbers, then fix any one from the $n-2$ numbers and choose a pair for it from the remaining $n-3$ ones and so on, see (1.3). Now if $\mathcal{L}$ is an arbitrary partition of $(1:n)$ and $\mathcal{K}^{\mathrm{II}} \in \mathcal{P}^{\mathrm{II}}_{(1:n)}$ then $(\mathcal{L}, \mathcal{K}^{\mathrm{II}})$ will be called a **diagram**. Actually $(\mathcal{L}, \mathcal{K}^{\mathrm{II}})$ can be regarded as a graph with vertices $M \in \mathcal{L}$ and edges $(k_1,k_2) \in \mathcal{K}^{\mathrm{II}}$. The edge (k_1,k_2) connects vertices M_1 and M_2 if $k_1 \in M_1$ and $k_2 \in M_2$.

The diagram $(\mathcal{L}, \mathcal{K}^{\mathrm{II}})$ will be called **closed** and **without loop** if the graph defined by it is closed and has no loop, respectively. A graph is called closed if any vertex can be reached from any other vertex. The diagram, Figure 1.1 according to the partitions $\mathcal{L} = \{L_1 = (1),\ L_2 = (2,3,4)\}$ and $\mathcal{K}^{\mathrm{II}} = \{(1,2),(3,4)\}$ of $(1:4)$ is closed and has a loop.

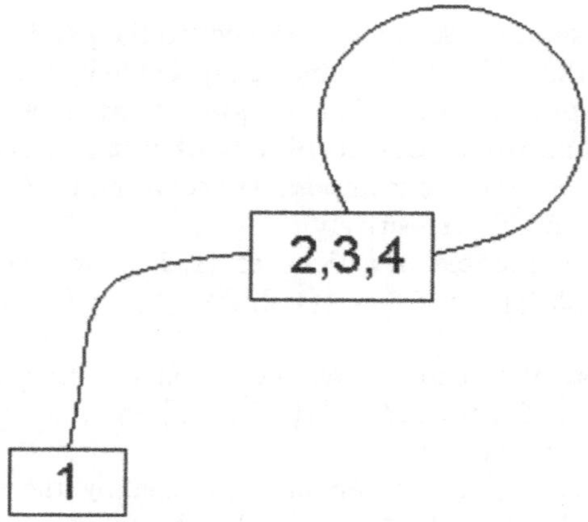

FIGURE 1.1. A closed diagram with a loop

1.4.2 Moments of Gaussian systems

Let us consider the higher order moments for a Gaussian system $\mathcal{X} = \{X_t, t \in\}$ with $\mathsf{E}X_t = 0$ and covariance $\sigma_{s,t} = \mathrm{Cov}(X_s, X_t)$, where $T \subseteq \mathbb{Z} = \{0, \pm 1, \pm 2, \dots\}$. We have seen in Example 5 that if n is even, then

$$\mathsf{E}X_1 \cdot X_2 \cdots X_n = \sum_{(n-1)!!} \prod_{l=1}^{n/2} \sigma_{j_l k_l},$$

where the summation is taken over all the possible partitions $\mathcal{K}^{\mathrm{II}}$ of $\mathcal{P}^{\mathrm{II}}_{(1:n)}$. One can put this into the form

$$\mathsf{E}X_1 \cdot X_2 \cdots X_n = \sum_{\mathcal{K}^{\mathrm{II}} \in \mathcal{P}^{\mathrm{II}}_{(1:n)}} \prod_{(j,k) \in \mathcal{K}^{\mathrm{II}}} \sigma_{jk}.$$

The first order moments of the Hermite polynomials are zero by the construction, see Recursion I, p. 7, the second order ones are

$$\mathsf{E}H_n(X_{t_{(1:n)}})H_m(X_{s_{(1:m)}}) = \delta_{\{n=m\}} \sum_{n!} \prod_{j=1}^{n} \sigma_{t_j s_{k_j}}, \tag{1.12}$$

where $t_{(1:n)} = (t_1, t_2, \dots, t_n)$, $s_{(1:m)} = (s_1, s_1, \dots, s_m)$ and the summation is taken for all the permutations $(s_{k_1}, s_{k_2} \dots, s_{k_m})$ of the indices $(s_1, s_2, \dots, s_m)$ while the order of $(t_1, t_2, \dots, t_n)$ is fixed. In particular

$$\mathsf{E}H_n(X_s)H_n(X_t) = n!\sigma^n_{t,s}.$$

Next take a partition $\mathcal{L} = (L_1, L_2, ..., L_p)$ of $(1:n)$ with $|L_j| = n_j$, i.e., the number of indices in the set L_j is n_j. Then

$$\mathsf{E}\prod_{j=1}^{p} H_{n_j}(X_{L_j}) = \sum_{\left\{ \mathcal{K}^{\mathrm{II}} \,\middle|\, \substack{(\mathcal{L},\mathcal{K}^{\mathrm{II}}) \\ \text{without loop}} \right\}} \prod_{(t_j,t_k)\in\mathcal{K}^{\mathrm{II}}} \sigma_{t_j t_k},$$

where the summation is taken over all diagrams $(\mathcal{L}, \mathcal{K}^{\mathrm{II}})$ without loops. If the set

$$\left\{ \mathcal{K}^{\mathrm{II}} \,\middle|\, \substack{(\mathcal{L},\mathcal{K}^{\mathrm{II}}) \\ \text{without loop}} \right\},$$

is empty, the expectation is zero. In particular if $p = 2$, i.e., $\mathcal{L} = (L_1, L_2)$, then the diagram $(\mathcal{L}, \mathcal{K}^{\mathrm{II}})$ without loop corresponds to a partition $\mathcal{K}^{\mathrm{II}}$ containing pairs (l_1, l_2) such that $l_1 \in L_1$ and $l_2 \in L_2$. One can reach all such types of partitions by fixing the first entries and permuting the second ones.

1.4.3 Cumulants for Hermite polynomials

We have seen in (1.11) that in general

$$\mathrm{Cum}(X_{\mathcal{L}}) = \sum_{\mathcal{L}\cup\mathcal{K}=\mathcal{I}} \prod_{M\in\mathcal{K}} \mathrm{Cum}(X_M), \tag{1.13}$$

remember that $X_{\mathcal{L}} = \left(\prod X_{L_1}, \prod X_{L_2}, \dots, \prod X_{L_p}\right)$.

Now because of the Gaussianity of the system all the cumulants are zero on the right hand side of (1.13) except for the second order ones. Therefore the summation is taken only for $\mathcal{K} \in \mathcal{P}^{\mathrm{II}}_{(1:n)}$. The assumption $\mathcal{L} \cup \mathcal{K}^{\mathrm{II}} = \mathcal{I}$, i.e., the partitions $\mathcal{L}$ and $\mathcal{K}^{\mathrm{II}}$ are indecomposable, is equivalent to the assumption that the graph $(\mathcal{L}, \mathcal{K}^{\mathrm{II}})$ is closed. We have thus the formula

$$\mathrm{Cum}(X_{\mathcal{L}}) = \sum_{\left\{ \mathcal{K}^{\mathrm{II}} \,\middle|\, \substack{(\mathcal{L},\mathcal{K}^{\mathrm{II}}) \\ \text{closed}} \right\}} \prod_{(t_i,t_j)\in\mathcal{K}^{\mathrm{II}}} \sigma_{t_i t_j}. \tag{1.14}$$

Take again a partition $\mathcal{L} = (L_1, L_2, ..., L_p)$ of $(1:n)$ such that $|L_j| = n_j$ then

$$\begin{aligned} &\mathrm{Cum}\left(H_{n_1}(X_{L_1}), H_{n_2}(X_{L_2}), ..., H_{n_p}(X_{L_p})\right) \\ &= \sum_{\left\{ \mathcal{K}^{\mathrm{II}} \,\middle|\, \substack{(\mathcal{L},\mathcal{K}^{\mathrm{II}}) \text{ closed,} \\ \text{without loop}} \right\}} \prod_{(t_j,t_k)\in\mathcal{K}^{\mathrm{II}}} \sigma_{t_j t_k}, \end{aligned} \tag{1.15}$$

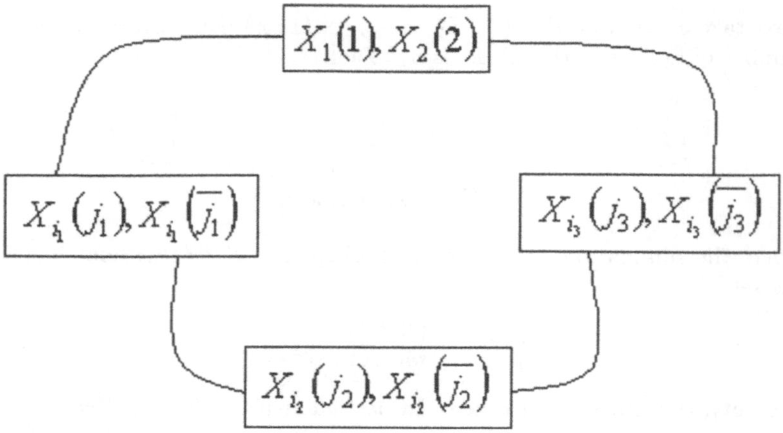

FIGURE 1.2. Diagram for $k = 4$

where the summation is over all closed diagrams $(\mathcal{L}, \mathcal{K}^{\mathrm{II}})$ without loops, see [79] for more details.

Suppose that all the variables below in the examples of this subsection are jointly Gaussian. The following equations will be useful for calculating higher order spectra for particular processes.

Example 10

$$\begin{aligned}
&\operatorname{Cum}\left(H_2\left(X_1(1), X_1(2)\right), H_2\left(X_2(1), X_2(2)\right), \ldots, H_2\left(X_k(1), X_k(2)\right)\right) \\
&\quad = \sum_{2^{k-1}(k-1)!} \operatorname{Cov}\left(X_1(1), X_{i_1}(j_1)\right) \operatorname{Cov}\left(X_{i_1}\left(\overline{j_1}\right), X_{i_2}(j_2)\right) \\
&\qquad \ldots \times \operatorname{Cov}\left(X_{i_2}\left(\overline{j_2}\right), X_{i_3}(j_3)\right) \operatorname{Cov}\left(X_{i_{k-1}}\left(\overline{j_{k-1}}\right), X_1(2)\right),
\end{aligned}$$

where the pair $\left(j_k, \overline{j_k}\right)$ *corresponds to a permutation of the numbers* $(1,2)$ *and the summation is taken over all the diagrams by the following construction. (See a particular case below for* $k = 4$*, Diagram 1.2.) Start from a pair,* $\left(X_1(1), X_1(2)\right)$ *say, and visit all the states* $\left(X_j(1), X_j(2)\right)$ *and arrive at* $\left(X_1(1), X_1(2)\right)$ *again. The number of such diagrams is* $2^{k-1}(k-1)!$ *because all possible orders of states are* $(k-1)!$ *and the connection between two states can be made by* 2×2 *different ways according to the permutations of their elements and the final step is determined. Therefore their number is* 2^{k-1}.

Example 11 *For any* $k \geq 1$

$$\operatorname{Cum}\left(H_2\left(X_1(1), X_1(2)\right), H_2\left(X_2(1), X_2(2)\right), \ldots, H_2\left(X_k(1), X_k(2)\right), Y_1\right) = 0.$$

Example 12

$$\begin{aligned}&\operatorname{Cum}\left(H_2\left(X_1(1), X_1(2)\right), \ldots, H_2\left(X_k(1), X_k(2)\right), Y_{(1:2)}\right)\\ &\quad=\sum_{2^k k!} \operatorname{Cov}\left(Y_1, X_{i_1}\left(j_1\right)\right) \operatorname{Cov}\left(X_{i_1}\left(\overline{j_1}\right), X_{i_2}\left(j_2\right)\right)\\ &\qquad\times \operatorname{Cov}\left(X_{i_2}\left(\overline{j_2}\right), X_{i_3}\left(j_3\right)\right) \ldots \operatorname{Cov}\left(X_{i_k}\left(\overline{j_k}\right), Y_2\right).\end{aligned}$$

The reasoning here is the same as in Example 10 and indices $\left(j_k, \overline{j_k}\right)$ are defined therein as well.

Example 13 *If $m \geq 3$ then*

$$\operatorname{Cum}\left(H_2(X_{(1,2)}), H_2(X_{(3,4)}), \ldots, H_2(X_{(k-1,k)}), Y_{(1:m)}\right)=0.$$

Example 14

$$\begin{aligned}&\operatorname{Cum}\left(H_3\left(Y_{(1:3)}\right), X_{(1:k)}\right)\\ &\qquad=\delta_{\{k=3\}} \sum_{3!} \operatorname{Cov}(Y_1, X_{i_1}) \operatorname{Cov}(Y_2, X_{i_2}) \operatorname{Cov}(Y_3, X_{i_3}).\end{aligned}$$

Example 15

$$\operatorname{Cum}\left(H_3\left(Y_{(1:3)}\right), H_2\left(X_{(1:2)}\right), H_1\left(X_3\right)\right)=\operatorname{Cum}\left(H_3\left(Y_{(1:3)}\right), X_{(1:3)}\right).$$

Example 16

$$\begin{aligned}&\operatorname{Cum}\left(H_3\left(Y_{(1:3)}\right), H_3\left(Z_{(1:3)}\right), X_{(1:2)}\right)\\ &\quad=\sum_{3!\cdot 2\cdot 3=36} \operatorname{Cov}(Y_1, Z_{i_1}) \operatorname{Cov}(Y_2, Z_{i_2}) \operatorname{Cov}(Y_3, X_{k_1}) \operatorname{Cov}(X_{k_2}, Z_{i_3}).\end{aligned}$$

The Figure 1.3 shows the structure of all possible closed diagrams without loops according to this example.

1.4.4 Products for Hermite polynomials

Remember that if $\mathcal{L}=(L_1, L_2, \ldots, L_p)$ is a partition of $(1:n)$, $X_{\mathcal{L}}$ denotes the vector of dimension p; $(\Pi X_{L_1}, \Pi X_{L_2}, \ldots, \Pi X_{L_p})$. Let us consider a partition $\mathcal{K}^{\mathrm{I,II}}$ with blocks having at most two elements. The diagram $(\mathcal{L}, \mathcal{K}^{\mathrm{I,II}})$ corresponds to a graph which might have free edges according to the blocs of $\mathcal{K}^{\mathrm{I,II}}$ with single elements. These free edges are called as arms of the diagram.$(\mathcal{L}, \mathcal{K}^{\mathrm{I,II}})$. Let $\mathcal{P}^{\mathrm{I,II}}_{(1:n)}$ denote the set of all partitions having one or two elements, i.e., $\mathcal{P}^{\mathrm{I,II}}_{(1:n)}=\left\{\mathcal{K}^{\mathrm{I,II}}\right\}$. The diagram, Figure 1.4 according to the partitions $\mathcal{L}=\{L_1=(1),\ L_2=(2,3,4)\}$ and $\mathcal{K}^{\mathrm{I,II}}=\{(1,2),(3),(4)\}$ of $(1:4)$ is closed and has two arms.

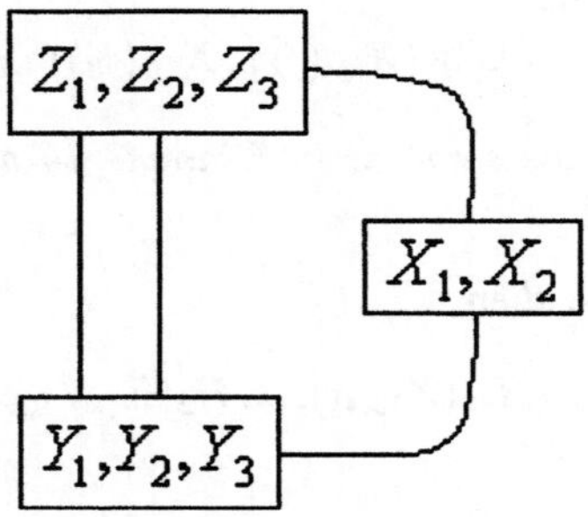

FIGURE 1.3. A closed diagram without loop

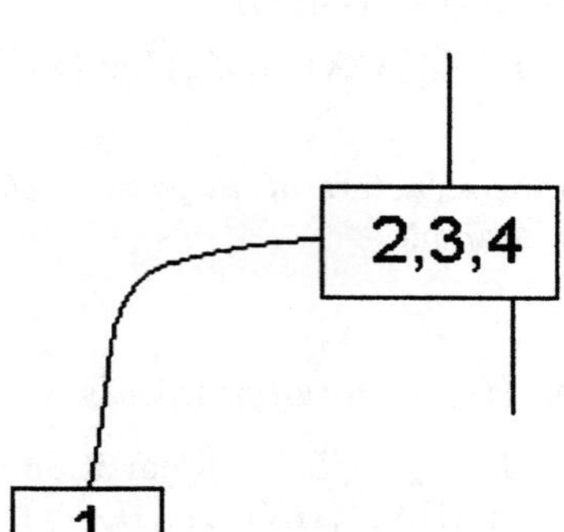

FIGURE 1.4. A closed diagram with two arms

The graph $(\mathcal{L}, \mathcal{K}^{\mathrm{I,II}})$, $\mathcal{K}^{\mathrm{I,II}} \in \mathcal{P}^{\mathrm{I,II}}_{(1:n)}$ has not only vertices $M \in \mathcal{L}$ and edges $(k_1, k_2) \in \mathcal{K}^{\mathrm{I,II}}$ but arms $(m) \in \mathcal{K}^{\mathrm{I,II}}$ as well. Put $d_{\mathcal{K}}$ for the number of arms in the partition $\mathcal{K}^{\mathrm{I,II}}$ and $D_{\mathcal{K}} = \{m \mid (m) \in \mathcal{K}^{\mathrm{I,II}}\}$ for the set of arms. Then

$$\Pi X_{(1:n)} = \sum_{\mathcal{K}^{\mathrm{I,II}}} \left(\prod_{(t_i,t_j)\in \mathcal{K}^{\mathrm{I,II}}} \sigma_{t_i t_j} \right) H_{d_{\mathcal{K}}}\left(X_{D_{\mathcal{K}}}\right), \tag{1.16}$$

where the sum is taken for all partition $\mathcal{K}^{\mathrm{I,II}} \in \mathcal{P}^{\mathrm{I,II}}_{(1:n)}$ and

$$X_{D_{\mathcal{K}}} = \{X_m \mid (m) \in \mathcal{K}^{\mathrm{I,II}}\}.$$

Note that it is necessary that the evenness of the number of arms $d_{\mathcal{K}}$ to be the same as the evenness of n, i.e., $d_{\mathcal{K}} \equiv n \bmod (2)$. Consider the case when $n = 4$. All the possible choices for $d_{\mathcal{K}}$ are 4, 2 and 0. If $d_{\mathcal{K}} = 4$, $\mathcal{K}^{\mathrm{I,II}}$ contains only arms, $\mathcal{K}^{\mathrm{I,II}} = \{(1), (2), (3), (4)\}$. If $d_{\mathcal{K}} = 2$, $\mathcal{K}^{\mathrm{I,II}}$ contains 2 arms and 1 pair, for example $\mathcal{K}^{\mathrm{I,II}} = \{(1), (2), (3,4)\}$. Now if $d_{\mathcal{K}} = 0$, $\mathcal{K}^{\mathrm{I,II}}$ contains only pairs, for example $\mathcal{K}^{\mathrm{I,II}} = \{(1,2), (3,4)\}$. We have

$$\begin{aligned}\prod X_{(1:4)} &= H_4\left(X_{(1:4)}\right) + \sigma_{12} H_2(X_3, X_4) + \sigma_{13} H_2(X_2, X_4) \\ &\quad + \sigma_{14} H_2(X_2, X_3) + \sigma_{23} H_2(X_1, X_4) + \sigma_{24} H_2(X_1, X_3) \\ &\quad + \sigma_{34} H_2(X_1, X_2) + \sigma_{12}\sigma_{34} + \sigma_{13}\sigma_{24} + \sigma_{14}\sigma_{23},\end{aligned}$$

and in particular,

$$X^4 = H_4(X) + 6\sigma^2 H_2(X) + 3\sigma^4.$$

It is worthwhile to mention the particular case when all the variables in (1.16) coincide, i.e., expressing X^n in terms of Hermite polynomials. We collect the coefficients according to $d_{\mathcal{K}}$, if $d_{\mathcal{K}} = k$ then there are $(n-k-1)!!$ partition available and $\binom{n}{k}$ possible choices. Therefore

$$\begin{aligned}X^n &= \sum_{k=0}^{n} (n-k-1)!! \binom{n}{k} H_k(X) \\ &= \sum_{d=0}^{\left[\frac{n}{2}\right]} \frac{n!}{(n-2d)!d!2^d} H_{n-2d}(X).\end{aligned} \tag{1.17}$$

This latter formula is usually referred to as the inversion formula for Hermite polynomials. We mention that $(n-k-1)!! = 0$ unless $n \equiv k, \bmod (2)$, so there are $\left[\frac{n}{2}\right]$ terms to add. These correspond to $n, n-2, \ldots, n-2d, \ldots, n-2\left[\frac{n}{2}\right]$, with coefficient

$$(2d-1)!! \binom{n}{n-2d} = \frac{n!}{(n-2d)!d!2^d}.$$

The Hermite polynomials $H_n(X_{(1:n)})$ can also be expressed in terms of the products $\prod X_{D_{\mathcal{K}}} = \prod_{(m)\in\mathcal{K}^{\mathrm{I,II}}} X_m$, i.e.,

$$H_n(X_{(1:n)}) = \sum_{(\mathcal{K}^{\mathrm{I,II}})} (-1)^{(n-d_{\mathcal{K}})/2} \left(\prod_{(t_i,t_j)\in\mathcal{K}^{\mathrm{I,II}}} \sigma_{t_i t_j} \right) \prod_{(m)\in\mathcal{K}^{\mathrm{I,II}}} X_m,$$

where the sum is taken for all the partitions $\mathcal{K}^{\mathrm{I,II}}$ for which the evenness of the number of arms $d_{\mathcal{K}}$ is the same as n, i.e., $d_{\mathcal{K}} \equiv n \bmod (2)$.

Take again a partition $\mathcal{L} = (L_1, L_2, ..., L_p)$ of $(1:n)$ with $|L_j| = n_j$. Then

$$\prod_{j=1}^{p} H_{n_j}(X_{L_j}) = \sum_{(\mathcal{L},\mathcal{K}^{\mathrm{I,II}})} \left(\prod_{(t_j,t_k)\in\mathcal{K}^{\mathrm{I,II}}} \sigma_{t_j t_k} \right) H_{d_{\mathcal{K}}}(X_{(m)\in\mathcal{K}^{\mathrm{I,II}}}), \tag{1.18}$$

where the summation is over all diagrams $(\mathcal{L}, \mathcal{K}^{\mathrm{I,II}})$ without loops. In particular if $p = 2$ and $L_1 = (1 : n-1)$, $L_2 = (n)$ then all the partitions of $\mathcal{P}^{\mathrm{I,II}}_{(1:n)}$ without loop are $\mathcal{K}^{\mathrm{I,II}} = \mathcal{O} = \{(1), (2), ..., (n)\}$, $\mathcal{K}^{\mathrm{I,II}} = \{(1), (2), ..., (j-1), (j,n), (j+1), ..., (n-1)\}$, $j = 1, 2, \ldots, n-1$, therefore

$$X_n H_{n-1}(X_{(1:n-1)}) = H_n(X_{(1:n)}) + \sum_{j=1}^{n-1} \sigma_{jn} H_{n-2}(X_{(1,2,\ldots,j-1,j+1,\ldots,n-1)}).$$

This formula has been considered as the recursion formula for the Hermite polynomials (see Recursion II, page 8).

We now present some Hermite equations to show the use of these formulae.

Example 17

$$H_2(X_1)H_2(X_2) = H_4(X_1, X_1, X_2, X_2) + 4\sigma_{12} H_2(X_1, X_2) + 2\sigma_{12}^2.$$

Example 18

$$\begin{aligned} H_4(X_1, X_1, X_2, X_2)H_2(X_3) = {} & H_6(X_1, X_1, X_2, X_2, X_3, X_3) \\ & + 2\sigma_{13} H_4(X_1, X_2, X_2, X_3) \\ & + 2\sigma_{23} H_4(X_1, X_1, X_2, X_3) \\ & + 2\sigma_{13}^2 H_2(X_2) + 2\sigma_{23}^2 H_2(X_1) \\ & + 8\sigma_{13}\sigma_{23} H_2(X_1, X_2). \end{aligned}$$

Example 19 *Remembering the notation* $X_{j[2]} = (X_j, X_j)$,

$$\begin{aligned}
H_6(X_{1[2]}, X_{2[2]}, X_{3[2]})H_2(X_4) &= H_8(X_{1[2]}, X_{2[2]}, X_{3[2]}, X_{4[2]}) \\
&+ 4\sigma_{14}H_6(X_1, X_{2[2]}, X_{3[2]}, X_4) \\
&+ 4\sigma_{24}H_6(X_{1[2]}, X_2, X_{3[2]}, X_4) \\
&+ 4\sigma_{34}H_6(X_{1[2]}, X_{2[2]}, X_3, X_4) \\
&+ 2\sigma_{14}^2 H_4(X_{2[2]}, X_{3[2]}) \\
&+ 2\sigma_{24}^2 H_4(X_{1[2]}, X_{3[2]}) \\
&+ 2\sigma_{34}^2 H_4(X_{1[2]}, X_{2[2]}) \\
&+ 8\sigma_{14}\sigma_{24}H_4(X_1, X_2, X_{3[2]}) \\
&+ 8\sigma_{14}\sigma_{34}H_4(X_1, X_{2[2]}, X_3) \\
&+ 8\sigma_{24}\sigma_{34}H_4(X_{1[2]}, X_2, X_3),
\end{aligned}$$

Example 20

$$\begin{aligned}
H_2(X_1)H_2(X_2)H_2(X_3) &= H_6(X_{1[2]}, X_{2[2]}, X_{3[2]}) \\
&+ 4\sigma_{23}H_4(X_{1[2]}, X_2, X_3) \\
&+ 4\sigma_{12}H_4(X_1, X_2, X_{3[2]}) \\
&+ 4\sigma_{13}H_4(X_1, X_{2[2]}, X_3) \\
&+ 8\sigma_{12}\sigma_{13}H_2(X_2, X_3) + 8\sigma_{13}\sigma_{23}H_2(X_1, X_2) \\
&+ 8\sigma_{12}\sigma_{23}H_2(X_1, X_3) + 2\sigma_{13}^2 H_2(X_2) \\
&+ 2\sigma_{23}^2 H_2(X_1) + 2\sigma_{12}^2 H_2(X_3) + 8\sigma_{12}\sigma_{13}\sigma_{23}.
\end{aligned}$$

Example 21

$$\begin{aligned}
\mathsf{E}H_2(X_1)H_2(X_2)H_2(X_3)H_2(X_4) &= 16\,[\sigma_{12}\sigma_{13}\sigma_{24}\sigma_{34} + \sigma_{12}\sigma_{23}\sigma_{14}\sigma_{24} \\
&+ \sigma_{23}\sigma_{13}\sigma_{24}\sigma_{14} + 4\left[\sigma_{13}^2\sigma_{24}^2 + \sigma_{23}^2\sigma_{14}^2 + \sigma_{12}^2\sigma_{34}^2\right].
\end{aligned}$$

This expectation corresponds to the term in (1.18) when $d_{\mathcal{K}} = 0$.

1.5 Stationary processes and spectra

1.5.1 Stochastic spectral representation

In this work a stochastic process X_t, $t \in \mathbb{Z}$, is called stationary if the expectation $\mathsf{E}X_t$ and the covariance function $\mathrm{Cov}(X_t, X_{t+s})$ are invariant under time shift, i.e., $\mathsf{E}X_t = \mathsf{E}X_0, \mathrm{Cov}(X_t, X_{t+s}) = \mathrm{Cov}(X_0, X_s)$. This type of stationarity is usually called weak stationarity or second order stationarity to make a difference between the cases when not only the first and second order moments are invariant but all the probability structure as well. This latter one is referred to as stationarity in the strict sense. If the process is Gaussian then these definitions of stationarity are equivalent. Let a stationary stochastic process X_t, with $\mathsf{E}X_t = 0$ and $\operatorname{Cov}(X_t, X_{t+s}) = C(s)$, be given. It is well known fact that the one has the representation of the covariance function as

$$C(s) = \int_0^1 e^{i2\pi s\omega}\Phi(d\omega),$$

and the process X_t as

$$X_t = \int_0^1 e^{i2\pi t\omega}W(d\omega),$$

for some spectral measure $\Phi(d\omega)$ and stochastic spectral measure $W(d\omega)$. The later is called the Cramér representation. These spectral measures are connected by the relationship

$$\mathsf{E}\,|W(d\omega)|^2 = \Phi(d\omega).$$

The spectral measure $\Phi(d\omega)$ is defined on the Borel sets of the real line. It is periodic, $\Phi(d\omega+1) = \Phi(d\omega)$, and symmetric about zero, $\Phi(-d\omega) = \Phi(d\omega)$, (only real valued processes are considered in this work). The stochastic spectral measure $W(d\omega)$ is a complex valued random measure on the Borel field $\mathcal{B}$ of the real line with the following properties:

- $\mathsf{E}W(A) = 0,\ A \in \mathcal{B}$,
- $\mathsf{E}W(A)\overline{W(B)} = \Phi(A \cap B)\ A, B \in \mathcal{B}$,
- $W(A) = \overline{W(-A)}$,
- $W(\sum_{k=1}^{\infty} A_k) = \sum_{k=1}^{\infty} W(A_k)$, if $A_1, A_2, \ldots, A_n \ldots$ are disjoint, the equality is meant by mean square sense.

A construction of $W(d\omega)$ from the process X_t is as follows, see [20], p. 101. Consider the discrete Fourier transform of X_t

$$d_{X,T}(\omega) = \sum_{t=-T}^{T} X_t e^{-i2\pi t\omega},$$

and

$$W_T([\omega_1, \omega_2)) = \int_{\omega_1}^{\omega_2} d_{X,T}(\omega)\,d\omega.$$

For simplicity put $W_T(\omega) = W_T([0,\omega))$, also $W(\omega) = W([0,\omega))$ then W_T tends to W in the mean of order ν for any $\nu > 0$, practiced the moments exist. This construction is reasonable because informally

$$W(d\omega) = \sum_{t=-\infty}^{\infty} X_t e^{-i2\pi t\omega}d\omega,$$

and the Cramér representation is straightforward.

The integration by the stochastic spectral measure $W(d\omega)$ is defined for all complex valued function g which are square integrable with respect to the spectral measure $\Phi(d\omega)$, i.e., for $g \in L^2_\Phi$. The integral

$$\int_0^1 g(\omega)W(d\omega),$$

is a real valued random variable if $g(\omega) = \overline{g(-\omega)}$. The relationships

$$\mathsf{E}\int_0^1 g(\omega)W(d\omega) = 0,$$

$$\mathsf{E}\int_0^1 g_1(\omega)W(d\omega)\overline{\int_0^1 g_2(\omega)W(d\omega)} = \int_0^1 g_1(\omega)\overline{g_2(\omega)}\Phi(d\omega),$$

hold true for this stochastic integral. The stochastic spectral measure $W(d\omega)$ according to a stationary process X_t is essentially unique. Starting with the stationary process X_t one can construct the stochastic spectral measure $W(d\omega)$ using linear operations only, see [20], [107], therefore if X_t is Gaussian, $W(d\omega)$ is Gaussian as well. Because $W(d\omega)$ is complex valued we mention, briefly, what is meant by a complex Gaussian system.

1.5.2 Complex Gaussian system

Consider a set $Z_{(1:n)}$ of random variables as a column vector $Z_{(1:n)} = (Z_1, Z_2, \ldots, Z_n)'$ where each component Z_k is complex valued, i.e., $Z_k = X_k + iY_k$. The mean

$$\mathsf{E}Z_{(1:n)} = \mathsf{E}X_{(1:n)} + i\mathsf{E}Y_{(1:n)},$$

and the covariance matrix

$$C_{ZZ} = \operatorname{Cov}(Z_{(1:n)}, Z_{(1:n)}) = \mathsf{E}\left(Z_{(1:n)} - \mathsf{E}Z_{(1:n)}\right)\overline{\left(Z_{(1:n)} - \mathsf{E}Z_{(1:n)}\right)'},$$

are complex valued when they exist. The vector $Z_{(1:n)}$ is called complex Gaussian if

- the real variables $\{X_{(1:n)}, Y_{(1:n)}\}$ form a Gaussian system
- the covariance matrix of the column vector $\left(X'_{(1:n)}, Y'_{(1:n)}\right)'$ is given by

$$\frac{1}{2}\begin{bmatrix} \operatorname{Re} C_{ZZ} & -\operatorname{Im} C_{ZZ} \\ \operatorname{Im} C_{ZZ} & \operatorname{Re} C_{ZZ} \end{bmatrix}.$$

This latter assumption leads, for example, to the following restriction: If $Z = X + iY$ is complex Gaussian random variable, the real part X and the imaginary part Y are independent with means $\operatorname{Re} \mathsf{E}Z$, $\operatorname{Im} \mathsf{E}Z$, respectively, and with same variance $C_{ZZ}/2$. Indeed $C_{ZZ} = \mathsf{E}\left(X^2 + Y^2\right)$ is real therefore X and Y are uncorrelated. It is easy to see also that for a complex Gaussian vector $Z_{(1:n)}$ it follows that

$$\mathsf{E}\left(Z_{(1:n)} - \mathsf{E}Z_{(1:n)}\right)\left(Z_{(1:n)} - \mathsf{E}Z_{(1:n)}\right)' = 0.$$

In general, a set of random variables with complex values will be called a complex Gaussian system if every finite dimensional subset forms a complex Gaussian vector. Two complex Gaussian vectors $Z_{(1:n)}$ and $V_{(1:m)}$ will be called independent if the $2n$ dimensional $\left(\operatorname{Re} Z'_{(1:n)}, \operatorname{Im} Z'_{(1:n)}\right)$ and the $2m$ dimensional $\left(\operatorname{Re} V'_{(1:m)}, \operatorname{Im} V'_{(1:m)}\right)$ Gaussian vectors are independent.

Some further properties of the stochastic Gaussian spectral measure $W(d\omega)$ include:

- for each $A \in \mathcal{B}$, $\operatorname{Re} W(A)$ and $\operatorname{Im} W(A)$ are independent with the same variance $\Phi(A)/2$,
- if $A_1 \cup -A_1, A_2 \cup -A_2, \dots A_n \cup -A_n$, are disjoint, then $W(A_1), W(A_2), \dots, W(A_n)$ are independent,
- if $A \cap -A = \emptyset$, then $W(A)$ and $W(-A)$ are independent.

1.5.3 *Spectra*

Suppose we have a stationary Gaussian stochastic process which is strictly stationary and has all the higher order moments; then a large class of processes transformed from this process retain these properties. If a strictly stationary process X_t has third order moments, then not only the covariances $\operatorname{Cov}(X_t, X_{t+s}) = C(s)$ are invariant with respect to the time-shift but the third order cumulants $\operatorname{Cum}(X_t, X_{t+r}, X_{t+s}) = C_3(r, s)$ as well.

Definition 22 *A stationary process X_t will be called stationary of kth order if all the cumulants exist up to k^{th} order and are invariant with respect to the time shift, i.e., for all $t \in T$, $l = 1, 2, \dots k$, and $t_{(1:l)}$, $t_j \in T$,*

$$\operatorname{Cum}\left(X_{t\vec{1}+t_{(1:l)}}\right) = \operatorname{Cum}\left(X_{t_{(1:l)}}\right).$$

For a stationary process of kth order the cumulants of lth $(l \le k)$ order depend on $l - 1$ variables only

$$\operatorname{Cum}\left(X_{t_{(1:l)}}\right) = \operatorname{Cum}\left(X_0, X_{t_{(2:l)} - t_1 \vec{1}}\right).$$

In particular the expectation is constant and the covariance depends only on the modulus of the difference of the time points

$$\mathrm{Cov}\left(X_t, X_{t+s}\right) = \mathrm{Cov}\left(X_0, X_{|s|}\right).$$

Assumption 23 *Besides the stationarity of kth order, we shall assume in the sequel that $\mathsf{E}X_t = 0$ and*

$$\sum_{\substack{t_j=-\infty \\ j=1,2,\ldots k-1}}^{\infty} \left|\mathrm{Cum}\left(X_0, X_{t_{(1:k-1)}}\right)\right| < \infty. \tag{1.19}$$

Assumption (1.19) implies that the Fourier transform of the cumulants exists is bounded and uniformly continuous. It will be convenient to use the notations $z = e^{i2\pi\omega}$, $z_k = e^{i2\pi\omega_k}$ such that $z_{(1:k)}^{-t_{(1:k)}} = e^{-i2\pi\sum t_j\omega_j}$.

Definition 24 *When it exists the Fourier transform*

$$S_k\left(\omega_{(1:k-1)}\right) = \sum_{\substack{t_j=-\infty \\ j=1,2,\ldots k-1}}^{\infty} \mathrm{Cum}\left(X_0, X_{t_{(1:k-1)}}\right) z_{(1:k-1)}^{-t_{(1:k-1)}}, \tag{1.20}$$

of the cumulant of a kth order stationary process X_t is called a kth order cumulant spectrum (or simply just a kth order spectrum).

In particular, the second order spectrum is usually simply called a spectrum; it is the spectral density of the process because under assumption (1.19) the spectral measure $\Phi(d\omega)$ of the process is absolute continuous with respect to the Lebesgue measure and S_2 is the corresponding Radon-Nikodym derivative of Φ.

Note that a kth order spectrum is a function of $k-1$ variables. There are some reasons to consider the kth order spectrum as a symmetric function of k variables and consider it on the hyperplane $\sum \omega_{(1:k)} = 0 \bmod (1)$,see [20] p. 26, [25] and [24], ($\bmod (1)$ is defined on page xix). Indeed, consider a linear filter of the form

$$Y_t = \sum_{m=-\infty}^{\infty} a_m X_{t-m},$$

such that along with the process X_t the process Y_t is k^{th} order stationary also. When it exists define the transfer function of this linear operator by

$$\alpha(z) = \sum_{m=-\infty}^{\infty} a_m z^{-m},$$

then the spectrum $S_{2,Y}$ of Y_t is given by

$$S_{2,Y}(z) = |\alpha(z)|^2 S_{2,X}(z).$$

In particular, when the X_t independent and identically distributed (i.i.d.) all spectra of the process X_t are constant and they are $\operatorname{Cum}(X_{0[k]})$. So

$$S_{2,Y}(z) = \sigma_X^2 |\alpha(z)|^2. \tag{1.21}$$

Clearly

$$\delta_1(\omega_1+\omega_2) S_{2,Y}(z_1,z_2) = \sigma_X^2 \delta_1(\omega_1+\omega_2)\alpha(z_1)\alpha(z_2),$$

is symmetric and equivalent to (1.21) where $\delta_1()$ is the Dirac comb. In general the kth order spectrum of a linear process has the form

$$\delta_1\left(\Sigma\omega_{(1:k)}\right) S_{k,Y}\left(z_{(1:k)}\right) = \operatorname{Cum}\left(X_{0[k]}\right)\delta_1\left(\Sigma\omega_{(1:k)}\right)\prod_{m=1}^{k}\alpha(z_m). \tag{1.22}$$

The symmetric versions of the spectra correspond to the symmetric property of the cumulants, i.e.,

$$\operatorname{Cum}\left(Y_{t\vec{1}+t_{(1:l)}}\right) = \operatorname{Cum}\left(Y_{t_{(1:l)}}\right),$$

for every $t \in T$ and the relationship

$$\operatorname{Cum}\left(Y_{t_{(1:l)}}\right) = \int_{[0,1]^l} z_{(1:l)}^{t_{(1:l)}}\delta_1\left(\Sigma\omega_{(1:l)}\right) S_{lY}\left(z_{(1:l)}\right) d\omega_{(1:l)},\ l = 1,2,\ldots k,$$

is valid.

It will generally not be confusing to denote the spectra defined by (1.20) either as a function of frequencies $\omega_{(1:k-1)}$ or function of exponents $z_{(1:k-1)}$ and either as a function of k or $k-1$ variables although these functions are not the same. It might be interesting to note that the symmetric versions of the spectra are not unique; for example the symmetric version $S_{2,Y}(z_1,z_2)$ of (1.21) can be either $\sigma_X^2\left(|\alpha(z_1)|^2+|\alpha(z_2)|^2\right)/2$ or $\sigma_X^2\alpha(z_1)\alpha(z_2)$. Both forms result in the same spectrum $S_{2,Y}(z)$ because they coincide on the hyperplane $\omega_1+\omega_2=0$. The first expression which can be obtained by the symmetrization of the spectrum, the second one follows from the property

$$\operatorname{Cum}(W(d\omega_1), W(d\omega_2)) = \delta_1(\omega_1+\omega_2)\sigma_X^2 d\omega_1 d\omega_2,$$

of the stochastic spectral measure $W(d\omega)$ of the independent series X_t. The higher order spectra can also be obtained this way, see [20] p. 100. Let $S_k\left(\omega_{(1:k-1)}\right)$ be the kth order spectra of X_t, then

$$\operatorname{Cum}(W(d\omega_1), W(d\omega_2), \ldots, W(d\omega_k)) = \delta_1\left(\Sigma\omega_{(1:k)}\right) S_k\left(z_{(1:k)}\right) d\omega_{(1:k)}.$$

If one starts with the symmetric version of the spectra then because of the presence of the Dirac comb

$$\begin{aligned} S_2(\omega_1) &= \delta_1(\omega_1+\omega_2) S_2(\omega_1,\omega_2) \\ &= S_2(\omega_1,-\omega_1) = S_2(-\omega_1,\omega_1) \\ &= S_2(\omega_1, 1-\omega_1) = S_2(1-\omega_1,\omega_1), \end{aligned}$$

and therefore

$$\begin{aligned} S_2(\omega) &= S_2(-\omega) \\ &= S_2(1-\omega), \end{aligned}$$

i.e., the spectrum is symmetric about *zero* and $1/2$, can be considered on the interval $[0,1/2]$. We shall frequently consider the third order spectrum S_3 called the bispectrum as well. While the spectrum is real and non-negative the bispectrum is generally complex valued. It has the following properties of symmetry,

$$\begin{aligned} S_3\left(\omega_{(1:2)}\right) &= \delta_1(\omega_1+\omega_2+\omega_3) S_3\left(\omega_{(1:3)}\right) = S_3(\omega_2,\omega_1) \\ &= S_3(\omega_1,-\omega_1-\omega_2) = S_3(-\omega_1-\omega_2,\omega_1) \\ &= S_3(-\omega_1-\omega_2,\omega_2) = S_3(\omega_2,-\omega_1-\omega_2) \\ &= \overline{S_3}\left(-\omega_{(1:2)}\right) \\ &= S_3(\omega_1+k,\omega_2+j), \quad k,j=1,2,\ldots . \end{aligned}$$

These equations imply that there are twelve triangles of frequencies in the plane, see Figure 1.5, each of which can be considered as the basic domain for the bispectrum for it is completely specified over the entire plane if it is determined over any one of the twelve triangles. We shall take the triangle with vertices $(0,0), (1/2,0), (1/3,1/3)$ as the basic domain for the bispectrum. The different shade of the triangles below means that the complex conjugation is necessary if one transform the bispectrum between them.

Example 25 *Brillinger [21] fitted an. AR(2) process to the data of Mississippi river run off, 1861-1880. The procedure of fitting will be considered later in Chapter 4, page 155. Here we give the bispectrum of a non-Gaussian AR(2) process with independent residuals and an estimate of the bispectrum. Specifically consider the process*

$$Y_t + \alpha_1 Y_{t-1} + \alpha_2 Y_{t-2} = e_t,$$

where e_t is an i.i.d. series. Suppose for example that $e_t = H_2(\mathrm{w}_t)$ with $\mathrm{w}_t \in \mathcal{N}(0,1)$ a Gaussian white noise series. Supposing $\alpha(z) \neq 0$, the bispectrum of Y_t is given by the general formula (1.22), i.e.,

$$S_{3,Y}(z_1,z_2) = \frac{\operatorname{Cum}(e_0,e_0,e_0)}{\alpha(z_1)\alpha(z_2)\alpha(z_3)}, \quad z_1 z_2 z_3 = 1.$$

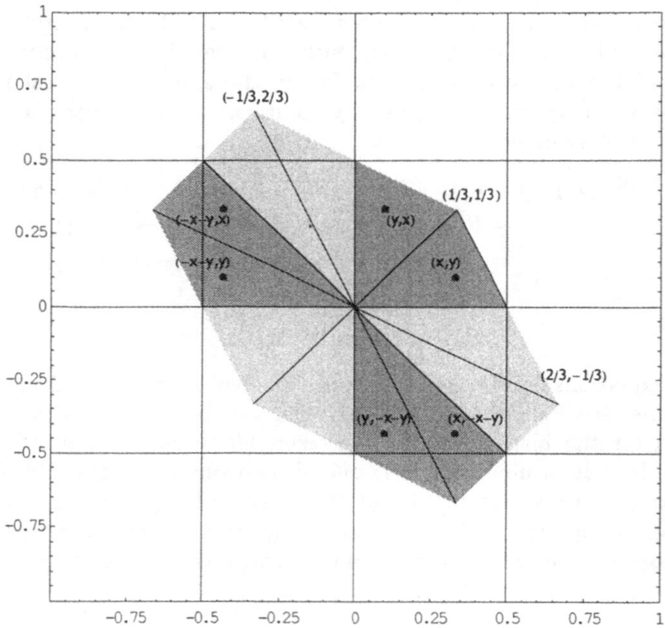

FIGURE 1.5. Twelve triangles of the bispectrum

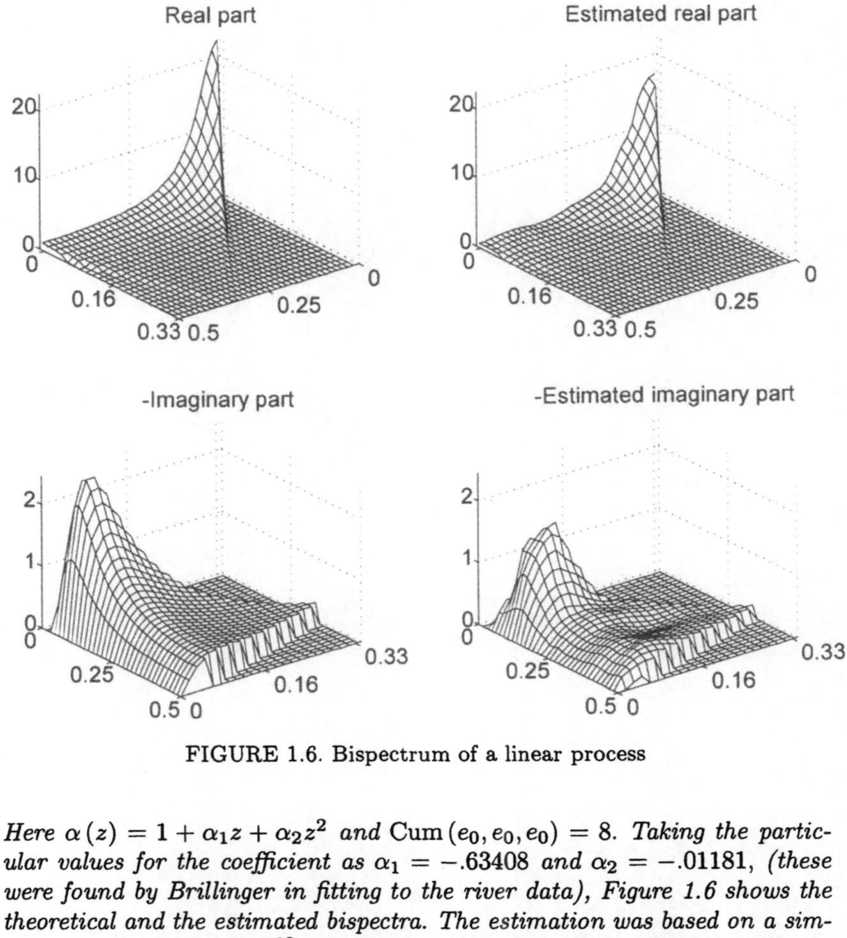

FIGURE 1.6. Bispectrum of a linear process

Here $\alpha(z) = 1 + \alpha_1 z + \alpha_2 z^2$ *and* $\operatorname{Cum}(e_0, e_0, e_0) = 8$. *Taking the particular values for the coefficient as* $\alpha_1 = -.63408$ *and* $\alpha_2 = -.01181$, *(these were found by Brillinger in fitting to the river data), Figure 1.6 shows the theoretical and the estimated bispectra. The estimation was based on a simulation of sample size* $2^{10} = 1024$ *and the frequencies are equidistant in interval* $[0, 1]$ *at spacing* 2^{-6}. *Note again that as the bispectrum is symmetric as above it is enough to consider the values on one of the twelve triangles. The values of the bispectrum are complex so we plot individually the real and the imaginary parts of* $S_{2,Y}(z_1, z_2) / \operatorname{Cum}(e_0, e_0, e_0)$. *For this particular case the triangle* $(0,0), (1/2,0), (1/3,1/3)$ *is chosen.*

We mention the works [92], [93] and [94] for the application of the cumulants and higher order spectra in signal processing.

2
The Multiple Wiener–Itô Integral

A basic tool of our investigations of nonlinear problems of time series analysis is the multiple Wiener–Itô integral. In the time domain Wiener started to investigate the stationary functionals of the Brownian motion processes in terms of higher order stochastic integrals. He developed the so called chaotic series representations [146]. The frequency domain analysis of stationary flows in the space of $\mathfrak{L}^2$ functionals of standard Wiener processes came into the picture when Rosenblatt's noncentral limit problem was studied in more general circumstances by Sinai[115], Dobrushin[33], Dobrushin and Major[34]. See in Major [78] for details. We give here some alternatives for the definition of multiple Wiener–Itô integral for a better understanding of this technique. The chaotic representation of a stationary subordinated processes will be considered as the generalization of the classic spectral representation of stationary random processes. Two particular cases are studied in detail. One is the closest possible process to the Gaussian stationary processes which has only first, i.e., Gaussian and second order multiple terms. The second is an appropriate function of a Gaussian stationary process.

2.1 Functions of spaces $\overline{L_\Phi^n}$ and $\widetilde{L_\Phi^n}$

The following Hilbert spaces will be important and are defined according to a spectral measure Φ. The correspondence of the spectral measure Φ to the stationary stochastic process X_t and to the stochastic spectral measure

W will be clear. Let $\overline{L_\Phi^n}$ denote the real Hilbert space of complex valued functions defined on $\mathbb{R}^n$ with the properties

- $f_n(\omega_1, \omega_2, \dots, \omega_n) = \overline{f_n(-\omega_1, -\omega_2, \dots, -\omega_n)}$,
- f_n is periodic with mod (1) in each variable, $f_n(\omega_1, \omega_2, \dots, \omega_k \pm 1, \dots, \omega_n) = f_n(\omega_1, \omega_2, \dots, \omega_k, \dots, \omega_n)$, $k = 1, 2, \dots, n$.
- $\|f_n\|_\Phi^2 = \int_0^1 |f_n(\omega_1, \omega_2, \dots, \omega_n)|^2 \Phi(d\omega_1)\Phi(d\omega_2)\dots\Phi(d\omega_n) < \infty$.

Roughly speaking, the functions in $\overline{L_\Phi^n}$ are Fourier transforms of real valued L_2 functions. The subspace $\widetilde{L_\Phi^n} \subset \overline{L_\Phi^n}$ consists of those functions $f_n \in \overline{L_\Phi^n}$ which are symmetric, i.e.,

- $f_n(\omega_1, \omega_2, \dots, \omega_n) = f_n\left(\omega_{\mathfrak{p}(1)}, \omega_{\mathfrak{p}(2)}, \dots, \omega_{\mathfrak{p}(n)}\right)$, for all $\mathfrak{p} \in \mathfrak{P}_n$, where $\mathfrak{P}_n$ denotes the set of all permutations of the numbers $(1:n) = (1, 2, \dots, n)$.

For every function $f_n \in \overline{L_\Phi^n}$ there corresponds an $\widetilde{f_n} \in \widetilde{L_\Phi^n}$, (symmetrized) version of f_n, defined by

$$\widetilde{f_n}(\omega_1, \omega_2, \dots, \omega_n) = \frac{1}{n!} \sum_{\mathfrak{p} \in \mathfrak{P}_n} f_n\left(\omega_{\mathfrak{p}(1)}, \omega_{\mathfrak{p}(2)}, \dots, \omega_{\mathfrak{p}(n)}\right).$$

It is denoted also by $f_r^{\mathfrak{P}}$. Sometimes we shall need to symmetrize a function f_n of variables $\omega_{(1:n)}$ by the variables $\omega_{(1:m)}$ where $n \le m$; in such cases it will be denoted by $\operatorname*{sym}_{\omega_{(1:m)}} f_n$, it belongs to $\widetilde{L_m^2}$ and is defined by

$$\operatorname*{sym}_{\omega_{(1:m)}} f_n(\omega_1, \omega_2, \dots, \omega_n) = \frac{1}{m!} \sum_{\mathfrak{p} \in \mathfrak{P}_m} f_n\left(\omega_{\mathfrak{p}(1)}, \omega_{\mathfrak{p}(2)}, \dots, \omega_{\mathfrak{p}(n)}\right).$$

It is easy to see that

$$\left\|\widetilde{f_n}\right\|_\Phi^2 \le \|f_n\|_\Phi^2 \text{ and } \left\|\operatorname*{sym}_{\omega_{(1:m)}} f_n\right\|_\Phi^2 \le c\|f_n\|_\Phi^2,$$

where constant c depends on the spectral measure Φ and n. Note that the spectral measure Φ is taken to correspond to a second order stationary process X_t and therefore $\operatorname{var}(X_t) = \Phi([0,1]) < \infty$.

For each fixed n we shall define the stochastic integral, as. a linear bounded (continuous) map from the Hilbert space $\overline{L_\Phi^n}$ into the Hilbert space of L_2-functionals of a stationary Gaussian process X_t. L_2-functionals of a process X_t are those random variables which are measurable with respect to the σ-algebra generated by the process X_t and have a finite variance.

2.2 The multiple Wiener–Itô integral of second order

If $n = 1$, then the stochastic integral for $g \in \overline{L_\Phi^1}$ is given by the formula

$$\mathrm{I}_1(g) = \int_0^1 g(\omega)W(d\omega).$$

$\mathrm{I}_1(g)$ is a real Gaussian variable and belongs to the Hilbert space $L_1^2(\mathcal{X})$ defined by all linear combination of X_t. Actually, $\mathrm{I}_1(\cdot)$ is an isometry from $\overline{L_\Phi^1}$ into $L_1^2(\mathcal{X})$, i.e.,

$$\|g(\omega)\|_\Phi^2 = \int_0^1 |g(\omega)|^2 \Phi(d\omega) = \mathrm{E}\,\mathrm{I}_1^2(g) = \|\mathrm{I}_1(g)\|^2 < \infty.$$

The linear transformation according to $g(\omega)$ can be derived directly. Set

$$g(\omega) = \sum_{k=-\infty}^{\infty} g_k e^{2\pi i k\omega},$$

then

$$\mathrm{I}_1(g) = \sum_{k=-\infty}^{\infty} g_k X_k,$$

where the convergence of the series is taken in the sense of L_2.

Consideration now turns to the case $n = 2$.

2.2.1 *Definition I*

Let a Gaussian stochastic spectral measure W with spectral measure Φ be given. In the space, $\overline{L_\Phi^2}$ the subspace of all finite linear combination of functions $\left\{g_{1k}(\omega_1)g_{1m}(\omega_2),\ g_{1k}, g_{1m} \in \overline{L_\Phi^1}\right\}$ is dense, i.e., every function $g_2(\omega_1, \omega_2) \in \overline{L_\Phi^2}$ can be approximated in L_2 norm by some linear combination of the functions of this type. Therefore it is enough to define the stochastic integral for a linear combination of the functions of the type $g_{1k}(\omega_1)g_{1m}(\omega_2)$. In other words the integral $\mathrm{I}_2\left(\sum_{k,m=1}^{K,M} a_{km} g_{1k} g_{1m}\right)$ is defined by

$$\begin{aligned}
&\mathrm{I}_2\left(\sum_{k,m=1}^{K,M} a_{km} g_{1k}(\omega_1) g_{1m}(\omega_2)\right) \\
&\quad = \sum_{k,m=1}^{K,M} a_{km} H_2\left(\int_0^1 g_{1k}(\omega)W(d\omega), \int_0^1 g_{1m}(\omega)W(d\omega)\right), \qquad (2.1)
\end{aligned}$$

where H_2 denotes the Hermite polynomial of second order. Note that it would be enough to define the integral above on the set of all finite linear combination of functions
$\{g_{1k}(\omega_1)g_{1m}(\omega_2) = e^{i2\pi(k\omega_1+m\omega_2)},\ k,m = \pm1,\pm2,\dots\}$ because this collection is also dense in $\overline{L_\Phi^2}$. We recall that H_2 depends on the covariances of the variables, specifically

$$\begin{aligned} H_2\left(\int_0^1 g_{11}(\omega)W(d\omega), \int_0^1 g_{12}(\omega)W(d\omega)\right) &= H_2\left(\mathrm{I}_1\left(g_{11}\right), \mathrm{I}_1\left(g_{12}\right)\right) \\ &= \mathrm{I}_1\left(g_{11}\right)\mathrm{I}_1\left(g_{12}\right) - \operatorname{Cov}\left(\mathrm{I}_1\left(g_{11}\right), \mathrm{I}_1\left(g_{12}\right)\right) \\ &= \mathrm{I}_1\left(g_{11}\right)\mathrm{I}_1\left(g_{12}\right) - \int_0^1 g_{11}(\omega)\overline{g_{12}(\omega)}\Phi(d\omega). \end{aligned}$$

This expression highlights the difference between the integral I_2 of product $g_{11}g_{12}$ and the ordinary product of I_1 integrals of functions g_{11} and g_{12}. As we have seen, the Hermite polynomial H_2 is bilinear, i.e.,

$$\begin{aligned} H_2\left(aX,Y\right) = H_2\left(X,aY\right) &= aH_2\left(X,Y\right), \\ H_2\left(X+Y,Z\right) &= H_2\left(X,Z\right) + H_2\left(Y,Z\right), \end{aligned}$$

showing how the integral I_2 follows for the linear combinations as well. Definition (2.1) implies that if

$$g_2\left(\omega_1,\omega_2\right) = \sum_{k,m=-\infty}^{\infty} g_{km}e^{i2\pi(k\omega_1+m\omega_2)},$$

then

$$\mathrm{I}_2\left(g_2\right) = \sum_{k,m=-\infty}^{\infty} g_{km}H_2\left(X_k,X_m\right). \tag{2.2}$$

Notice further that H_2 is symmetrical, therefore by definition (2.1)

$$\mathrm{I}_2\left(g_{11}(\omega_1)g_{12}(\omega_2)\right) = \mathrm{I}_2\left(g_{11}(\omega_2)g_{12}(\omega_1)\right) = \mathrm{I}_2\left(\operatorname{sym}\left(g_{11}(\omega_1)g_{12}(\omega_2)\right)\right).$$

The integral has zero mean by the definition (2.1) and the variance of the integral follows from (1.12)

$$\begin{aligned} \mathrm{E}\,\mathrm{I}_2^2\left(g_{11}g_{12}\right) &= \mathrm{E}\,H_2^2\left(\mathrm{I}_1\left(g_{11}\right), \mathrm{I}_1\left(g_{12}\right)\right) \\ &= \|g_{11}\|_\Phi^2\,\|g_{12}\|_\Phi^2 + \left(g_{11},g_{12}\right)_\Phi^2 \\ &= 2!\,\|\operatorname{sym}\left(g_{11}g_{12}\right)\|_\Phi^2 \\ &\le 2!\,\|g_{11}g_{12}\|_\Phi^2\,. \end{aligned}$$

The basic properties of the stochastic integral of second order for $g_2 \in \overline{L_\Phi^2}$ may now be summarized as for $g_2, h_2 \in \overline{L_\Phi^2}$,

1. $\mathrm{I}_2(g_2)$ is real valued,
2. $\mathrm{E}\,\mathrm{I}_2(g_2) = 0$,
3. $\mathrm{I}_2(g_2) = \mathrm{I}_2(\operatorname{sym} g_2)$,
4. $\mathrm{E}\,\mathrm{I}_2^2(g_2) = 2!\,\|g_2\|_\Phi$ for $g_2 \in \widetilde{L_\Phi^2}$ and $\mathrm{E}\,\mathrm{I}_2^2(g_2) \le 2!\,\|g_2\|_\Phi$ for $g_2 \in \overline{L_\Phi^2}$,
5. $\mathrm{E}\,\mathrm{I}_1(g_1)\mathrm{I}_2(g_2) = 0$ for any $g_1 \in \overline{L_\Phi^1}$ and $g_2 \in \overline{L_\Phi^2}$,
6.

$$\mathrm{E}\,\mathrm{I}_2(g_2)\,\mathrm{I}_2(h_2) = 2! \int_{[0,1]^2} \operatorname{sym} g_2(\omega_{(1:2)})\overline{\operatorname{sym} h_2(\omega_{(1:2)})} \times S_2(\omega_1)\, S_2(\omega_2)\, d\omega_1 d\omega_2$$

$$= 2! \int_{[0,1]^2} g_2(\omega_{(1:2)})\overline{\operatorname{sym} h_2(\omega_{(1:2)})} S_2(\omega_1)\, S_2(\omega_2)\, d\omega_1 d\omega_2.$$

We shall use also the notation

$$\mathrm{I}_2(g_2) = \int_{[0,1]^2} g_2(\omega_1, \omega_2) W(d\omega_1, d\omega_2).$$

To understand the connection between the stochastic Gaussian spectral measure $W(d\omega)$ and this $W(d\omega_1, d\omega_2)$ consider the following particular case. Suppose $A_1 \in \mathcal{B}$ and $A_2 \in \mathcal{B}$ are two subsets of the interval $[0,1]$, with $A_{-j} = 1 - A_j$, $j = 1, 2$, and such that $A_j \cap A_k = \emptyset$, if $j \neq \pm k$, $\quad j, k = \pm 1$, ± 2. Put

$$g_2(\omega_1, \omega_2) = g_{1,2}\chi_{A_1}(\omega_1)\chi_{A_2}(\omega_2) + g_{-1,-2}\chi_{A_{-1}}(\omega_1)\chi_{A_{-2}}(\omega_2) + g_{1,-2}\chi_{A_1}(\omega_1)\chi_{A_{-2}}(\omega_2) + g_{-1,2}\chi_{A_{-1}}(\omega_1)\chi_{A_2}(\omega_2),$$

where $g_{j,k} = a_j a_k$ and $\chi_A(\omega)$ is defined by $\operatorname{mod}(1)$, i.e.,

$$\chi_A(\omega) = \begin{cases} 1 & \text{if } \omega \in A, \operatorname{mod}(1), \\ 0 & \text{otherwise.} \end{cases},$$

($\operatorname{mod}(1)$ is defined on page xix). If we assume that $g_2 \in \overline{L_\Phi^2}$, then

$$g_2(\omega_1, \omega_2) = \overline{g_2(-\omega_1, -\omega_2)},$$

therefore $g_{j,k} = \overline{g_{-j,-k}}$, $\quad j \neq \pm k$, $\quad j, k = \pm 1, \pm 2$. As I_2 is additive one has

$$\mathrm{I}_2(g_2) = \int_{[0,1]^2} \left(a_1\chi_{A_1}(\omega_1) + a_{-1}\chi_{A_{-1}}(\omega_1)\right) \times \left(a_2\chi_{A_2}(\omega_2) + a_{-2}\chi_{A_{-2}}(\omega_2)\right) W(d\omega_1, d\omega_2)$$

$$= H_2\left(a_1 W(A_1) + a_{-1} W(A_{-1}), a_2 W(A_2) + a_{-2} W(A_{-2})\right)$$
$$= \left(a_1 W(A_1) + a_{-1} W(A_{-1})\right)\left(a_2 W(A_2) + a_{-2} W(A_{-2})\right),$$

since $a_1 W(A_1) + a_{-1} W(A_{-1})$ and $a_2 W(A_2) + a_{-2} W(A_{-2})$ are real valued and uncorrelated.

2.2.2 Definition II

Now we begin by defining a particular dense subclass $\widehat{\overline{L_\Phi^2}}$ of $\overline{L_\Phi^2}$. Let $\{A_j, j = \pm 1, \pm 2, \ldots \pm n\}$ be a finite set of intervals in $\mathbb{R}$. It is called a regular system if $A_j = -A_{-j}$, and $A_j \cap A_k = \emptyset$ when $j \neq k$ for all $j, k = \pm 1, \pm 2, \ldots \pm n$. A function $h \in \overline{L_\Phi^2}$ is adapted to this system if it is constant on the sets $A_j \times A_k$, $j, k = \pm 1, \pm 2, \ldots \pm n$ and vanishes outside these sets and on the sets $A_j \times A_k$, $j = \pm k$. A function $h \in \widehat{\overline{L_\Phi^2}}$ if $h \in \overline{L_\Phi^2}$ and there exists a regular system such that h is adapted to it. The definition of the stochastic integral I_2 for $h \in \widehat{\overline{L_\Phi^2}}$ is then the following

$$\begin{aligned} \mathrm{I}_2(h) &= \int h(\omega_1, \omega_2) W(d\omega_1, d\omega_2) \\ &= \sum_{j,k=\pm 1, \pm 2, \ldots \pm n} h(\omega_j, \omega_k) W(A_j) W(A_k), \end{aligned}$$

where $\omega_l \in A_l$, $l = \pm 1, \pm 2, \ldots \pm n$, see Major [78] for more details. The restriction of this integral to the square $[0, 1]^2$ gives the same result as the previous definition. Definition I and Definition II are connected by the so called Itô formula , which we shall give later.

2.2.3 Definition III

In the construction of $W(d\omega_1, d\omega_2)$ from the process X_t we might follow the previous one dimensional case. Consider the discrete Fourier transform of $H_2(X_t, X_s)$

$$d_{XX,T_1T_2}(\omega_1, \omega_2) = \sum_{t_1=-T_1}^{T_1} \sum_{t_2=-T_2}^{T_2} H_2(X_s, X_t) e^{-i2\pi(s\omega_1 + t\omega_2)},$$

and

$$W_{T_1T_2}(\omega_1, \omega_2) = \int_0^{\omega_1} \int_0^{\omega_2} d_{XX,T_1T_2}(\omega_1, \omega_2) d\omega_1 d\omega_2,$$

where $W_{T_1T_2}(\omega_1, \omega_2)$ is used for $W_{T_1T_2}([0, \omega_1) \times [0, \omega_2))$. Now the cumulant

$$\operatorname{Cum}(W_{T_1T_2}(\omega_1, \omega_2), W_{T_1T_2}(\lambda_1, \lambda_2)),$$

has the symbolic limit

$$\int_0^{\omega_1}\int_0^{\omega_2}\int_0^{\lambda_1}\int_0^{\lambda_2} \delta_1(\omega_1+\lambda_1)\,\delta_1(\omega_2+\lambda_2)+\delta_1(\omega_1+\lambda_2)$$

$$\times\,\delta_1(\omega_2+\lambda_1)\,S_2(\lambda_1)\,S_2(\lambda_2)\,d\omega_1 d\omega_2 d\lambda_1 d\lambda_2, \tag{2.3}$$

as $T_1, T_2 \to \infty$, where $S_2(\lambda)$ is the spectral density of the process X_t. From (2.3) it follows easily that $W_{T_1T_2}$ is a Cauchy sequence and therefore converges in the mean square to some stochastic measure say W_0. Some algebra leads to the Cramér representation of $H_2(X_t, X_s)$, i.e.,

$$H_2(X_{t_1}, X_{t_2}) = \int_{[0,1]^2} e^{i2\pi(t_1\omega_1+t_2\omega_2)} W_0(d\omega_1, d\omega_2). \tag{2.4}$$

The two stochastic integrals (2.1) (2.4) coincide on a dense subset of functions so $W(d\omega_1, d\omega_2)$ can be represented by $W_0(d\omega_1, d\omega_2)$. The expression (2.3) gives another explanation of the necessity of symmetrization in evaluating covariances of multiple integrals. In fact, from (2.3) it follows that

$$\begin{aligned}\mathrm{E}\left(W(d\omega_1, d\omega_2)\overline{W(d\lambda_1, d\lambda_2)}\right) &= \left[\delta_1(\omega_1-\lambda_1)\,\delta_1(\omega_2-\lambda_2)+\delta_1(\omega_1-\lambda_2)\,\delta_1(\omega_2-\lambda_1)\right]\\ &\qquad\times S_2(\lambda_1)\,S_2(\lambda_2)\,d\omega_1 d\omega_2 d\lambda_1 d\lambda_2.\end{aligned}$$

Taking the expectation of the product of integrals

$$\mathrm{E}\,\mathrm{I}_2(g_2)\,\overline{\mathrm{I}_2(h_2)},$$

one obtains the expression

$$\begin{aligned}\int_{[0,1]^4} & g_2(\omega_1, \omega_2)\,\overline{h_2(\lambda_1, \lambda_2)}\\ &\times\left[\delta_1(\omega_1-\lambda_1)\,\delta_1(\omega_2-\lambda_2)+\delta_1(\omega_1-\lambda_2)\,\delta_1(\omega_2-\lambda_1)\right]\\ &\times S_2(\lambda_1)\,S_2(\lambda_2)\,d\omega_1 d\omega_2 d\lambda_1 d\lambda_2,\end{aligned}$$

to be evaluated. The result is

$$\begin{aligned}\int_{[0,1]^2} & \left[g_2(\lambda_1, \lambda_2)\,\overline{h_2(\lambda_1, \lambda_2)}+g_2(\lambda_2, \lambda_1)\,\overline{h_2(\lambda_1, \lambda_2)}\right]\\ &\qquad\times S_2(\lambda_1)\,S_2(\lambda_2)\,d\lambda_1 d\lambda_2\\ &= 2!\int_{[0,1]^2}\operatorname{sym}\left[g_2(\lambda_1, \lambda_2)\right]\overline{h_2(\lambda_1, \lambda_2)}S_2(\lambda_1)\,S_2(\lambda_2)\,d\lambda_1 d\lambda_2,\end{aligned}$$

which is just the same as

$$2!\int_{[0,1]^2}\operatorname{sym} g_2(\lambda_1, \lambda_2)\,\overline{\operatorname{sym} h_2(\lambda_1, \lambda_2)}S_2(\lambda_1)\,S_2(\lambda_2)\,d\lambda_1 d\lambda_2.$$

Note here that it is the symmetry of the spectral measure $S_2(\lambda_1)\,S_2(\lambda_2)\,d\lambda_1 d\lambda_2$ that has made this last step possible.

2.3 The multiple Wiener–Itô integral of order n

We proceed with the general case. Functions of the space $\overline{L^n_\Phi}$ are approximated by linear combinations of the products $\prod g_{1k_{(1:n)}}$, where each $g_{1k_j} \in \overline{L^1_\Phi}$. The multiple Wiener–Itô integral of order n, $\mathrm{I}_n\left(\sum_{k_{(1:n)}=1}^{K_{(1:n)}} a_{k_{(1:n)}} \prod g_{1k_{(1:n)}}\right)$, is defined by the equation

$$\mathrm{I}_n\left(\sum_{k_{(1:n)}=\vec{1}}^{K_{(1:n)}} a_{k_{(1:n)}} \prod_{j=1}^{n} g_{1k_j}(\omega_j)\right) = \sum_{k_{(1:n)}=\vec{1}}^{K_{(1:n)}} a_{k_{(1:n)}} \tag{2.5}$$

$$\times H_n\left(\int_0^1 g_{1k_1}(\omega)W(d\omega), \int_0^1 g_{1k_2}(\omega)W(d\omega), \ldots, \int_0^1 g_{1k_n}(\omega)W(d\omega)\right),$$

where H_n is the Hermite polynomial of order n. In particular, if we put

$$\prod_{j=1}^{n} g_{1k_j}(\omega_j) = e^{i2\pi\left(\sum_j k_j\omega_j\right)},$$

we get

$$\mathrm{I}_n\left(e^{i2\pi\left(\sum_j k_j\omega_j\right)}\right) = H_n\left(X_{k_{(1:n)}}\right).$$

The notation

$$\mathrm{I}_n(f_n) = \int_{[0,1]^n} f_n(\omega_1, \omega_2, \ldots, \omega_n)\, W(d\omega_1, d\omega_2, \ldots, d\omega_n),$$

for the nth order integral will also be used. The Itô formula is also a special case of the definition. Let $h_1, h_2, \ldots, h_m$, $h_j \in \overline{L^2_1}$, $j = 1, 2, \ldots, m$ be an orthogonal system and let a fixed vector of integers $k_{(1:n)} \in \mathbb{Z}^n$ be given. Put $\sum k_{(1:n)} = K$. The random variables $Y_j = \int_0^1 h_j(\omega)W(d\omega)$, $j = 1, 2, \ldots, m$ are orthogonal and Gaussian therefore

$$H_K\left(Y_{1[k_1]}, Y_{2[k_2]}, \ldots, Y_{m[k_m]}\right) = \prod_{j=1}^{m} H_{k_j}\left(Y_{j[k_j]}\right).$$

Here we used the notation $Y_{j[k_j]}$ for the repeated variable, i.e., $Y_{j[k_j]} = \underbrace{(Y_j, Y_j, \ldots, Y_j)}_{k_j}$ is a k_j dimensional vector. Now by the definition of integral I_n we get

$$\prod_{j=1}^{m} H_{k_j}\left(Y_{j[k_j]}\right) = \mathrm{I}_K\left(\prod_{l=1}^{K} g_l(\omega_l)\right),$$

where $g_l = h_j$, for $\sum_1^{j-1} k_i < l \leq \sum_1^j k_i$. If H_{k_j} denotes the Hermite polynomial with one variable of order k_j, we get the Itô formula in its original form

$$\prod_{j=1}^{m} H_{k_j}\left(\int_0^1 h_j(\omega)W(d\omega)\right) = \mathrm{I}_K\left(\prod_{l=1}^{K} g_l(\omega_l)\right)$$
$$= \mathrm{I}_K\left(\operatorname{sym}\prod_{l=1}^{K} g_l(\omega_l)\right).$$

2.3.1 Properties

Let $g_n \in \overline{L^n_\Phi}$ then

1. $\mathrm{I}_n(g_n)$ is real valued,
2. $\mathrm{E}\,\mathrm{I}_n(g_n) = 0$,
3. $\mathrm{I}_n(g_n) = \mathrm{I}_n(\operatorname{sym} g_n)$,
4. $\mathrm{E}\,\mathrm{I}_n^2(g_n) = n!\,\|g_n\|$ for $g_n \in \widetilde{L^n_\Phi}$ and $\mathrm{E}\,\mathrm{I}_n^2(g_n) \leq n!\,\|g_n\|$ for $g_n \in \overline{L^n_\Phi}$,
5. If $m \neq n$ then $\mathrm{E}\,\mathrm{I}_m(g_m)\mathrm{I}_n(g_n) = 0$ for any $g_m \in \overline{L^2_m}$ and $g_n \in \overline{L^n_\Phi}$,
6. for $g_n, h_n \in \overline{L^n_\Phi}$,

$$\mathrm{E}\,\mathrm{I}_n(g_n)\,\mathrm{I}_n(h_n) = n!\int_{[0,1]^n} \operatorname{sym} g_n(\omega_{(1:n)})\overline{\operatorname{sym} h_n(\omega_{(1:n)})} \times \prod_{i=1}^{n} S_2(\omega_i)\,d\omega_{(1:n)}$$
$$= n!\int_{[0,1]^n} g_n(\omega_{(1:n)})\overline{\operatorname{sym} h_n(\omega_{(1:n)})} \times \prod_{i=1}^{n} S_2(\omega_i)\,d\omega_{(1:n)}.$$

2.3.2 Diagram formula

One of the basic rules of manipulating with multiple Wiener–Itô integral called Diagram formula. It originates from the product of Hermite polynomials. We have seen that the product of Hermite polynomials is expressed by the linear combination of Hermite polynomials, again see (1.18), which is very natural, but the explicit expression of the coefficients of this linear combination is more interesting. Start with the simplest case of taking the product of the H_1 and H_{n-1}.

$$H_1\left(X_{k_n}\right) H_{n-1}(X_{k_{(1:n-1)}})$$
$$= H_n(X_{k_{(1:n)}}) + \sum_{j=1}^{n-1} \sigma_{k_j k_n} H_{n-2}(X_{k_{(1,2,\ldots,j-1,j+1,\ldots,n-1)}}),$$

where $\sigma_{k_j k_n} = \operatorname{Cov}\left(X_{k_j}, X_{k_n}\right) = C\left(|k_j - k_n|\right)$ because of the stationarity of X_t. This may be rewritten by multiple integrals as

$$\int_0^1 e^{i2\pi k_n \omega} W(d\omega) \int_{[0,1]^{n-1}} e^{i2\pi\left(\sum_{l=1}^{n-1} k_l \omega_l\right)} W(d\omega_{(1:n-1)})$$
$$= \int_{[0,1]^n} e^{i2\pi\left(\sum_{l=1}^{n} k_l \omega_l\right)} W(d\omega_{(1:n)}) + \sum_{j=1}^{n-1} \int_{[0,1]^{n-2}} \int_0^1 e^{i2\pi\left(\sum_{l=1}^{n-1} k_l \omega_l\right)}$$
$$\times e^{-i2\pi k_n \omega_j} \Phi\left(d\omega_j\right) W(d\omega_{(1:j-1)}, d\omega_{(j+1:n-1)}). \tag{2.6}$$

We introduce a function corresponding to the deterministic integral of the right hand side in (2.6). Let $h_1 \in \overline{L^1_\Phi}$, $g_{n-1} \in \overline{L^{n-1}_\Phi}$ and define the function $h_1 \boxtimes_j g_{n-1}$ of $n-2$ variables by

$$\left[h_1 \boxtimes_j g_{n-1}\right]\left(\omega_{(1:j-1)}, \omega_{(j+1:n-1)}\right) = \int_0^1 g_{n-1}\left(\omega_{(1:n-1)}\right) \overline{h_1\left(\omega_j\right)} \Phi\left(d\omega_j\right).$$

It is easy to see that $h_1 \boxtimes_j g_{n-1} \in \overline{L^{n-2}_\Phi}$. The formula for the product $\mathrm{I}_1\left(h_1\right) \mathrm{I}_{n-1}\left(g_{n-1}\right)$ is now straightforward consequence of (2.6)

$$\int_0^1 h_1(\omega) W(d\omega) \int_{[0,1]^{n-1}} g_{n-1}\left(\omega_{(1:n-1)}\right) W(d\omega_{(1:n-1)})$$
$$= \int_{[0,1]^n} h_1\left(\omega_n\right) g_{n-1}\left(\omega_{(1:n-1)}\right) W(d\omega_{(1:n)})$$
$$+ \sum_{j=1}^{n-1} \int_{[0,1]^{n-2}} \left[h_1 \boxtimes_j g_{n-1}\right]\left(\omega_{(1:j-1)}, \omega_{(j+1:n-1)}\right)$$
$$\times W(d\omega_{(1:j-1)}, d\omega_{(j+1:n-1)}). \tag{2.7}$$

Note that $h_1 g_{n-1} \in \overline{L^n_\Phi}$.

Some more formulation is necessary for the general diagram formula. Consider a partition $\mathcal{L} = (L_1, L_2, \ldots, L_p)$ of $(1:n)$ with $|L_j| = n_j$ and $\mathcal{K}^{\mathrm{I,II}}$ a partition with blocks having at most two elements, see Section 1.4.1, p. 15. Denote the number of arms $d_{\mathcal{K}}$ according to the blocks $\{(k_i), i = 1, 2, \ldots, d_{\mathcal{K}}\}$ with one element in the diagram $\left(\mathcal{L}, \mathcal{K}^{\mathrm{I,II}}\right)$ without a loop such that the number of edges (k_j, k_m) is $N = (n - d_{\mathcal{K}})/2$. Let $h_l \in \overline{L_\Phi^{n_l}}$, $l = 1, 2, \dots, p$ and define the function

$$\left[\boxtimes_{(\mathcal{L},\mathcal{K}^{\mathrm{I},\mathrm{II}})} h_l\right] \left(\omega_{k_i}, i = 1, 2, \dots, d_{\mathcal{K}}\right)$$
$$= \int_{[0,1]^{n-d_{\mathcal{K}}}} \prod_{i=1}^{p} h_i\left(\omega_{L_i}\right) \prod_{(k_j,k_m)\in\mathcal{K}^{\mathrm{I},\mathrm{II}}} \delta_1\left(\omega_{k_j}+\omega_{k_m}\right) \Phi\left(d\omega_{k_j}\right) \Phi\left(d\omega_{k_m}\right),$$

where $\delta_1()$ is the periodic Dirac delta. Then put

$$\lambda_{(1:d_{\mathcal{K}})} = \left(\omega_{k_i}, i = 1, 2, \dots, d_{\mathcal{K}}\right),$$

so

$$\prod_{i=1}^{p} \int_{[0,1]^{n_i}} h_i\left(\omega_{L_i}\right) W\left(d\omega_{L_i}\right) \tag{2.8}$$
$$= \sum_{\mathcal{K}^{\mathrm{I},\mathrm{II}}} \int_{[0,1]^{d_{\mathcal{K}}}} \left[\boxtimes_{(\mathcal{L},\mathcal{K}^{\mathrm{I},\mathrm{II}})} h_l\right] \left(\lambda_{(1:d_{\mathcal{K}})}\right) W\left(\lambda_{(1:d_{\mathcal{K}})}\right),$$

where the summation is taken for all partitions $\mathcal{K}^{\mathrm{I},\mathrm{II}}$ such that the diagram $(\mathcal{L}, \mathcal{K}^{\mathrm{I},\mathrm{II}})$ has no loops. Formula (2.8) is called the diagram formula. It is a generalization of (2.7) and the formula (1.18) concerning the product for Hermite polynomials. If $d_{\mathcal{K}} = 0$ then the integrand is constant and this can only happen if n is even. In this case $\mathcal{K}^{\mathrm{I},\mathrm{II}}$ includes only pairs and the product $\boxtimes_{(\mathcal{L},\mathcal{K}^{\mathrm{I},\mathrm{II}})} h_l$ is constant. Therefore we introduce the notation

$$\left[\boxtimes_{\mathcal{K}^{\mathrm{II}}} h_l\right] = \int_{[0,1]^{n/2}} \prod_{i=1}^{p} h_i\left(\omega_{L_i}\right) \prod_{(k_j,k_m)\in\mathcal{K}^{\mathrm{II}}} \delta_1\left(\omega_{k_j} + \omega_{k_m}\right) \Phi\left(d\omega_{k_j}\right).$$

A consequence of the diagram formula is that the expected value of the left hand side is zero if n is odd, otherwise

$$\mathrm{E} \prod_{i=1}^{p} \int_{[0,1]^{n_i}} h_i\left(\omega_{L_i}\right) W\left(d\omega_{L_i}\right) = \sum_{\left\{ \mathcal{K}^{\mathrm{II}} \,\middle|\, \substack{(\mathcal{L},\mathcal{K}^{\mathrm{II}}) \\ \text{without loop}} \right\}} \left[\boxtimes_{\mathcal{K}^{\mathrm{II}}} h_l\right],$$

where the summation is taken by all partitions $\mathcal{K}^{\mathrm{II}}$ of pairs from $(1:n)$ such that the diagram $(\mathcal{L}, \mathcal{K}^{\mathrm{II}})$ has no loops.

The cumulant of the multiple integrals has fewer of terms because the summation is taken only by the diagrams $(\mathcal{L}, \mathcal{K}^{\mathrm{II}})$ which are closed and have no loops

$$\operatorname{Cum}\left(\left\{\int_{[0,1]^{n_i}} h_i\left(\omega_{L_i}\right) W\left(d\omega_{L_i}\right), L_i \in \mathcal{L}\right\}\right) \tag{2.9}$$

$$= \sum_{\left\{ \mathcal{K}^{\mathrm{II}} \,\middle|\, \begin{array}{l}(L,K^{\mathrm{II}})\text{ closed,}\\ \text{without loop}\end{array} \right\}} \left[\boxtimes_{(\mathcal{L},\mathcal{K}^{\mathrm{II}})} h_l\right].$$

2.3.3 Fock space

We have seen that for symmetric functions $g_n \in \widetilde{L_\Phi^n}$ the $\mathrm{E}\,\mathrm{I}_n(g_n) = 0$ and $\mathrm{E}\,\mathrm{I}_n^2(g_n) = n!\,\|g_n\|_\Phi^2$. So the integral $(n!)^{-1/2}\,\mathrm{I}_n$ provides a unitary transformation between the space $\widetilde{L_\Phi^n}$ and the linear Hilbert space $L_n^2(\mathcal{X})$ generated by all possible Hermite polynomials with order n of the system $\mathcal{X} = \{X_t,\, t \in \mathbb{Z}\}$. Recall that $\{X_t\}$ denotes a Gaussian stationary process with spectral measure Φ and stochastic spectral measure W. The space of the series $g = (g_0, g_1, g_2, \ldots, g_n, \ldots)$ of functions $g_n \in \widetilde{L_\Phi^n}$ is called Fock space if

$$\|g\|^2 = \sum_{n=0}^{\infty} n!\,\|g_n\|_\Phi^2 < \infty,$$

where g_0 is constant and $\widetilde{L_\Phi^0}$ contains only constants. The Fock space is denoted by $\operatorname{Exp}\left(\widetilde{L_\Phi^2}\right)$. For each $g \in \operatorname{Exp}\left(\widetilde{L_\Phi^2}\right)$ define the multiple integral

$$\mathrm{I}(g) = \sum_{n=0}^{\infty} \mathrm{I}_n(g_n).$$

This integral is a unitary transformation from Fock space $\operatorname{Exp}\left(\widetilde{L_\Phi^2}\right)$ into the Hilbert space $\mathfrak{L}^2(\mathcal{X})$ of all random variables with finite variances and measurable with respect to the process $\mathcal{X} = \{X_t\}$. It follows from the definitions that if $g \in \operatorname{Exp}\left(\widetilde{L_\Phi^2}\right)$, $\mathrm{I}(g) \in \mathfrak{L}^2(\mathcal{X})$. The converse of this statement is more interesting. If $Y \in \mathfrak{L}^2(\mathcal{X})$, it can be put into the form of a multiple integral

$$Y = \sum_{n=0}^{\infty} \mathrm{I}_n(g_n),$$

and the corresponding functions $g_n \in \widetilde{L_\Phi^n}$ are unique, see Major [78] for more detail.

2.3.4 Stratonovich integral in frequency domain and the Hu–Meyer formula

An n-fold stochastic measure can be obtained by standard measure theory methods by taking the n times product of stochastic spectral measures $W(d\omega)$ with spectral measure Φ. Formally one may start with the integral of the functions of type $e^{i2\pi(\sum_j k_j\omega_j)}$, i.e.,

$$\mathrm{I}_n^S\left(e^{i2\pi(\sum_j k_j\omega_j)}\right) = \prod X_{k_{(1:n)}},$$

where

$$X_t = \int_0^1 e^{i2\pi t\omega} W(d\omega),$$

as usual. Notice that the right hand side is symmetric therefore the integrand can also be symmetrized. Now define I_n^S, for all functions $g_n \in \widetilde{L_\Phi^n}$, by the L_2 limit

$$\mathrm{I}_n^S(g_n) = \int_{[0,1]^n} g_n(\omega_1, \omega_2, \ldots, \omega_n) W(d\omega_1) W(d\omega_2) \cdots W(d\omega_n).$$

This type of stochastic integration is usually referred as a Stratonovich stochastic integral. It is different from the n-fold Wiener–Itô integral defined above. The main difference is the contribution of the diagonals to the integrals. Formula (1.16) gives the connection between the Stratonovich and Wiener–Itô integrals. If $g_n \in \widetilde{L_\Phi^n}$,

$$\mathrm{I}_n^S(g_n) = \sum_{d=0}^{[\frac{n}{2}]} \frac{n!}{(n-2d)!d!2^d} \mathrm{I}_{n-2d}(\mathsf{Tr}_d[g_n]), \tag{2.10}$$

where the operator $\mathsf{Tr}_d(g_n)$ is the trace of order d of the transfer function g_n. The coincidence of d pairs of coordinates according to the spectral measure Φ gives rise to the trace of order d of the function g_n, which is

$$\mathsf{Tr}_d[g_n](\lambda_{(1:n-2d)})$$
$$= \int_{[0,1]^{2d}} g_n(\omega_{(1:2d)}, \lambda_{(1:n-2d)}) \prod_{i=1}^d \delta_1(\omega_{2i-1} + \omega_{2i}) \Phi(d\omega_{2i}).$$

Clearly $\mathsf{Tr}_d[g_n] \in \widetilde{L_\Phi^{n-2d}}$. In time domain analysis, formula (2.10) is called the Hu–Meyer formula. It is important in work with the Feynman integral, see [65] for more details.

2.4 Chaotic representation of stationary processes

2.4.1 Subordinated functionals of Gaussian processes

Consider a time shift operator U_s for a stationary Gaussian process X_t, i.e., $\mathsf{U}_s X_t = X_{t+s}$ for every s, t (p. xix). The process X_t is strictly stationary therefore the time shift U_s shifts the realizations of the process without affecting the distributions. Fix the time points u, v and define a process Y_t by

$$Y_t = H_2\left(X_{u+t}, X_{v+t}\right).$$

This process is strictly stationary with the same time shift operator U_s and, starting from time zero,

$$Y_t = \mathsf{U}_t Y_0 = H_2\left(\mathsf{U}_t X_u, \mathsf{U}_t X_v\right),$$

and so

$$Y_t = \int_{[0,1]^2} e^{i2\pi t(\omega_1+\omega_2)} e^{i2\pi(u\omega_1+v\omega_2)} W(d\omega_1, d\omega_2).$$

In general if we shift, by U_t, any element $\mathrm{I}_2\left(g_2\right)$ of the space $L_2^2\left(\mathcal{X}\right)$, we shall get a srtictly (and weakly) stationary process with multiple spectral representation

$$Y_t = \mathsf{U}_t \mathrm{I}_2\left(g_2\right) = \int_{[0,1]^2} e^{i2\pi t(\omega_1+\omega_2)} g_2(\omega_1, \omega_2) W(d\omega_1, d\omega_2).$$

This spectral representation is different from the Cramér representation given above. First of all the multiple spectral representation contains all the information with respect to the process, while the Cramer representation contains only the second order ones. The Cramér representation exists for a very general class of stationary processes, this so called multiple spectral representation is available only in some very particular cases.

Let X_t be a stationary Gaussian first order autoregressive process with parameters a and σ^2 $\left(\sigma^2 > 0\right)$, i.e.,

$$X_t = aX_{t-1} + \mathrm{w}_t,$$

where w_t is a Gaussian white noise series with the stochastic spectral representation

$$\mathrm{w}_t = \int_0^1 e^{i2\pi t\omega} W\left(d\omega\right).$$

Suppose that $|a| < 1$, i.e., X_t depends only on the past of w_t. Fix the indices u and v and put

$$\begin{aligned} Y_t &= H_2\left(X_{u+t}, X_{v+t}\right) \\ &= X_{u+t}X_{v+t} - \frac{\sigma^2 a^{|u-v|}}{1-a^2} \\ &= \int_{[0,1]^2} e^{i2\pi t(\omega_1+\omega_2)} \frac{e^{i2\pi(u\omega_1+v\omega_2)}}{\left(1-ae^{i2\pi\omega_1}\right)\left(1-ae^{i2\pi\omega_2}\right)} W(d\omega_1, d\omega_2). \end{aligned}$$

Then Y_t is a stationary process with the covariance structure

$$\operatorname{Cov}\left(Y_t, Y_{t+s}\right) = \frac{\sigma^4}{\left(1-a^2\right)^2}\left(a^{2|s|} + a^{|s+u-v|}a^{|s+v-u|}\right),$$

and with the spectral density

$$\begin{aligned} S_2(\omega) &= \frac{\sigma^4}{\left|1-a^2e^{i2\pi\omega}\right|^2} \\ &+ \frac{\sigma^4 a^{2|u-v|}}{1-a^2}\left[\frac{e^{i2\pi\omega(u-v)}}{\left(1-a^2e^{-i2\pi\omega}\right)\left(1-e^{i2\pi\omega}\right)} + \frac{e^{i2\pi\omega(v-u)}}{\left(1-a^2e^{i2\pi\omega}\right)\left(1-e^{-i2\pi\omega}\right)}\right]. \end{aligned}$$

In general, if $g \in \mathsf{Exp}\left(\widetilde{L_\Phi^2}\right)$, we may apply the time shift operator U_s to the random variable

$$Y = \sum_{n=0}^{\infty} \mathrm{I}_n\left(g_n\right),$$

and arrive at the stationary process

$$\begin{aligned} Y_t &= \mathsf{U}_t Y \\ &= \sum_{n=0}^{\infty} \mathsf{U}_t \mathrm{I}_n\left(g_n\right) \\ &= \sum_{n=0}^{\infty} \int_{[0,1]^n} e^{i2\pi t \Sigma\omega_{(1:n)}} g_n\left(\omega_{(1:n)}\right) W\left(d\omega_{(1:n)}\right). \end{aligned} \tag{2.11}$$

The interesting result of Dobrushin is that a large class of stationary processes in the space $\mathfrak{L}^2(\mathcal{X})$, $\mathcal{X} = \{X_t\}$ can be put into the form (2.11). It is natural to introduce the following definition.

Definition 26 *The stationary process $\{Y_t,\ t \in \mathbb{Z}\}$ in the space $\mathfrak{L}^2(\mathcal{X})$, $\mathcal{X} = \{X_t\}$ (statinary and Gaussian) is said to be subordinated to $\{X_t,\ t \in \mathbb{Z}\}$ if the time shift operator U_s of X_t is the time shift operator of Y_t as well.*

The process defined by (2.11) is subordinated to X_t.

Theorem 27 (Dobrushin) *Let $\mathcal{X} = \{X_t,\ t \in \mathbb{Z}\}$ denote some stationary Gaussian process. Every stationary process $\{Y_t,\ t \in \mathbb{Z}\}$ in the space $\mathfrak{L}^2(\mathcal{X})$ which is subordinated to X_t can be describe by the representation*

$$Y_t = \sum_{n=0}^{\infty} \int_{[0,1]^n} e^{i2\pi t \Sigma \omega_{(1:n)}} g_n(\omega_{(1:n)}) W(d\omega_{(1:n)}), \tag{2.12}$$

where the transfer function $g = (g_0, g_1, g_2, \dots, g_n, \dots) \in \mathsf{Exp}\left(\widetilde{L_\Phi^2}\right)$ is essentially unique.

It is easy to get the covariance function from (2.12) because the stochastic integrals of different order are orthogonal. Specifically

$$\operatorname{Cov}(Y_t, Y_{t+s}) = \sum_{n=1}^{\infty} n! \int_{[0,1]^n} e^{i2\pi s \Sigma \omega_{(1:n)}} \left|g_n(\omega_{(1:n)})\right|^2 \prod_{k=1}^{n} \Phi(d\omega_k),$$

where Φ denotes the spectral measure of X_t. If the spectral density $S_{2,X}$ of X_t exists then one can derive, from the covariance, the fact that the spectral density $S_{2,Y}$ of Y_t is

$$S_{2,Y}(\omega) = |g_1(\omega)|^2 S_{2,X}(\omega) + \sum_{n=2}^{\infty} n! \int_{[0,1]^n} \delta_1\left(\omega - \Sigma\omega_{(1:n)}\right) \left|g_n\left(\omega_{(1:n)}\right)\right|^2 \prod_{k=1}^{n} S_{2,X}(\omega_k)\, d\omega_{(1:n)}.$$

Note that

$$\begin{aligned}\int_{[0,1]^n} \delta_1\left(\omega - \Sigma\omega_{(1:n)}\right) \left|g_n\left(\omega_{(1:n)}\right)\right|^2 \prod_{k=1}^{n} S_{2,X}(\omega_k)\, d\omega_{(1:n)} \\ = \int_{[0,1]^{n-1}} \left|g_n\left(\left(\omega, \omega_{(2:n)}\right) - \left(\omega_{(2:n)}, 0\right)\right)\right|^2 S_{2,X}(\omega - \omega_2) \\ \times \prod_{k=2}^{n-1} S_{2,X}(\omega_k - \omega_{k+1})\, S_{2,X}(\omega_n)\, d\omega_{(2:n)}.\end{aligned}$$

In particular put $n = 3$ then

$$\begin{aligned}\int_{[0,1]^3} \delta_1\left(\omega - \Sigma\omega_{(1:3)}\right) \left|g_3\left(\omega_{(1:3)}\right)\right|^2 \prod_{k=1}^{3} S_{2,X}(\omega_k)\, d\omega_{(1:3)} \\ = \int_{[0,1]^2} |g_3(\omega - \omega_2, \omega_2 - \omega_3, \omega_3)|^2 \\ \times S_{2,X}(\omega - \omega_2)\, S_{2,X}(\omega_2 - \omega_3)\, S_{2,X}(\omega_3)\, d\omega_1 d\omega_2.\end{aligned}$$

If g_n is constant then the result is n^{th} convolution of $S_{2,X}$. Indeed, in that case

$$Y_t = \sum_{n=1}^{\infty} g_n H_n (X_t),$$

and

$$S_{2,Y}(\omega) = \sum_{n=1}^{\infty} n! \sigma_X^{2n} g_n^2 \breve{S}_{2,X}^{*n}(\omega),$$

where $\breve{S}_{2,X}^{*n}$ denotes the nth order convolution of $S_{2,X}/\sigma_X^2$, i.e.,

$$\breve{S}_{2,X}^{*n}(\omega)$$
$$= \frac{1}{\sigma_X^{2n}} \int_{[0,1]^{n-1}}^{2} S_{2,X}(\omega - \omega_2) \prod_{k=2}^{n-1} S_{2,X}(\omega_l - \omega_{k+1}) S_{2,X}(\omega_n) d\omega_{(2:n)}.$$

See section 2.4.3 below for details.

2.4.2 Spectra for processes with Hermite degree-2

Assume again that $\mathrm{w}_t \in \mathcal{N}(0, \sigma^2)$ is a Gaussian white noise series with the stochastic spectral representation

$$\mathrm{w}_t = \int_0^1 e^{i2\pi t\omega} W(d\omega).$$

Put

$$Y_t = \sum_{r=0}^{\infty} \int_{\mathcal{D}^r} z_{(1:r)}^{t\vec{1}} g_r(z_{(1:r)}) W(d\omega_{(1:r)}),$$

where $\mathbf{z}_{(r)} = (z_1, z_2, \ldots, z_r)$, $z_k = \exp(i2\pi\omega_k)$, $\vec{1} = (1, 1, \ldots, 1)$, and $z_{(1:r)}^{t\vec{1}} = \prod_{k=1}^{r} z_k^t$.

Definition 28 *The Hermite process of degree two, with* $\mathsf{E}Y_t = 0$, *is given by*

$$Y_t = \sum_{r=1}^{2} \int_{[0,1]^r} z_{(1:r)}^{t\vec{1}} g_r(z_{(1:r)}) W(d\omega_{(1:r)}),$$

i.e., all the transfer functions but the first and second ones are zero. It is supposed that $g_2(z_{(1:2)})$ *is symmetric.*

We have considered the cumulants of second and first order Hermite polynomials in examples 10-13. The spectra of the Hermite process of degree two, Y_t, are based on those formulae and are easy to derive. We give the proof of the fourth order spectrum S_4, called the trispectrum, only as a demonstration of the technique.

Some spectra of Y_t follow:

Spectrum

$$S_2(z_1) = \sigma^2 |g_1(z_1)|^2 + 2\sigma^4 \int_0^1 \left|g_2(v, z_1 v^{-1})\right|^2 d\lambda, \tag{2.13}$$

where v denotes the $\exp(i2\pi\lambda)$.

Bispectrum

$$S_3(z_{(1:2)}) = 6\sigma^4 \operatorname*{sym}_{z_{(3)}} \left[g_1(z_1) g_2(z_1^{-1}, z_2^{-1}) g_1(z_2)\right]$$

$$+ 8\sigma^6 \operatorname*{sym}_{z_{(3)}} \left[\int_0^1 g_2(v, z_1 v^{-1}) g_2(v z_1^{-1}, z_2^{-1} v^{-1}) g_2(v z_2, v^{-1}) d\lambda\right], \tag{2.14}$$

where $\prod_{k=1}^3 z_k = 1$, i.e.,, put $z_3 = \prod_{k=1}^2 z_k^{-1}$. This form of $S_3\left(\omega_{(1:2)}\right)$ is the discrete analog of the formula given by [16] in the continuous time case.

Trispectrum

$$S_4(z_{(1:3)}) = 48\sigma^6 \operatorname*{sym}_{z_{(4)}} \left[g_1(z_1) g_2(z_1^{-1}, z_1 z_3) g_2((z_1 z_3)^{-1}, z_2^{-1}) g_1(z_2)\right]$$
$$+ 48\sigma^8 \operatorname*{sym}_{z_{(4)}} \left[\int_0^1 g_2(v, z_1 v^{-1}) g_2(v z_1^{-1}, z_1 z_4 v^{-1})\right.$$
$$\left.\times g_2(v (z_1 z_4)^{-1}, v^{-1} z_2^{-1}) g_2(v z_2, v^{-1}) d\lambda\right],$$

where $\prod_{k=1}^4 z_k = 1$, i.e., put $z_4 = \prod_{k=1}^3 z_k^{-1}$.

Quadrospectrum

$$S_5(z_{(1:4)}) = 480\sigma^8 \operatorname*{sym}_{z_{(5)}} \left[g_1(z_1) g_2(z_1^{-1}, z_1 z_3) g_2((z_1 z_3)^{-1}, z_1 z_3 z_4)\right.$$
$$\left. g_2((z_1 z_3 z_4)^{-1}, z_2^{-1}) g_1(z_2)\right]$$
$$+ 384\sigma^{10} \operatorname*{sym}_{z_{(5)}} \left[\int_0^1 g_2(v, z_1 v^{-1}) g_2(v z_1^{-1}, z_1 z_5 v^{-1})\right.$$
$$\left.\times g_2(v (z_1 z_5)^{-1}, v^{-1} z_1 z_4 z_5) g_2(v (z_1 z_4 z_5)^{-1}, v^{-1} z_2^{-1}) g_2(v z_2, v^{-1}) d\lambda\right],$$

where $\prod_{k=1}^5 z_k = 1$ i.e.,, put $z_5 = \prod_{k=1}^4 z_k^{-1}$.

The pentaspectrum S_6 is given in the same manner. The proof of the spectrum (2.13) can be found on page 143 below.

Proof. For the trispectrum S_4. The process $Y_t = Y_{1,t} + Y_{2,t}$ where $Y_{1,t}$ is Gaussian process and $Y_{2,t}$ is given as a homogeneous chaos of second order:

$$Y_{1,t} = \int_{[0,1]} z_1^t g_1(z_1) W(d\omega_1),$$
$$Y_{2,t} = \int_{[0,1]^r} z_{(1:2)}^{t\vec{1}} g_2(z_{(1:2)}) W(d\omega_{(1:2)}).$$

The fourth order cumulant is

$$\operatorname{Cum}(Y_{t_1}, Y_{t_2}, Y_{t_3}, Y_{t_4}) = \sum \operatorname{Cum}(Y_{1,t_i}, Y_{1,t_j}, Y_{2,t_k}, Y_{2,t_l}) + \sum \operatorname{Cum}(Y_{2,t_1}, Y_{2,t_2}, Y_{2,t_3}, Y_{2,t_4}), \tag{2.15}$$

because all the other types of cumulants are zero, see Examples 10-13 above and the formula (2.9) for cumulant of multiple Wiener–Itô integrals. Consider $\operatorname{Cum}(Y_{1,t_1}, Y_{1,t_2}, Y_{2,t_3}, Y_{2,t_4})$. Take the series expansion of $Y_{1,t}$ and $Y_{2,t}$ in terms of the Hermite polynomials see (2.2). The cumulant is multilinear so the typical term to evaluate is

$$\operatorname{Cum}(H_1(\mathrm{w}_{t_1-k}), H_1(\mathrm{w}_{t_2-l}), H_2(\mathrm{w}_{t_3-m_1}, \mathrm{w}_{t_3-m_2}), H_2(\mathrm{w}_{t_4-n_1}, \mathrm{w}_{t_4-n_2}))$$

according to the product $g_{1,k}g_{1,l}g_{2,m_1,m_2}g_{2,n_1,n_2}$, where $g_{1,k}$ and g_{2,m_1,m_2} are the Fourier coefficients of the functions $g_1(z_1)$ and $g_2(z_{(1:2)})$ respectively. We have seen in Example 12 that this cumulant is the sum of the products of second order cumulants like

$$\begin{aligned}
&\operatorname{Cov}(\mathrm{w}_{t_1-k}, \mathrm{w}_{t_3-m_1}) \operatorname{Cov}(\mathrm{w}_{t_3-m_2}, \mathrm{w}_{t_4-n_1}) \operatorname{Cov}(\mathrm{w}_{t_4-n_2}, \mathrm{w}_{t_2-l}) \\
&= \sigma^6 \int_{[0,1]^3} e^{i2\pi[(t_1-k-t_3+m_1)\omega_1 + (t_3-m_2-t_4+n_1)\omega_2]} \\
&\qquad \times e^{i2\pi(t_4-n_2-t_2+l)\omega_3} d\omega_1 d\omega_2 d\omega_3 \\
&= \sigma^6 \int_{[0,1]^4} e^{i2\pi[(t_1\lambda_1+t_3\lambda_2+t_4\lambda_3+t_2\lambda_4)-(k\lambda_1-m_1\lambda_1+m_2(\lambda_1+\lambda_2))]} \\
&\qquad \times e^{i2\pi[-n_1(\lambda_1+\lambda_2)-n_2\lambda_4+l\lambda_4]} \delta_1(\Sigma\lambda_{(1:4)}) \, d\lambda_1 d\lambda_2 d\lambda_3 d\lambda_4.
\end{aligned}$$

This integral will not change if we take any other term of covariances according to Example 12, therefore

$$\begin{aligned}
&\operatorname{Cum}(H_1(\mathrm{w}_{t_1-k}), H_1(\mathrm{w}_{t_2-l}), H_2(\mathrm{w}_{t_3-m_1}, \mathrm{w}_{t_3-m_2}), H_2(\mathrm{w}_{t_4-n_1}, \mathrm{w}_{t_4-n_2})) \\
&\qquad = \sigma^6 2^2 2! \int_{[0,1]^4} e^{i2\pi(t_1\lambda_1+t_3\lambda_2+t_4\lambda_3+t_2\lambda_4-k\lambda_1+m_1\lambda_1)} \\
&\times e^{-i2\pi(m_2(\lambda_1+\lambda_2)-n_1(\lambda_1+\lambda_2)-n_2\lambda_4+l\lambda_4)} \delta_1(\Sigma\lambda_{(1:4)}) \, d\lambda_1 d\lambda_2 d\lambda_3 d\lambda_4.
\end{aligned}$$

Inserting $g_{1,k}g_{1,l}g_{2,m_1,m_2}g_{2,n_1,n_2}$ and summing up by the indices k, l, m_1, m_2, n_1, n_2, we have

$$\operatorname{Cum}\left(Y_{1,t_1}, Y_{1,t_2}, Y_{2,t_3}, Y_{2,t_4}\right) = \sigma^6 2^2 2! \int_{[0,1]^4} e^{i2\pi(t_1\omega_1+t_3\omega_2+t_4\omega_3+t_2\omega_4)}$$
$$\times g_1(z_1) g_2(z_1^{-1}, z_1 z_2) g_2((z_1 z_2)^{-1}, z_4^{-1}) g_1(z_4) \delta_1\left(\Sigma\omega_{(1:4)}\right) d\omega_{(1:4)}.$$

The first sum of equation (2.15) contains all the possible choices of times for the first and for the second degree processes, therefore the sum is symmetric and contains $\binom{4}{2} = 6$ terms, so the number of symmetrized functions is $2^2 2! 6 = 48$. The calculation for the second term of equation (2.15) is simpler; the number of symmetrized functions in this case will be the same as in Example 10, i.e., $2^3 3! = 48$. ■

2.4.3 The process $F(X_t)$

Let $Y_t = F(X_t)$ where $F \in L_G^2$, i.e.,

$$\int_{-\infty}^{\infty} F^2(x)\, dG(x) < \infty,$$

with G is the Gaussian distribution function according to X_0,

$$G(x) = \frac{1}{\sigma_X \sqrt{2\pi}} \int_{-\infty}^{x} e^{-\frac{(u-\mu_X)^2}{2\sigma_X^2}}\, du.$$

This Y_t is a stationary process in $\mathfrak{L}^2(\mathcal{X})$. Put F into the Hermite series expansion

$$F(X_0) = \sum_{n=1}^{\infty} h_n H_n(X_0),$$

where the Hermite coefficients h_n are defined by

$$\begin{aligned} h_n &= \frac{1}{\sigma_X^{2n} n!} \operatorname{E} F(X_0) H_n(X_0) \\ &= \frac{1}{\sigma_X^{2n} n!} \int_{-\infty}^{\infty} F(x) H_n(x)\, dG(x) \\ &= \frac{1}{\sigma_X^{2n} n!} f_n. \end{aligned}$$

The chaotic multiple representation of Y_t is now

$$Y_t = \sum_{n=0}^{\infty} h_n \int_{[0,1]^n} e^{i2\pi t \Sigma\omega_{(1:n)}} W_X(d\omega_{(1:n)}), \tag{2.16}$$

with $W_X(d\omega)$ the stochastic spectral measure of X_t. The transfer functions of the representation (2.12) are constant in this case.

An approximation of the spectral density of $F(X_t)$

One has

$$\begin{aligned}\operatorname{Cov}(Y_t, Y_{t+s}) &= \sum_{n=1}^{\infty} n! h_n^2 \int_{[0,1]^n} e^{i2\pi s \Sigma\omega_{(1:n)}} \prod_{k=1}^{n} \Phi(d\omega_k) \\ &= \sum_{n=1}^{\infty} n! h_n^2 C_X^n(s) \\ &= \sum_{n=1}^{\infty} n! \sigma_X^{2n} h_n^2 \frac{C_X^n(s)}{C_X^n(0)},\end{aligned} \tag{2.17}$$

with $C_X(s) = \operatorname{Cov}(X_t, X_{t+s})$ and $\sigma_X^2 = C_X(0)$. The spectral density of Y_t may also be expressed in terms of the spectral density $S_{2,X}(\omega)$ of X_t. Take the Fourier transform of both sides of (2.17) to get

$$S_{2,Y}(\omega) = \sum_{n=1}^{\infty} n! \sigma_X^{2n} h_n^2 \breve{S}_{2,X}^{*n}(\omega),$$

where $\breve{S}_{2,X}^{*n}(\omega)$ is the nth convolution power, see page xx or 49, of the $\breve{S}_{2,X}(\omega) = S_{2,X}(\omega)/C_X(0)$. In this case

$$\int_{[0,1]} \breve{S}_{2,X}(\omega)\, d\omega = 1.$$

Grenander and Rosenblatt [44] suggested the following approximation for $S_{2,Y}$. First observe that if $s \neq 0$,

$$\left| \frac{C_X^n(s)}{C_X^n(0)} \right| \ll 1,$$

in particular, when n is large and the process X_t has a short memory, i.e., its covariance decays exponentially. In that case

$$\breve{S}_{2,X}^{*n}(\omega) = \sum_{s=-\infty}^{\infty} \frac{C_X^n(s)}{C_X^n(0)} e^{-i2\pi s\omega} \sim \frac{C_X^n(0)}{C_X^n(0)} = 1.$$

Therefore

$$\begin{aligned}
S_{2,Y}^{appr}(\omega) &= \sum_{k=1}^{n} k!\sigma_X^{2k} h_k^2 \breve{S}_{2,X}^{*k}(\omega) + \sum_{k=n+1}^{\infty} k!\sigma_X^{2k} h_k^2 \\
&= \sum_{k=1}^{n} k!\sigma_X^{2k} h_k^2 \left[\breve{S}_{2,X}^{*k}(\omega) - 1\right] + \sum_{k=0}^{\infty} k!\sigma_X^{2k} h_k^2 - h_0^2 \\
&= \sum_{k=1}^{n} k!\sigma_X^{2k} h_k^2 \left[\breve{S}_{2,X}^{*k}(\omega) - 1\right] + \mathrm{E}\, F^2(X_0) - (\mathrm{E}\, F(X_0))^2 \\
&= \sum_{k=1}^{n} k!\sigma_X^{2k} h_k^2 \left[\breve{S}_{2,X}^{*k}(\omega) - 1\right] + \operatorname{var}(Y_0) \\
&= \sum_{k=1}^{n} k! h_k^2 \left[S_{2,X}^{*k}(\omega) - \sigma_X^{2k}\right] + \operatorname{var}(Y_0).
\end{aligned}$$

In some particular cases this approximation can work well even for $n = 1$, or 2. The same reasoning applies for the process

$$Y_t = \sum_{n=0}^{\infty} \int_{[0,1]^n} e^{i2\pi t \Sigma\omega_{(1:n)}} g_n\left(\Sigma\omega_{(1:n)}\right) W_X(d\omega_{(1:n)}). \tag{2.18}$$

Indeed,

$$\operatorname{Cov}(Y_t, Y_{t+s}) = \sum_{n=1}^{\infty} n! \int_{[0,1]^n} e^{i2\pi s \Sigma\omega_{(1:n)}} \left|g_n\left(\Sigma\omega_{(1:n)}\right)\right|^2 \prod_{k=1}^{n} \Phi(d\omega_k),$$

changing variables in this integral by putting $\lambda_k = \Sigma\omega_{(1:k)}$, if $1 \leq k < n$ and $\omega = \Sigma\omega_{(1:n)}$ we have

$$\operatorname{Cov}(Y_t, Y_{t+s}) = \sum_{n=1}^{\infty} n!\sigma_X^{2n} \int_{[0,1]} e^{i2\pi s\omega} \left|g_n(\omega)\right|^2 \breve{S}_{2,X}^{*n}(\omega)\, d\omega.$$

Therefore the spectrum of Y_t can be approximated by

$$S_{2,Y}^{appr}(\omega) = \sum_{k=1}^{n} k! \left|g_k(\omega)\right|^2 \left[S_{2,X}^{*k}(\omega) - \sigma_X^{2k}\right] + \sum_{k=1}^{\infty} k!\sigma_X^{2k} \left|g_k(\omega)\right|^2.$$

The second term here can be considered as the spectrum of Y_t with some Gaussian white noise input w_t with $\operatorname{var}(\mathrm{w}_t) = \sigma_X^2$.

An approximation of the bispectrum of $F(X_t)$

The idea employed in the previous section will be useful for the approximation of the higher order spectra as well. As the process (2.16) is a particular case of (2.18), we shall consider the approximation of the bispectrum for the process (2.18) first and then give the approximation of the bispectrum of $F(X_t)$. Suppose that if $s \neq 0$ and $n > 2$ then

$$\left| \frac{C_X^n(s)}{C_X^n(0)} \right| \ll 1, \tag{2.19}$$

as above. For the third order cumulants $\mathrm{Cum}(Y_s, Y_t, Y_u)$ it would be necessary to consider all the possible triplets of higher order integrals from (2.18). Instead of doing this, begin with considering some particular cases. The cumulant

$$\mathrm{Cum}(H_k(X_s), H_m(X_t), H_n(X_u)) = \mathsf{E} H_k(X_s) H_m(X_t) H_n(X_u), \tag{2.20}$$

can be evaluated either by the formula (1.15), page 17, or in two steps. First express $H_k(X_s) H_m(X_t)$ in terms of Hermite polynomials, see (1.18) page 22, and then take the expectation of the product with $H_n(X_u)$. Because of the symmetry of cumulants, it can be assumed that $k \leq m \leq n$. Now (2.20) will be different from zero if $k+m+n$ is not only even but also $n = k+m-2N$ with some $N = 0, 1, 2 \ldots$. If it is nonzero, it contains only products of covariances $C_X(t-s)$, $C_X(u-s)$ and $C_X(u-t)$ in some appropriate powers. The assumption (2.19) indicates that we are interested in products of type

$$C_X^p(t-s) C_X^r(u-s) C_X^l(u-t),$$

for which $p \leq r \leq l$ and either $p \leq 2$ or $r \leq 2$ or $l \leq 2$. These cases will be considered below. In general, put

$$\mathrm{I}_n\left(z_{(1:n)}^{t\vec{1}} g_n\right) = \int_{[0,1]^n} e^{i2\pi t \Sigma\omega_{(1:n)}} g_n\left(\Sigma\omega_{(1:n)}\right) W_X(d\omega_{(1:n)}),$$

where

$$z_{(1:n)}^{t\vec{1}} = e^{i2\pi t \Sigma\omega_{(1:n)}}.$$

Consider now the cumulants containing the first order term I_1. If $1 \le m \le n$,

$$\begin{aligned}
&\operatorname{Cum}\left(\mathrm{I}_1\left(z_1^s g_1\right), \mathrm{I}_m\left(z_{(1:m)}^{t\vec{1}} g_m\right), \mathrm{I}_n\left(z_{(1:n)}^{u\vec{1}} g_n\right)\right) \\
&= \delta_{m=n-1} n! \int_{[0,1]^n} e^{i2\pi\left((s-u)\omega_1-(t-u)\Sigma\omega_{(2:n)}\right)} g_1\left(\omega_1\right) g_{n-1}\left(\Sigma\omega_{(2:n)}\right) \\
&\qquad \times \overline{g_n}\left(\Sigma\omega_{(1:n)}\right) \prod_{k=1}^{n} \Phi\left(d\omega_k\right) \\
&= \delta_{m=n-1} n! \int_{[0,1]^3} e^{i2\pi(s\lambda_1+t\lambda_2+u\lambda_3)} g_1\left(\lambda_1\right) g_{n-1}\left(\lambda_2\right) g_n\left(\lambda_3\right) S_{2,X}\left(\lambda_1\right) \\
&\times \int_{[0,1]^{n-2}} S_{2,X}\left(\lambda_2-\omega_3\right) \prod_{k=3}^{n-1} S_{2,X}\left(\omega_k-\omega_{k+1}\right) S_{2,X}\left(\omega_n\right) d\omega_{(3:n)} \\
&\qquad \times \delta_1\left(\Sigma\lambda_{(1:3)}\right) d\lambda_{(1:3)} \\
&= \delta_{m=n-1} n! \int_{[0,1]^3} e^{i2\pi(s\lambda_1+t\lambda_2+u\lambda_3)} g_1\left(\lambda_1\right) g_{n-1}\left(\lambda_2\right) g_n\left(\lambda_3\right) \\
&\qquad \times S_{2,X}\left(\lambda_1\right) S_{2,X}^{*(n-1)}\left(\lambda_2\right) \delta_1\left(\Sigma\lambda_{(1:3)}\right) d\lambda_{(1:3)}.
\end{aligned}$$

The contribution of this to the bispectrum is the symmetrization

$$3! \operatorname*{sym}_{\lambda_{(1:3)}}\left(g_1\left(\lambda_1\right) g_{n-1}\left(\lambda_2\right) g_n\left(\lambda_3\right) S_{2,X}\left(\lambda_1\right) S_{2,X}^{*(n-1)}\left(\lambda_2\right)\right).$$

The assumption (2.19) implies that $\breve{S}_{2,X}^{*(n-2)} \sim 1$ if $n > 4$ therefore the approximation of the bispectrum will contain

$$\begin{aligned}
&3! \operatorname*{sym}_{\lambda_{(1:3)}}\left(g_1\left(\lambda_1\right) \breve{S}_{2,X}\left(\lambda_1\right)\left\{2!\sigma_X^4 g_1\left(\lambda_2\right) g_2\left(\lambda_3\right) \breve{S}_{2,X}\left(\lambda_2\right)\right.\right. \\
&\left.\left.+ 3!\sigma_X^6 g_2\left(\lambda_2\right) g_3\left(\lambda_3\right) \breve{S}_{2,X}\left(\lambda_1\right) \breve{S}_{2,X}^{*2}\left(\lambda_2\right) + \sum_{n=4}^{\infty} n!\sigma_X^{2n} g_{n-1}\left(\lambda_2\right) g_n\left(\lambda_3\right)\right\}\right).
\end{aligned}$$

In particular,

$$\begin{aligned}
&\operatorname{Cum}\left(\mathrm{I}_1\left(z_1^s\right), \mathrm{I}_m\left(z_{(1:m)}^{t\vec{1}}\right), \mathrm{I}_n\left(z_{(1:n)}^{u\vec{1}}\right)\right) = \delta_{m=n-1} n! \\
&\quad \times \int_{[0,1]^2} e^{i2\pi(s\lambda_1+t\lambda_2+u\lambda_3)} S_{2,X}\left(\lambda_1\right) S_{2,X}^{*(n-1)}\left(\lambda_2\right) \delta_1\left(\Sigma\lambda_{(1:3)}\right) d\lambda_{(1:3)}.
\end{aligned}$$

Moreover if the cumulant contains the second order term I_2, there are two cases to be considered. One is

$$\begin{aligned}
&\operatorname{Cum}\left(\mathrm{I}_2\left(z_{(1:2)}^{s\vec{1}} g_2\right), \mathrm{I}_{n-2}\left(z_{(1:n-2)}^{t\vec{1}} g_{n-2}\right), \mathrm{I}_n\left(z_{(1:n)}^{u\vec{1}} g_n\right)\right)\\
&= n! \int_{[0,1]^n} e^{i2\pi\left(s\Sigma\omega_{(1:2)}+t\Sigma\omega_{(3:n)}-u\Sigma\omega_{(1:n)}\right)} g_2\left(\Sigma\omega_{(1:2)}\right) g_{n-2}\left(\Sigma\omega_{(3:n)}\right)\\
&\quad\times \overline{g_n}\left(\Sigma\omega_{(1:n)}\right) \prod_{k=1}^{n} \Phi\left(d\omega_k\right)\\
&= n! \int_{[0,1]^3} e^{i2\pi(s\lambda_1+t\lambda_2+u\lambda_3)} g_2\left(\lambda_1\right) g_{n-2}\left(\lambda_2\right) g_n\left(\lambda_3\right)\\
&\int_{[0,1]^{n-2}} S_{2,X}\left(\lambda_1-\omega_2\right) S_{2,X}\left(\omega_2\right) d\omega_2 S_{2,X}\left(\lambda_2-\omega_4\right)\\
&\quad\times \prod_{k=4}^{n-1} S_{2,X}\left(\omega_k-\omega_{k+1}\right) S_{2,X}\left(\omega_n\right) d\omega_2 d\omega_{(4:n)} \delta_1\left(\Sigma\lambda_{(1:3)}\right) d\lambda_{(1:3)}\\
&= n! \int_{[0,1]^3} e^{i2\pi(s\lambda_1+t\lambda_2+u\lambda_3)} g_2\left(\lambda_1\right) g_{n-2}\left(\lambda_2\right) g_n\left(\lambda_3\right) S_{2,X}^{*2}\left(\lambda_1\right)\\
&\quad\times S_{2,X}^{*(n-2)}\left(\lambda_2\right) \delta_1\left(\Sigma\lambda_{(1:3)}\right) d\lambda_{(1:3)}.
\end{aligned}$$

In particular,

$$\begin{aligned}
&\operatorname{Cum}\left(\mathrm{I}_2\left(z_{(1:2)}^{(t+s)\vec{1}}\right), \mathrm{I}_{n-2}\left(z_{(1:n-2)}^{(t+r)\vec{1}}\right), \mathrm{I}_n\left(z_{(1:n)}^{t\vec{1}}\right)\right)\\
&\quad= n! \int_{[0,1]^3} e^{i2\pi(s\lambda_1+r\lambda_2)} S_{2,X}^{*2}\left(\lambda_1\right) S_{2,X}^{*(n-2)}\left(\lambda_2\right) \delta_1\left(\Sigma\lambda_{(1:3)}\right) d\lambda_{(1:3)}.
\end{aligned}$$

The second case is

$$\begin{aligned}
&\operatorname{Cum}\left(\mathrm{I}_2\left(z_{(1:2)}^{s\vec{1}} g_2\right), \mathrm{I}_{n-1}\left(z_{(1:n-1)}^{t\vec{1}} g_{n-1}\right), \mathrm{I}_{n-1}\left(z_{(1:n-1)}^{u\vec{1}} g_{n-1}\right)\right)\\
&\quad= 2\left(n-1\right)^2\left(n-2\right)! \int_{[0,1]^n} e^{i2\pi\left(s(\omega_1-\omega_2)+t\Sigma\omega_{(2:n)}-u\left(\omega_1+\Sigma\omega_{(3:n)}\right)\right)}\\
&\qquad\times g_2\left(\omega_1-\omega_2\right) g_{n-1}\left(\Sigma\omega_{(2:n)}\right) \overline{g_{n-1}}\left(\omega_1+\Sigma\omega_{(3:n)}\right) \prod_{k=1}^{n} \Phi\left(d\omega_k\right)
\end{aligned}$$

$$= 2(n-1)(n-1)! \int_{[0,1]^3} e^{i2\pi(s\lambda_1+t\lambda_2+u\lambda_3)} g_2(\lambda_1) g_{n-1}(\lambda_2) g_{n-1}(\lambda_3)$$
$$\times \int_{[0,1]^{n-2}} S_{2,X}(\lambda_1+\lambda_2-\omega_3) S_{2,X}(\lambda_2-\omega_3) \prod_{k=3}^{n-1} S_{2,X}(\omega_k-\omega_{k+1})$$
$$\times S_{2,X}(\omega_n) d\omega_{(3:n)} \delta_1 \left(\Sigma\lambda_{(1:3)}\right) d\lambda_{(1:3)}$$
$$= 2(n-1)(n-1)! \int_{[0,1]^3} e^{i2\pi(s\lambda_1+t\lambda_2+u\lambda_3)} g_2(\lambda_1) g_{n-1}(\lambda_2) g_{n-1}(\lambda_3)$$
$$\times \int_{[0,1]} S_{2,X}(\lambda_1+\lambda_2-\omega_3) S_{2,X}(\lambda_2-\omega_3) S_{2,X}^{*(n-2)}(\omega_3) d\omega_3$$
$$\times \delta_1 \left(\Sigma\lambda_{(1:3)}\right) d\lambda_{(1:3)}.$$

In particular,

$$\operatorname{Cum}\left(\mathrm{I}_2\left(z_{(1:2)}^{s\vec{1}}\right), \mathrm{I}_{n-1}\left(z_{(1:n-1)}^{t\vec{1}}\right), \mathrm{I}_{n-1}\left(z_{(1:n-1)}^{u\vec{1}}\right)\right) = 2(n-1)(n-1)!$$
$$\times \int_{[0,1]^3} e^{i2\pi(s\lambda_1+t\lambda_2+u\lambda_3)} \int_{[0,1]} S_{2,X}(\lambda_1+\lambda_2-\omega_3)$$
$$\times S_{2,X}(\lambda_2-\omega_3) S_{2,X}^{*(n-2)}(\omega_3) d\omega_3 \delta_1 \left(\Sigma\lambda_{(1:3)}\right) d\lambda_{(1:3)}.$$

Let us apply assumption (2.19) again. We see that an approximation of the bispectrum can be based on terms like

$$S_{2,X}^{*2}(\lambda_1) S_{2,X}(\lambda_2), \qquad S_{2,X}^{*2}(\lambda_1) S_{2,X}^{*2}(\lambda_2), \text{ and } S_{2,X}^{*2}(\lambda_1),$$

by the first case and terms like

$$\int_{[0,1]} S_{2,X}(\lambda_1+\lambda_2-\omega_3) S_{2,X}(\lambda_2-\omega_3) S_{2,X}(\omega_3) d\omega_3,$$
$$\int_{[0,1]} S_{2,X}(\lambda_1+\lambda_2-\omega_3) S_{2,X}(\lambda_2-\omega_3) S_{2,X}^{*2}(\omega_3) d\omega_3, \text{ and}$$
$$\int_{[0,1]} S_{2,X}(\lambda_1+\lambda_2-\omega_3) S_{2,X}(\lambda_2-\omega_3) d\omega_3 = S_{2,X}^{*2}(\lambda_1),$$

by the second case.

Next consider

$$\operatorname{Cum}\left(\mathrm{I}_3\left(z_{(1:3)}^{s\mathbf{1}}g_3\right), \mathrm{I}_{n-2}\left(z_{(1:n-2)}^{t\mathbf{1}}g_{n-2}\right), \mathrm{I}_{n+1}\left(z_{(1:n-1)}^{u\mathbf{1}}g_{n-1}\right)\right)$$

$$= 3!n\binom{n+1}{2}(n-1)!\int_{[0,1]^n} e^{i2\pi\left(-s\Sigma\omega_{(1:3)}-t\Sigma\omega_{(3:n)}+u\left(\Sigma\omega_{(1:2)}+\Sigma\omega_{(4:n)}\right)\right)}$$

$$\times g_3\left(-\Sigma\omega_{(1:3)}\right)g_{n-2}\left(-\Sigma\omega_{(3:n)}\right)g_{n-1}\left(\Sigma\omega_{(1:2)}+\Sigma\omega_{(4:n)}\right)\prod_{k=1}^{n}\Phi\left(d\omega_k\right)$$

$$= 3!n!\binom{n+1}{2}\int_{[0,1]^3} e^{i2\pi(s\lambda_1+t\lambda_2+u\lambda_3)}g_3\left(\lambda_1\right)g_{n-2}\left(\lambda_2\right)g_{n-1}\left(\lambda_3\right)$$

$$\times\int_{[0,1]^{n-2}} S_{2,X}^{*2}\left(\lambda_1+\lambda_2-\omega_3\right)S_{2,X}\left(\lambda_2+\omega_3\right)$$

$$\times\prod_{k=4}^{n-1}S_{2,X}\left(\omega_k-\omega_{k+1}\right)S_{2,X}\left(\omega_n\right)d\omega_{(3:n)}\delta_1\left(\Sigma\lambda_{(1:3)}\right)d\lambda_{(1:3)}$$

$$= 3!n!\binom{n+1}{2}\int_{[0,1]^3} e^{i2\pi(s\lambda_1+t\lambda_2+u\lambda_3)}g_3\left(\lambda_1\right)g_{n-2}\left(\lambda_2\right)g_{n-1}\left(\lambda_3\right)$$

$$\times\int_{[0,1]} S_{2,X}^{*2}\left(\lambda_1+\lambda_2-\omega_3\right)S_{2,X}\left(\lambda_2+\omega_3\right)$$

$$\times S_{2,X}^{*(n-3)}\left(\omega_3\right)d\omega_3\delta_1\left(\Sigma\lambda_{(1:3)}\right)d\lambda_{(1:3)}.$$

Finally consider

$$\operatorname{Cum}\left(\mathrm{I}_4\left(z_{(1:4)}^{s\mathbf{1}}g_4\right), \mathrm{I}_{n-2}\left(z_{(1:n-2)}^{t\mathbf{1}}g_{n-2}\right), \mathrm{I}_{n-2}\left(z_{(1:n-2)}^{u\mathbf{1}}g_{n-2}\right)\right)$$

$$= 4!\binom{n-2}{2}^2(n-4)!\int_{[0,1]^n} e^{i2\pi\left(-s\Sigma\omega_{(1:4)}+t\left(\Sigma\omega_{(1:2)}-\Sigma\omega_{(5:n)}\right)+u\Sigma\omega_{(3:n)}\right)}$$

$$\times g_4\left(-\Sigma\omega_{(1:4)}\right)g_{n-2}\left(\Sigma\omega_{(1:2)}-\Sigma\omega_{(5:n)}\right)g_{n-2}\left(\Sigma\omega_{(3:n)}\right)\prod_{k=1}^{n}\Phi\left(d\omega_k\right)$$

$$= 3!\left(n-2\right)\left(n-3\right)\left(n-2\right)!\int_{[0,1]^3} e^{i2\pi(s\lambda_1+t\lambda_2+u\lambda_3)}g_4\left(\lambda_1\right)g_{n-2}\left(\lambda_2\right)$$

$$\times g_{n-2}\left(\lambda_3\right)\int_{[0,1]^{n-2}} S_{2,X}^{*2}\left(\lambda_3+\omega_3\right)S_{2,X}^{*2}\left(\lambda_2-\omega_3\right)S_{2,X}\left(\omega_n\right)$$

$$\times\prod_{k=5}^{n-1}S_{2,X}\left(\omega_k-\omega_{k+1}\right)d\omega_{(3:n)}\delta_1\left(\Sigma\lambda_{(1:3)}\right)d\lambda_{(1:3)}$$

$$= 3!\left(n-2\right)\left(n-3\right)\left(n-2\right)!\int_{[0,1]^3} e^{i2\pi(s\lambda_1+t\lambda_2+u\lambda_3)}g_3\left(\lambda_1\right)g_{n-2}\left(\lambda_2\right)$$

$$\times g_{n-1}\left(\lambda_3\right)\int_{[0,1]} S_{2,X}^{*2}\left(\lambda_3+\omega_3\right)S_{2,X}^{*2}\left(\lambda_2-\omega_3\right)$$

$$\times S_{2,X}^{*(n-4)}\left(\omega_3\right)d\omega_3\delta_1\left(\Sigma\lambda_{(1:3)}\right)d\lambda_{(1:3)}.$$

Let us now summarize the cases necessary for the approximation. Cumulants producing products

$$C_X^p(t-s)\, C_X^r(u-s)\, C_X^l(u-t),$$

with $p, r \leq 2$ have already been considered. The product

$$C_X(t-s)\, C_X^r(u-s)\, C_X^l(u-t),$$

occurs with cumulants

$$\operatorname{Cum}\left(H_{r+1}(X_s), H_{l+1}(X_t), H_{r+l}(X_u)\right),$$

where we may suppose that $r \leq l$, otherwise there is no restriction. The cases $r = 1, 2$ are included in the previous cumulants. If $3 \leq r \leq l$,

$$\left|\frac{C_X^r(u-s)}{C_X^r(0)}\right| \ll 1, \text{ and } \left|\frac{C_X^l(u-t)}{C_X^l(0)}\right| \ll 1,$$

unless $u = s$ and $u = t$, therefore

$$\sum_{s,t,u} \frac{C_X(t-s)\, C_X^r(u-s)\, C_X^l(u-t)}{C_X^{r+l+1}(0)} z_1^{-s} z_2^{-t} z_3^{-u} \sim 1.$$

The situation for the product

$$C_X^2(t-s)\, C_X^r(u-s)\, C_X^l(u-t),$$

with cumulant

$$\operatorname{Cum}\left(H_{r+2}(X_s), H_{l+2}(X_t), H_{r+l}(X_u)\right),$$

is clearly the same.

The approximation for the bispectrum $S_{3Y}(\lambda_1, \lambda_2)$ starts with the normalization of cumulants, i.e.,

$$\frac{1}{\sigma_X^{k+m+n}} \operatorname{Cum}\left(\mathrm{I}_k\left(z_{(1:k)}^{s\vec{1}}\right), \mathrm{I}_m\left(z_{(1:m)}^{t\vec{1}}\right), \mathrm{I}_n\left(z_{(1:n)}^{u\vec{1}}\right)\right).$$

Then, using assumption (2.19), we collect first those terms which contain $S_{2,X}$, then all the rest are summarized in an additional term

$$
\begin{aligned}
&S_{3Y}^{appr}(\lambda_1,\lambda_2)\\
&= 3!\delta_1\left(\Sigma\lambda_{(1:3)}\right)\operatorname*{sym}_{\lambda_{(1:3)}}\left[g_1(\lambda_1)\left(S_{2,X}(\lambda_1)-\sigma_X^2\right)\right.\\
&\times\{2!g_1(\lambda_2)g_2(\lambda_3)\left(S_{2,X}(\lambda_2)-\sigma_X^2\right)+3!g_2(\lambda_2)g_3(\lambda_3)\\
&\times\left(S_{2,X}(\lambda_1)-\sigma_X^2\right)\left(S_{2,X}^{*2}(\lambda_2)-\sigma_X^4\right)+\sum_{n=4}^{\infty}n!g_{n-1}(\lambda_2)g_n(\lambda_3)\Bigg\}\\
&+g_2(\lambda_1)\left(S_{2,X}^{*2}(\lambda_1)-\sigma_X^4\right)\{4!g_2(\lambda_2)g_4(\lambda_3)\\
&\times\left(S_{2,X}^{*2}(\lambda_2)-\sigma_X^4\right)+\sum_{n=4}^{\infty}n!g_{n-2}(\lambda_2)g_n(\lambda_3)\Bigg\}\\
&+g_2(\lambda_1)\{8g_2(\lambda_2)g_2(\lambda_3)\\
&\times\int_{[0,1]}\left(S_{2,X}(\lambda_3+\omega)-\sigma_X^2\right)\left(S_{2,X}(\lambda_2-\omega)-\sigma_X^2\right)\\
&\times\left(S_{2,X}(\omega)-\sigma_X^2\right)d\omega\\
&+36g_3(\lambda_2)g_3(\lambda_3)\int_{[0,1]}\left(S_{2,X}(\lambda_3+\omega)-\sigma_X^2\right)\left(S_{2,X}(\lambda_2-\omega)-\sigma_X^2\right)\\
&\times\left(S_{2,X}^{*2}(\omega)-\sigma_X^4\right)d\omega+\sum_{n=5}^{\infty}2(n-1)(n-1)!g_{n-1}(\lambda_2)g_{n-1}(\lambda_3)\\
&\times\left(S_{2,X}^{*2}(-\lambda_1)-\sigma_X^4\right)\}+6!15g_3(\lambda_1)g_3(\lambda_2)g_4(\lambda_3)\\
&\times\int_{[0,1]}\left(S_{2,X}(\lambda_3+\omega)-\sigma_X^2\right)\left(S_{2,X}(\lambda_2-\omega)-\sigma_X^2\right)\\
&\times\left(S_{2,X}^{*2}(\omega)-\sigma_X^4\right)d\omega\\
&+5!4!g_3(\lambda_1)\sum_{n=6}^{\infty}g_{n-2}(\lambda_2)g_{n-1}(\lambda_3)\int_{[0,1]}\left(S_{2,X}^{*2}(\lambda_3+\omega)-\sigma_X^2\right)\\
&\times\left.\left(S_{2,X}^{*2}(\lambda_2-\omega)-\sigma_X^2\right)\left(S_{2,X}^{*2}(\omega)-\sigma_X^4\right)d\omega\right]+S_{3Y}^{w}(\lambda_1,\lambda_2).
\end{aligned}
$$

Here $S_{3Y}^{w}(\lambda_1,\lambda_2)$ denotes the bispectrum of Y_t with white noise input. The reason for this is that if the input is white noise, the spectrum $S_{2,X}$ is constant, σ_X^2, and $S_{3,Y}^{appr}=S_{3,Y}^{w}$. Note that

$$
\operatorname{Cum}(Y_0,Y_0,Y_0)=\operatorname{E}\left(Y_0-\operatorname{E}Y_0\right)^3.
$$

It is now worthwhile to put down the approximation for the process Y_t given by representation (2.16)

$$
\begin{aligned}
S_{3,Y}^{appr}\left(\lambda_1,\lambda_2\right) &= 3!\delta_1\left(\Sigma\lambda_{(1:3)}\right)\operatorname*{sym}_{\lambda_{(1:3)}}\left[h_1\left(S_{2,X}\left(\lambda_1\right)-\sigma_X^2\right)\right.\\
&\times\left\{2!h_1h_2\left(S_{2,X}\left(\lambda_2\right)-\sigma_X^2\right)+3!h_2h_3\left(S_{2,X}\left(\lambda_1\right)-\sigma_X^2\right)\right.\\
&\times\left.\left(S_{2,X}^{*2}\left(\lambda_2\right)-\sigma_X^4\right)+\sum_{n=4}^{\infty}n!h_{n-1}h_n\right\}+h_2\left(S_{2,X}^{*2}\left(\lambda_1\right)-\sigma_X^4\right)\\
&\times\left\{4!h_2h_4\left(S_{2,X}^{*2}\left(\lambda_2\right)-\sigma_X^4\right)+\sum_{n=4}^{\infty}n!h_{n-2}h_n\right\}\\
&+h_2\left\{8h_2^2\int_{[0,1]}\left(S_{2,X}\left(\lambda_3+\omega\right)-\sigma_X^2\right)\left(S_{2,X}\left(\lambda_2-\omega\right)-\sigma_X^2\right)\right.\\
&\times\left(S_{2,X}\left(\omega\right)-\sigma_X^2\right)d\omega+36h_3^2\int_{[0,1]}\left(S_{2,X}\left(\lambda_3+\omega\right)-\sigma_X^2\right)\\
&\times\left(S_{2,X}\left(\lambda_2-\omega\right)-\sigma_X^2\right)\left(S_{2,X}^{*2}\left(\omega\right)-\sigma_X^4\right)d\omega\\
&+\left.\sum_{n=5}^{\infty}2\left(n-1\right)\left(n-1\right)!h_{n-1}^2\left(S_{2,X}^{*2}\left(-\lambda_1\right)-\sigma_X^4\right)\right\}\\
&+6!15h_3^2h_4\int_{[0,1]}\left(S_{2,X}\left(\lambda_3+\omega\right)-\sigma_X^2\right)\\
&\times\left(S_{2,X}\left(\lambda_2-\omega\right)-\sigma_X^2\right)\left(S_{2,X}^{*2}\left(\omega\right)-\sigma_X^4\right)d\omega\\
&+5!4!h_3\sum_{n=6}^{\infty}h_{n-2}h_{n-1}\int_{[0,1]}\left(S_{2,X}^{*2}\left(\lambda_3+\omega\right)-\sigma_X^2\right)\\
&\times\left.\left(S_{2,X}^{*2}\left(\lambda_2-\omega\right)-\sigma_X^2\right)\left(S_{2,X}^{*2}\left(\omega\right)-\sigma_X^4\right)d\omega\right]+\operatorname{Cum}\left(Y_0,Y_0,Y_0\right).
\end{aligned}
$$

3

Stationary Bilinear Models

Over the past fifteen years or so a great deal of attention has been paid to a particular class of nonlinear systems, namely to the bilinear ones. One can mention the lecture notes of Subba Rao and Gabr [122] and the paper of Tuan [96] by time series side and the works [4] and [88] by system theory side. Papers discussing bilinear time series include: [2], [22], [30], [40], [47], [46], [45], [51], [62], [64], [69], [76], [75], [81], [83], [88], [95], [102], [103], [105], [101], [98], [99], [112], [77], [113], [111], [117], [116], [119], [120], [62], [126], [139], [128], [132], [141] and [124].

In this chapter we consider bilinear processes with Gaussian white noise input. First we show that each bilinear process has some particular state space representation, and then we obtain expressions for higher-order transfer functions associated with the bilinear model. An important aspect of this expansion is that for these models there is a recursive relation between transfer functions of various orders. The recursive nature of the transfer functions and the properties of the norms of the symmetric transfer functions may be used to obtain the necessary and sufficient conditions for the second-order stationarity. This technique allows us to get a necessary and sufficient condition for the existence of the $2n^{th}$ moment not only in the first order case but also in the higher order case, see also Terdik and Ispány [134]. The exact spectrum for bilinear processes is determined by using the method of Wiener–Itô chaotic spectral representation. The spectrum of the bilinear process itself does not contain information about the bilinear parameters in general, therefore it is necessary to use the third order properties, i.e., the bispectrum for the identification. It turns out that all the information is given either by the spectrum and the bispectrum or by the spectrum and the cross-spectrum between the process and its square. Therefore the non-Gaussian method of estimation based on the second and third order spectra, see Brillinger[21], will be used for the estimation of the parameters of the bilinear process in Chapter 4. A section is devoted to the identification of multiple-bilinear models and the state space realization of bilinear processes. The realization procedures for the deterministic bilinear systems are based on the generalization of the methods used in linear cases, see Rugh[109]. Isidori's paper, Isidori[63], has had much influence on the works concerning the stochastic realization theory of bilinear systems because of the generalization of the Hankel matrix. The investigations have up to now assumed the various moments and cross-moments of the output and the input, Desai[32], Frazho[38], which is an unreal assumption for a bilinear time series. On the other hand there are more natural quantities than the cross-moments for the realization of a bilinear input-output system, namely the Hermite polynomials of the input should be considered as the starting point for an efficient analysis, see Terdik[129].

3.1 Definition of bilinear models

The bilinear model is defined by the following equation:

$$\sum_{m=0}^{P} a_m Y_{t-m} = \sum_{m=0}^{Q} b_m \mathrm{w}_{t-m} + \sum_{k=1}^{R} \sum_{l=0}^{S} d_{k,l} Y_{t-l} \mathrm{w}_{t-k},$$

where $a_0 = b_0 = 1$, w_t is a Gaussian white noise series and some $d_{k,l} \neq 0$. Y_t is called a bilinear process. The model is linear in the series Y_t and w_t separately but not in both. In particular, if $d_{k,l} = 0$ unless $k \leq l$, the bilinear model is called lower triangular. This means that the multiplicative terms contain such pairs of noise and observation that the observation Y_{t-l} occurs only before the noise factor w_{t-k}, therefore they are independent. (A physically realizable observation does not depend on the future.) This fact makes the analysis easier and more realistic, though the technique we apply for the analysis would work for more general model, (see Section 3.6, page 137). From now on we shall consider the lower triangular bilinear model only and it will be simply called "bilinear".

Remark 29 *Concerning the stationarity of the bilinear model, the assumption of lower triangularity is only technical. There has been shown by Terdik [130] that the method of investigation of stationarity, which we follow bellow, works in general situation also. Moreover, this method allows putting not only linear, i.e., $MA(Q)$ part, but any finite degree polynomial of the noise series into the bilinear model.*

We express the back shifting of the observation according to the noise more explicitly by the following notation:

$$\sum_{m=0}^{P} a_m Y_{t-m} = \sum_{m=0}^{Q} b_m \mathrm{w}_{t-m} + \sum_{m=1}^{R}\sum_{n=0}^{S} d_{m,m+n} Y_{t-m-n} \mathrm{w}_{t-m}. \tag{3.1}$$

We define the model (3.1) as a bilinear time series model $BL(P, Q, R, S)$. If the bilinear process Y_t is second order stationary and subordinated to w_t, one can put it into the Wiener–Itô expansion, see Theorem 27, page 48, as

$$Y_t = \sum_{r=0}^{\infty} \int_{\mathcal{D}^r} g_r(\omega_{(1:r)}) e^{i2\pi t \Sigma\omega_{(1:r)}} W(d\omega_{(1:r)}) \quad ; \; \mathcal{D}^r = [0,1]^r,$$

where the integrals are r-fold Wiener–Itô stochastic integrals with respect to the Gaussian stochastic measure $W(d\omega)$, $(\mathsf{E}W(d\omega) = 0, E|W(d\omega)|^2 = \sigma^2 d\omega)$ of w_t, i.e.,

$$\mathrm{w}_t = \int_{\mathcal{D}} e^{i2\pi t\omega} W(d\omega).$$

$\omega_{(1:r)}$ denotes the vector $(\omega_1, \omega_2, \cdots, \omega_r)$; $\omega_k \in [0,1]$ and the $\Sigma\omega_{(1:r)} = \sum_{k=1}^{r} \omega_k$. The transfer functions g_r can be derived in the same way as those (3.7) on page 68. We shall see later that they are given by the recursive formula

$$g_1(\omega_1) = \frac{\beta(\omega_1)}{\alpha(\omega_1)};$$

$$g_r(\omega_{(1:r)}) = \frac{\gamma(\Sigma\omega_{(1:r-1)}, \Sigma\omega_{(1:r)})}{\alpha(\Sigma\omega_{(1:r)})} g_{r-1}(\omega_{(1:r-1)}); \quad r \geq 2,$$

where the polynomials are determined by the coefficients of the bilinear equation (3.1), i.e.

$$\alpha(\omega) = \sum_{m=0}^{P} a_m e^{-i2\pi m\omega} \; ; \qquad \beta(\omega) = \sum_{k=0}^{Q} b_k e^{-i2\pi k\omega};$$

$$\begin{aligned}\gamma(\omega, \omega+\lambda) &= \sum_{m=1}^{R}\sum_{n=0}^{S} d_{m,m+n} e^{-i2\pi((m+n)\omega + m\lambda)} \\ &= \sum_{m=1}^{R}\sum_{n=0}^{S} d_{m,m+n} e^{-i2\pi(n\omega + m(\omega+\lambda))}.\end{aligned}$$

In the case $g_2(\omega_{(1:2)}) \neq 0$, i.e., Y_t is not purely linear, the recursive formula shows that it has infinitely many nonzero transfer functions, therefore its degree is considered to be infinite. (See also e.g., Terdik and Subba Rao [139].)

3.2 Identification of a bilinear model with scalar states

3.2.1 Multiple spectral representation and stationarity

Bilinear model BL(1,0,1,0) with zero mean

We start the investigation with a particular case which shows up many of the problems one faces in the general case. Let $\{\mathrm{w}_t, t \in \mathbb{Z}\}$, $\mathbb{Z} = \{0, \pm 1, \pm 2, \dots\}$ be Gaussian white noise, with mean zero and variance σ^2. Let the spectral representation of w_t be

$$\mathrm{w}_t = \int_{\mathcal{D}} \exp(i2\pi t\omega) W(d\omega), \qquad t \in \mathbb{Z},$$

where $\mathcal{D} = [0,1]$ and $W(d\omega)$ is a Gaussian stochastic measure with $E(W(d\omega)) = 0$ and $E|W(d\omega)|^2 = \Phi(d\omega) = \sigma^2 d\omega$.

Suppose that $\{Y_t\,;\, t \in \mathbf{Z}\}$ fulfils the bilinear equation

$$Y_t = aY_{t-1} + dY_{t-1}\mathrm{w}_{t-1} + \mathrm{w}_t + f_0, \qquad t \in \mathbb{Z}, \tag{3.2}$$

where $a, d \in \mathbf{R}$ and $f_0 = -\sigma^2 d$ is added to keep $\mathsf{E}Y_t = 0$ for a stationary solution. One reason for supposing $\mathsf{E}Y_t = 0$ is that during a preliminary analysis of real data it often happens that the sample mean is removed, for example in removing seasonality, trends etc. We shall investigate some properties of a solution of equation (3.2) satisfying the following properties.

Assumption 30 *$\{Y_t\,,\, t \in \mathbb{Z}\}$ is*

- *second order stationary* process,
- *physically realizable* in terms of $\{\mathrm{w}_t\,,\, t \in \mathbb{Z}\}$, that is Y_t is measurable with respect to $\sigma\{\mathrm{w}_s\,,\, s \le t\}$ for all $t \in \mathbb{Z}$ and
- *subordinated* to $\{\mathrm{w}_t\,,\, t \in \mathbb{Z}\}$, that is (beside stationarity) if T_s is a shift transformation for w_t, $T_s\mathrm{w}_t = \mathrm{w}_{t+s}$ then it is also one for Y_t, i.e., $T_sY_t = Y_{t+s}$ for all $s, t \in \mathbb{Z}$, see Section 2.4.1, page 46.

We shall refer to $\{Y_t\,,\, t \in \mathbb{Z}\}$ as a *bilinear process* BL(1,0,1,0) *with scalar states.* Let $X_t = Y_t - \mathrm{w}_t$. Then $\mathsf{E}X_t = 0$ and $\{X_t\,,\, t \in \mathbb{Z}\}$ satisfies the equation

$$\begin{aligned} X_t &= aX_{t-1} + dX_{t-1}\mathrm{w}_{t-1} + b_1 H_1(\mathrm{w}_{t-1}) + dH_2(\mathrm{w}_{t-1}), \\ Y_t &= X_t + \mathrm{w}_t, \qquad t \in \mathbb{Z}, \end{aligned} \tag{3.3}$$

where $b_1 = a$ and H_n denotes the nth Hermite polynomial (with leading coefficient 1, see Section 1.2.2, page 7). It is seen that X_t does not depend on w_t because of physical realizability, therefore it is independent of w_t. Since

$\{X_t, t \in \mathbf{Z}\}$ is also physically realizable, subordinated to $\{w_t, t \in \mathbb{Z}\}$ and second order stationary, it has a Wiener–Itô representation, see Theorem 27, page 48,

$$X_t = \sum_{r=1}^{\infty} \int_{\mathcal{D}^r} \exp\left(i2\pi t \sum_{j=1}^{r} \omega_j\right) f_r(\omega_{(1:r)}) W(d\omega_{(1:r)}), \qquad t \in \mathbb{Z},$$

where $f_n \in \overline{L^n_\Phi}$. The integrals are r–fold Wiener–Itô stochastic integrals with respect to the Gaussian stochastic measure $W(d\omega)$ and

$$\sum_{r=0}^{\infty} r! \| \operatorname{sym} f_r \|_\Phi^2 < \infty. \tag{3.4}$$

The quantity $\|f_r\|_\Phi^2$ was defined earlier, page 34

It is well–known that f_r is not uniquely determined for $r \geq 2$ in a Wiener–Itô representation rather only up to symmetrization. This fact makes it possible to choose the transfer function f_r to make the calculations easier. The product $X_t w_t$ has a multiple stochastic spectral representation and can be calculated by the help of the diagram formula, see (2.7), page 42,

$$\begin{aligned}
X_t w_t &= \sum_{r=1}^{\infty} \int_{\mathcal{D}} \exp(i2\pi t\omega) W(d\omega) \\
&\quad \times \int_{\mathcal{D}^r} \exp\left(i2\pi t \sum_{j=1}^{r} \omega_j\right) f_r(\omega_{(1:r)}) W(d\omega_{(1:r)}) \\
&= \sum_{r=1}^{\infty} \int_{\mathcal{D}^{r+1}} \exp(i2\pi t \sum_{j=1}^{r+1} \omega_j) f_r(\omega_{(1:r)}) W(d\omega_{(1:r+1)}) \\
&+ \sum_{r=1}^{\infty} \int_{\mathcal{D}^{r-1}} \exp(i2\pi t \sum_{j=1}^{r-1} \omega_j) \sigma^2 \sum_{k=1}^{r} \int_{\mathcal{D}} f_r(\omega_{(1:r)}) d\omega_k W(d\omega_{(1:r-1)}).
\end{aligned} \tag{3.5}$$

Observe that the constant term appears only when $r = 1$. It is

$$\sigma^2 \int_{\mathcal{D}} f_1(\omega_1) d\omega_1 = \mathsf{E} X_t w_t = 0.$$

The process X_t depends on $w_{t-1}, w_{t-2} \ldots$ and if $\varphi_{k_{(1:r)}}$ denotes the Fourier coefficients of $f_r(\omega_{(1:r)})$ expanding via $\exp(-i2\pi \sum_{j=1}^{r} k_j \omega_j)$, then $\varphi_{k_{(1:r)}}$ is the coefficient of $H_r\left(w_{(t\vec{1}-k_{(1:r)})}\right)$ in the Hermite expansion of X_t, see formula (2.5), page 40. Therefore in general

$$\sigma^2 \int_{\mathcal{D}} f_r(\omega_{(1:r)}) d\omega_k = \varphi_{(0,0,\ldots,0)} = 0,$$

because the smallest index of the Fourier coefficients of $f_r(\omega_{(1:r)})$ is 1. That is. if at least one entry of $k_{(1:r)}$ is less then 1, then $\varphi_{k_{(1:r)}} = 0$. As the second term in (3.5) is zero we have

$$X_t \mathrm{w}_t = \sum_{r=1}^{\infty} \int_{\mathcal{D}^{r+1}} \exp(i2\pi t \sum_{j=1}^{r+1} \omega_j) f_r(\omega_{(1:r)}) W(d\omega_{(1:r+1)}). \tag{3.6}$$

Now if we compare the multiple stochastic spectral representation of both sides of equation (3.3), using (3.6) for a chaotic representation of $X_{t-1}\mathrm{w}_{t-1}$, we get the transfer functions

$$\begin{aligned} f_0 &= 0, \\ f_1(\omega_1) &= (\exp(i2\pi\omega_1) - a)^{-1} b_1, \\ f_2(\omega_{(1:2)}) &= (\exp(i2\pi(\omega_1 + \omega_2)) - a)^{-1} d(f_1(\omega_1) + 1), \\ f_3(\omega_{(1:3)}) &= \left(\exp\left(i2\pi \sum_{j=1}^{3} \omega_j\right) - a\right)^{-1} df_2(\omega_{(1:2)}) \end{aligned}$$

and

$$f_r(\omega_{(1:r)}) = \left(\exp\left(i2\pi \sum_{j=1}^{r} \omega_j\right) - a\right)^{-1} df_{r-1}(\omega_{(1:r-1)}),\ r \geq 3. \tag{3.7}$$

Although the recursion for f_r is valid only if $r \geq 3$, this will not cause any problem to prove the convergence of (3.4). First we show that if one take two different permutations $\mathfrak{p}, \mathfrak{q} \in \mathfrak{P}_r$ where $\mathfrak{p}(1:r) = (\mathfrak{p}(1), \mathfrak{p}(2), \dots, \mathfrak{p}(r))$ and $\mathfrak{q}(1:r) = (\mathfrak{q}(1), \mathfrak{q}(2), \dots, \mathfrak{q}(r))$, then $f_r(\omega_{\mathfrak{p}(1:r)})$ and $f_r(\omega_{\mathfrak{q}(1:r)})$ are orthogonal, i.e.

$$\int_{\mathcal{D}^r} f_r(\omega_{\mathfrak{p}(1:r)}) \overline{f_r(\omega_{\mathfrak{q}(1:r)})} d\omega_{(1:r)} = 0, \tag{3.8}$$

unless $\mathfrak{q}(1:r) = (\mathfrak{p}(2), \mathfrak{p}(1), \mathfrak{p}(3), \dots, \mathfrak{p}(r))$. In other words unless the permutation $\mathfrak{q}(1:r)$ comes from $\mathfrak{p}(1:r)$ by changing the first two entries. It is enough to prove that this is true for $f_r(\omega_{(1:r)})$ and $f_r(\omega_{\mathfrak{p}(1:r)})$ since for every permutation $\mathfrak{p}$, $\mathfrak{q} \in \mathfrak{P}_r$ there exists a $\mathfrak{p}_1 \in \mathfrak{P}_r$ such that

$$\int_{\mathcal{D}^r} f_r(\omega_{\mathfrak{p}(1:r)}) \overline{f_r(\omega_{\mathfrak{q}(1:r)})} d\omega_{(1:r)} = \int_{\mathcal{D}^r} f_r(\omega_{(1:r)}) \overline{f_r(\omega_{\mathfrak{p}_1(1:r)})} d\omega_{(1:r)}. \tag{3.9}$$

Notice that f_2 is a sum of two orthogonal functions

$$\begin{aligned} f_2(\omega_{(1:2)}) = {} & (\exp(i2\pi(\omega_1 + \omega_2)) - a)^{-1} df_1(\omega_1) \\ & + d(\exp(i2\pi(\omega_1 + \omega_2)) - a)^{-1}. \end{aligned}$$

Let us separate the transfer functions according to these terms, i.e.

$$f_r(\omega_{(1:r)}) = g_r(\omega_{(1:r)}) + h_r(\omega_{(1:r)}).$$

This results in the two systems of functions

$$\begin{aligned}
g_1(\omega_{(1)}) &= f_1(\omega_{(1)}),\\
h_1(\omega_{(1)}) &= 0,\\
g_2(\omega_{(1:2)}) &= (\exp(i2\pi(\omega_1+\omega_2) - a)^{-1} d f_1(\omega_1),\\
h_2(\omega_{(1:2)}) &= (\exp(i2\pi(\omega_1+\omega_2) - a)^{-1} d,\\
g_r(\omega_{(1:r)}) &= \left(\exp\left(i2\pi\sum_{j=1}^{r}\omega_j\right) - a\right)^{-1} d g_{r-1}(\omega_{(1:r-1)}), \qquad r \geq 3,\\
h_r(\omega_{(1:r)}) &= \left(\exp\left(i2\pi\sum_{j=1}^{r}\omega_j\right) - a\right)^{-1} d h_{r-1}(\omega_{(1:r-1)}), \qquad r \geq 3,
\end{aligned}$$

i.e., the same recursion as (3.7) is valid for both g_r and h_r. It is clear that g_r and h_r are orthogonal and h_r behaves as g_{r-1} considering that h_2 depends on $\omega_1 + \omega_2$ only. That is a function of one variable only. It is easy to see that there is no constant term in the series expansion of f_1, i.e.

$$f_1(\omega_1) = b_1 \sum_{k=1}^{\infty} a^k \exp(-i2\pi k\omega_1).$$

Now the series expansion of g_2 is

$$g_2(\omega_{(1:2)}) = b_1 d \sum_{k,l=1}^{\infty} a^{k+l} \exp\left(-i2\pi\left((k+l)\,\omega_1 + l\omega_2\right)\right).$$

This has the important property that the coefficient of ω_1, $k+l$, is always greater than the coefficient of ω_2, l. This property implies that $g_2(\omega_{(1:2)})$ and $g_2(\omega_{\mathfrak{p}(1:2)})$ where $\omega_{\mathfrak{p}(1:2)} = (\omega_2, \omega_1)$, are orthogonal. In general the transfer function g_r depends on ω_1, $\omega_1 + \omega_2$, $\ldots$, $\sum_{j=1}^{r}\omega_j$ and the only nonzero Fourier coefficients $\psi_{-k_{(1:r)}}$ of g_r are those for which the entries of indices $k_{(1:r)}$ fulfil $0 < k_r < k_{r-1} < k_{r-2} < \ldots < k_1$, see (3.7). If one takes a permutation $\omega_{\mathfrak{p}(1:r)}$ of $\omega_{(1:r)}$, then the $\exp\left(-i2\pi\sum_{j=1}^{r} k_j\omega_{\mathfrak{p}(j)}\right)$ will be orthogonal for all $\exp(-i2\pi\sum_{j=1}^{r} l_j\omega_j)$ with $0 < l_r < l_{r-1} < l_{r-2} < \ldots < l_1$. The same is true for h_r changing ω_1 and ω_2. Thus the orthogonality (3.8) is proved. Now because of the orthogonality

$$\begin{aligned}
r!\|\operatorname{sym} g_r\|_\Phi^2 &= r!\sigma^{2r}\int_{\mathcal{D}^r} g_r(\omega_{(1:r)})\operatorname{sym}\overline{g_r(\omega_{(1:r)})}\,d\omega_{(1:r)}\\
&= \frac{\sigma^{2r} d^{2(r-1)} b_1^2}{(1-a^2)^r}, \; r \geq 1,
\end{aligned}$$

and because of the symmetry in ω_1 and ω_2

$$\begin{aligned} r!\| \operatorname{sym} h_r \|_\Phi^2 &= r!\sigma^{2r} \int_{\mathcal{D}^r} h_r(\omega_{(1:r)}) \operatorname{sym} \overline{h_r(\omega_{(1:r)})} d\omega_{(1:r)} \\ &= \frac{2\sigma^{2r} d^{2(r-1)}}{(1-a^2)^{r-1}}, \; r \geq 2. \end{aligned}$$

The explicit form of the variance of X_t is

$$\begin{aligned} \operatorname{var}(X_t) &= \sum_{r=0}^{\infty} r! \| \operatorname{sym} f_r \|_\Phi^2 \\ &= \sum_{r=0}^{\infty} r! \left(\| \operatorname{sym} g_r \|_\Phi^2 + \| \operatorname{sym} h_r \|_\Phi^2 \right) \\ &= \sigma^2 \frac{a^2 + 2\sigma^2 d^2}{1 - a^2 - \sigma^2 d^2}. \end{aligned}$$

The series is convergent if and only if $a^2 + \sigma^2 d^2 < 1$. The situation when $a^2 + \sigma^2 d^2$ is close to *one* has some interesting consequencies, see [71] in linear case. We have proved the following theorem.

Theorem 31 *The necessary and sufficient condition for the existence of a stationary, physically realizable and subordinated to* w_t *solution (Assumption 30) of the bilinear model*

$$Y_t = aY_{t-1} + dY_{t-1}\mathrm{w}_{t-1} + \mathrm{w}_t + f_0, \qquad t \in \mathbb{Z},$$

where $a, d \in \mathbb{R}$ *and* $f_0 = -\sigma^2 d$ *is*

$$a^2 + \sigma^2 d^2 < 1.$$

In that case the transfer functions of $Y_t - \mathrm{w}_t$ *are given by (3.7). In particular*

$$\operatorname{var}(Y_t) = \sigma^2 \frac{1 + \sigma^2 d^2}{1 - a^2 - \sigma^2 d^2},$$

while the covariance of $X_t = Y_t - \mathrm{w}_t$ *has the form*

$$\operatorname{cov}(X_t, X_{t+s}) = \int_{\mathcal{D}} \exp(i2\pi s\omega) \frac{a^2 + 2\sigma^2 d^2}{1 - a^2 - \sigma^2 d^2} \frac{\sigma^2 (1 - a^2)}{|\exp(i2\pi\omega) - a|^2} d\omega,$$

and therefore the spectrum of Y_t *is*

$$S_{Y,Y}(\omega) = \frac{a^2 + 2\sigma^2 d^2}{1 - a^2 - \sigma^2 d^2} \frac{\sigma^2 (1 - a^2)}{|\exp(i2\pi\omega) - a|^2} + \sigma^2.$$

The transfer functions of the bilinear model (3.2) in time domain was given by Priestley [98] and in frequency domain by Terdik [126].

Bilinear model with nonzero mean

There might be a situation in which the process fulfils the bilinear equation

$$Y_t = aY_{t-1} + dY_{t-1}\mathrm{w}_{t-1} + \mathrm{w}_t, \tag{3.10}$$

i.e., in distinction to the process considered earlier in equation (3.2), the mean of Y_t is not zero. The mean is

$$\mathsf{E}Y_t = \frac{\sigma^2 d}{1-a}.$$

This case has been considered by Subba Rao and Gabr [122], for example. The contribution of our treatment here, as compared to the results available, is that the higher order spectra of the process (3.10) will be given clearly enough to base parameter estimation upon. Assume that $\mathrm{w}_t \in \mathcal{N}(0,1)$ otherwise standardize by w_t/σ, Y_t/σ and σd. It is direct that (3.10) is equivalent to the system of equations

$$\begin{aligned} X_t &= aX_{t-1} + dX_{t-1}\mathrm{w}_{t-1} + b_1 H_1(\mathrm{w}_{t-1}) + dH_2(\mathrm{w}_{t-1}), \\ Y_t &= X_t + \mathrm{w}_t + \frac{d}{1-a}, \end{aligned} \tag{3.11}$$

where

$$b_1 = a + d^2/(1-a).$$

One has

$$\mathsf{E}X_t = 0, \ .$$

Notice that the difference between (3.3) and (3.11) is only in the coefficient b_1 which has changed by $d^2/(1-a)$. The transfer function system $\{f_0, f_1, \dots, f_n, \dots\}$ for X_t, where f_0 is zero, is given by the recursion formulae, see (3.7),

$$\begin{aligned} f_1(z_1) &= \frac{b_1}{z_1 - a}, \\ f_2(z_{(1:2)}) &= \frac{d}{z_{(1:2)}^{\mathbf{1}} - a}\left(f_1(z_1) + 1\right), \\ f_r(z_{(1:r)}) &= \frac{d}{z_{(1:r)}^{\mathbf{1}} - a} f_{r-1}(z_{(1:r-1)}),\ r > 2. \end{aligned} \tag{3.12}$$

The consideration that each transfer function has nonzero Fourier coefficients with negative indexes only , i.e.

$$f_r(z_{(1:r)}) = \sum_{k_j \geq 1} f_{k_{(1:r)}} z_{(1:r)}^{-k_{(1:r)}}, \quad k_{(1:r)} = (k_1, k_2, \dots, k_r), \tag{3.13}$$

will be very useful later on. If $f \in \overline{L_\Phi^r}$ and $g \in \overline{L_\Phi^r}$ where $\Phi(d\omega) = d\omega$ then

$$\int_{\mathcal{D}^r} f_r(z_{(1:r)}) W(d\omega_{(1:r)}), \quad \int_{\mathcal{D}^r} g_r(z_{(1:r)}) W(d\omega_{(1:r)}),$$

are well defined. We introduce the inner product

$$\begin{aligned}(f,g)_r^{\mathfrak{P}} &= r! \int_{\mathcal{D}^r} f_r^{\mathfrak{P}}(z_{(1:r)}) g_r^{\mathfrak{P}}(\overline{z_{(1:r)}}) \sigma^{2r} d\omega_{(1:r)} \\ &= \mathsf{E} \int_{\mathcal{D}^r} f_r(z_{(1:r)}) W(d\omega_{(1:r)}) \int_{\mathcal{D}^r} g_r(z_{(1:r)}) W(d\omega_{(1:r)}).\end{aligned}$$

Here $f_r^{\mathfrak{P}}$ denotes the symmetrized version for f, i.e.

$$f_r^{\mathfrak{P}}(z_{(1:r)}) = \mathsf{sym}_{z_{(1:r)}} f(z_{(1:r)}) = \frac{1}{r!} \sum_{\mathfrak{p} \in \mathfrak{P}_r} f(z_{\mathfrak{p}(1:r)}),$$

where $\mathfrak{P}_r$ is the group of permutations for the integers $1, 2, \ldots, r$, see also on page 34. Sometimes when we would like to point out the number of variables to be symmetrized by and will use the notation $[f]_r^{\mathfrak{P}}$ as well. The proof of the following lemma is quite evident.

Lemma 32 *If* $f, g \in \overline{L_\Phi^{r-1}}$, $a(z)$, $b(z) \in \overline{L_\Phi^1}$, *all these functions have property (3.13) and*

$$\int_0^1 f(z_{(1:r-1)}) d\omega_k = \int_0^1 g(z_{(1:r-1)}) d\omega_k = 0, \ k = 1, 2, .., r-1,$$

Then

$$\left(a(z_{(1:r)}^{\vec{1}}) f(z_{(1:r-1)}), b(z_{(1:r)}^{\vec{1}}) g(z_{(1:r-1)}) \right)_r^{\mathfrak{P}} = (a(z), b(z))_1 \, (f, g)_{r-1}^{\mathfrak{P}}.$$

Proof. As $a(z_{(1:r)}^{\vec{1}})$ and $b(z_{(1:r)}^{\vec{1}})$ are symmetric therefore

$$\begin{aligned}\mathsf{sym}_{z_{(1:r)}} \left(a(z_{(1:r)}^{\vec{1}}) f(z_{(1:r-1)}) \right) &= a(z_{(1:r)}^{\vec{1}}) \mathsf{sym}_{z_{(1:r)}} f(z_{(1:r-1)}) \\ &= a(z_{(1:r)}^{\vec{1}}) \frac{1}{r} \sum_{k=1}^{r} [f]_{r-1}^{\mathfrak{P}}(z_{(1:r)\setminus\{k\}}),\end{aligned}$$

where $z_{(1:r)\setminus\{k\}}$ denotes the vector $z_{(1:r)}$ with the kth entry missing. Now it is enough to show that $a \times [f]_{r-1}^{\mathfrak{P}}$ and $b \times [g]_{r-1}^{\mathfrak{P}}$ are orthogonal unless their variables are the same, i.e., for $k \neq r$

$$\int_0^1 a(z_{(1:r)}^{\vec{1}}) f(z_{(1:r-1)}) b(\overline{z}_{(1:r)}^{\vec{1}}) g(\overline{z}_{(1:r)\setminus\{k\}}) d\omega_k = 0,$$

Changing the variables $z_{(1:r)}^{\vec{1}} = v_k$ and $z_j = v_j$, $j \neq k$ then integrating by ω_r one gets the desired result. ■

3.2.2 *Spectra*

Spectrum

Now it is straightforward to derive the spectrum for the state process X_t of equation (3.11). Start with the covariances

$$\begin{aligned}\mathsf{E}X_tX_{t+s} &= \sum_{r=1}^{\infty}\left(z_{(1:r)}^{s\vec{1}}f_r(z_{(1:r)}), f_r(z_{(1:r)})\right)_r^{\mathfrak{P}} \\ &= \int_0^1 z^s\left|\frac{b_1}{z-a}\right|^2 d\omega + \int_0^1 z^s\left|\frac{d}{z-a}\right|^2 d\omega \\ \times &\left[\int_0^1 (|f_1(z)|^2+2)d\omega + \sum_{r=3}^{\infty}(f_{r-1}(z_{(1:r-1)}), f_{r-1}(z_{(1:r-1)}))_{r-1}^{\mathfrak{P}}\right] \\ &= \int_0^1 z^s\frac{const.d^2+b_1^2}{|z-a|^2}d\omega,\end{aligned}$$

which means that the spectrum of X_t has the form

$$S_{XX}(z) = \frac{const.d^2+b_1^2}{|z-a|^2}.$$

The constant can be easily determined because $\mathsf{E}X_t^2$ can be got directly from the equation (3.11) and from the spectrum also

$$\mathsf{E}X_t^2 = \frac{2d^2+b_1^2}{1-a^2-d^2} = \frac{const.d^2+b_1^2}{1-a^2}.$$

Therefore

$$S_{XX}(z) = \frac{2d^2+b_1^2}{|z-a|^2}\frac{1-a^2}{1-a^2-d^2}.$$

Now if we put $s=0$ in the covariances above, we see by the previous Lemma 32 that

$$(f_r(z_{(1:r)}), f_r(z_{(1:r)}))_r^{\mathfrak{P}} = \left[\frac{d^2}{1-a^2}\right]^{r-2}(f_2(z_{(2)}), f_2(z_{(2)}))_2^{\mathfrak{P}}, \tag{3.14}$$

therefore the necessary and sufficient condition for the convergence of the series

$$\sum_{r=1}^{\infty}(f_r(z_{(1:r)}), f_r(z_{(1:r)}))_r^{\mathfrak{P}},$$

i.e., for the second order stationarity of X_t, as we have shown earlier, is that $a^2+d^2<1$.

Quadratic Process

If the solution X_t of the equation (3.11) is stationary in second order, , it is stationary in the strict sense as well. In that case the quadratic process $Q_t = X_t^2$ is also strictly stationary and under condition to be given later in Theorem 37, page 88, it is second order stationary (Terdik [130]). Namely, as

$$(a + d\mathrm{w}_0)^2 = a^2 + d^2 + 2adH_1(\mathrm{w}_0) + d^2 H_2(\mathrm{w}_0),$$

the necessary and sufficient condition of stationarity of Q_t is that the second order moment of $(a + d\mathrm{w}_0)^2$ is less then one, i.e., $2d^4 + 4a^2d^2 + (a^2 + d^2)^2 < 1$. Put the Wiener–Itô representation for Q_t

$$Q_t = q_0 + \sum_{r=1}^{\infty} \int_{D^r} z_{(1:r)}^{t\tilde{1}} q_r(z_{(1:r)}) W(d\omega_{(1:r)}). \tag{3.15}$$

The transfer function system for Q is determined by the square of equation (3.11) as

$$\begin{aligned}Q_t &= Q_{t-1}\left[d^2 H_2(\mathrm{w}_{t-1}) + 2ad\mathrm{w}_{t-1} + a^2 + d^2\right] \\ &+ 2X_{t-1}\left[d^2 H_3(\mathrm{w}_{t-1}) + d(a+b_1)H_2(\mathrm{w}_{t-1}) + (2d^2 + ab_1)\mathrm{w}_{t-1} + b_1 d\right] \\ &+ d^2 H_4(\mathrm{w}_{t-1}) + 2b_1 d H_3(\mathrm{w}_{t-1}) + \left[4d^2 + b_1^2\right] H_2(\mathrm{w}_{t-1}) \\ &+ 4b_1 d\mathrm{w}_{t-1} + b_1^2 + 2d^2.\end{aligned} \tag{3.16}$$

The diagram formula of second order is used to put both sides of the above equation into Wiener–Itô spectral representation. If we compare the transfer functions of the same order, we find that the transfer functions $q_0, \ldots, q_4$, are given explicitly by

$$\begin{aligned}q_0 &= \frac{b_1^2 + 2d^2}{1 - a^2 - d^2} = \mathsf{E}X_t^2, \\ q_1(z_1) &= \frac{1}{z_1 - a^2 - d^2}\left(2b_1 d f_1(z_1) + 2adq_0 + 4b_1 d\right), \\ q_2(z_{(2)}) &= \frac{1}{z_{(2)}^{\tilde{1}} - a^2 - d^2}\Big(2adq_1(z_1) + d^2 q_0 \\ &\quad + 2b_1 d f_2(z_{(2)}) + 2(2d^2 + ab_1) f_1(z_1) + 4d^2 + b_1^2\Big), \\ q_3(z_{(1:3)}) &= \frac{1}{z_{(1:3)}^{\tilde{1}} - a^2 - d^2}\Big(2adq_2(z_{(2)}) + d^2 q_1(z_1) + 2b_1 d f_3(z_{(1:3)}) \\ &\quad + 2(2d^2 + ab_1) f_2(z_{(2)}) + 2d(a+b_1) f_1(z_1) + 2b_1 d\Big), \\ q_4(z_{(4)}) &= \frac{1}{z_{(1:r)}^{\tilde{1}} - a^2 - d^2}\Big(2adq_3(z_{(1:3)}) + d^2 q_2(z_{(2)}) + 2b_1 d f_4(z_{(4)}) \\ &\quad + 2(2d^2 + ad) f_3(z_{(1:3)}) + 2d(a+b_1) f_2(z_{(2)}) + 2d^2 f_1(z_1) + d^2\Big).\end{aligned} \tag{3.17}$$

If $r > 4$, the following recursive formula is applied:

$$q_r(z_{(1:r)}) = \frac{1}{z_{(1:r)}^{\vec{1}} - a^2 - d^2} \left(2adq_{r-1}(z_{(r-1)}) + d^2 q_{r-2}(z_{(r-2)})\right.$$
$$+ 2b_1 d f_r(z_{(1:r)}) + 2(2d^2 + ad) f_{r-1}(z_{(r-1)})$$
$$\left. + 2d(a+d) f_{r-2}(z_{(r-2)}) + 2d^2 f_{r-3}(z_{(r-3)})\right).$$

Cross-spectrum

We are going to consider the cross-spectrum between X_t and Q_t. The following inner products can be determined by the result of the Lemma 32. If $r > 4$,

$$\left(z_{(1:r)}^{s\vec{1}} q_r(z_{(1:r)}), f_r(z_{(1:r)})\right)_r^{\mathfrak{P}}$$
$$= \int_0^1 z^s \frac{2ad^2}{(\overline{z} - a)(z - a^2 - d^2)} d\omega \left(q_{r-1}(z_{(r-1)}), f_{r-1}(z_{(r-1)})\right)_{r-1}^{\mathfrak{P}}$$
$$+ 2d(2d^2 + ad) \int_0^1 z^s \frac{1}{(\overline{z} - a)(z - a^2 - d^2)} d\omega \left(f_{r-1}(z_{(r-1)}), f_{r-1}(z_{(r-1)})\right)_{r-1}^{\mathfrak{P}}$$
$$+ 2b_1 d^3 \int_0^1 z^s \frac{1}{|z - a|^2 (z - a^2 - d^2)} d\omega \left(f_{r-1}(z_{(r-1)}), f_{r-1}(z_{(r-1)})\right)_{r-1}^{\mathfrak{P}}.$$

The product $\left(f_{r-1}(z_{(r-1)}), f_{r-1}(z_{(r-1)})\right)_{r-1}^{\mathfrak{P}}$ has been determined before, see (3.14) and the cross-product $\left(q_{r-1}(z_{(r-1)}), f_{r-1}(z_{(r-1)})\right)_{r-1}^{\mathfrak{P}}$ fulfils the recursion

$$\left(q_{r-1}(z_{(r-1)}), f_{r-1}(z_{(r-1)})\right)_{r-1}^{\mathfrak{P}} = \frac{2ad^2 \left(q_{r-2}(z_{(r-2)}), f_{r-2}(z_{(r-2)})\right)_{r-2}^{\mathfrak{P}}}{1 - a(a^2 + d^2)}$$
$$+ \frac{\frac{2ab_1 d^2}{1 - a^2} + 2d(2d^2 + ab_1)}{1 - a(a^2 + d^2)}$$
$$\times \left(f_{r-2}(z_{(r-2)}), f_{r-2}(z_{(r-2)})\right)_{r-2}^{\mathfrak{P}}.$$

Similar expressions are valid for $r = 1, 2, 3, 4$ as well. The cross-spectrum follows as

$$S_{Q,X}(z) = C_1 \frac{1}{|z - a|^2 (z - a^2 - d^2)} + C_2 \frac{1}{(\overline{z} - a)(z - a^2 - d^2)},$$

where C_1 and C_2 are constants to be determined. The first consideration is that

$$\int_0^1 S_{Q,X}(z) d\omega = \mathsf{E} X_t^3.$$

Therefore

$$C_1 \frac{a}{(1-a^2)(1-a(a^2+d^2))} + C_2 \frac{1}{(1-a(a^2+d^2))} = \mathsf{E}X_t^3.$$

On the other hand using the equation for X_t and X_t^2 we get

$$\begin{aligned}\mathsf{E}X_t^3 &= E(aX_{t-1} + dX_{t-1}\mathrm{w}_{t-1} + b_1\mathrm{w}_{t-1} + dH_2(\mathrm{w}_{t-1}))X_t^2 \\ &= E(a(a^2+d^2)X_{t-1}^3 + 2ab_1dX_{t-1}^2 + 2ad^2X_{t-1}^3 \\ &+ 2d(2d^2+ab_1)X_{t-1}^2 + 2ab_1dX_{t-1}^2 + 4b_1^2d \\ &+ 2d^3X_{t-1}^2 + 2d(4d^2+b_1^2)),\end{aligned}$$

The expectation $\mathsf{E}X_t^3$ does not depend on t by the assumption of stationarity and $\mathsf{E}X_t^2$ has been determined above. So

$$\begin{aligned}\mathsf{E}X_t^3 &= \frac{(6ab_1d + 6d^3)\mathsf{E}X_t^2 + 2d(3b_1^2+4d^2)}{1-a(a^2+3d^2)} \\ &= 2d\frac{(3ab_1+3d^2)\mathsf{E}X_t^2 + 3b_1^2 + 4d^2}{1-a(a^2+3d^2)}.\end{aligned} \tag{3.18}$$

Constant C_1 can be determined directly from the inner products as the sum of the coefficients of the integral

$$\int_0^1 z^s \frac{1}{|z-a|^2(z-a^2-d^2)} d\omega,$$

i.e.

$$\begin{aligned}C_1 &= 2b_1^3d + 2b_1d^3(f_1(z), f_1(z)) + 4b_1d^3 \\ &+ 2b_1d^3\sum_{r=2}^{\infty}(f_r(z_{(1:r)}), f_r(z_{(1:r)}))_r^{\mathfrak{P}} \\ &= 2b_1^3d + 4b_1d^3 + 2b_1d^3\mathsf{E}X_t^2 \\ &= 2b_1d\left[1-a^2\right]\mathsf{E}X_t^2.\end{aligned}$$

Now we have

$$C_2 = (1-a(a^2+d^2))\mathsf{E}X_t^3 - 2ab_1d\mathsf{E}X_t^2.$$

Therefore the explicit expression for the cross-spectrum is

$$\begin{aligned}S_{Q,X}(z) = \frac{1-a(a^2+d^2)}{(z^{-1}-a)(z-a^2-d^2)} &\underset{XXX}{\mathrm{Cum}}(0,0) \\ &+ \frac{2b_1d}{(1-az^{-1})(z-a^2-d^2)}\underset{XX}{\mathrm{Cum}}(0).\end{aligned}$$

Note here that if a process X_t is k^{th} order stationary, its kth order cumulants depend on $k-1$ variables only, therefore we shall use the notation

$$\operatorname{Cum}\left(X_{t_{(1:k)}}\right) = \underset{\underbrace{XX\ldots X}_{k}}{\operatorname{Cum}}\left(t_{(2:k)} - t_1\right),$$

in particular,

$$\operatorname{Cum}\left(X_{t_{(1:3)}}\right) = \underset{XXX}{\operatorname{Cum}}\left(t_2 - t_1, t_3 - t_1\right).$$

Spectrum for Q

The same method we used to determine the cross-spectrum leads to the spectrum of Q as well. Consider the inner product for $r > 4$

$$\begin{aligned}(z_{(1:r)}^{s\vec{1}} q_r(z_{(1:r)}), q_r(z_r))_r^{\mathfrak{P}} &= const. \int_0^1 z^s \frac{1}{|z-a^2-d^2|^2} d\omega \\ &+ 4b_1^2 d^4 (f_{r-1}(z_{(r-1)}), f_{r-1}(z_{(r-1)}))_{r-1}^{\mathfrak{P}} \\ &\times \int_0^1 z^s \frac{1}{|z-a^2-d^2|^2 |z-a|^2} d\omega \\ &+ 8ab_1 d^3 (q_{r-1}(z_{(r-1)}), f_{r-1}(z_{(r-1)}))_{r-1}^{\mathfrak{P}} \\ &\times \operatorname{Re} \int_0^1 z^s \frac{1}{|z-a^2-d^2|^2 (z-a)} d\omega,\end{aligned}$$

where Re denotes the real part of the complex valued integral. The cases $r = 1, 2, 3, 4$ produce same terms as above with different coefficients of the integrals. Therefore the spectrum is given up to some constants.

$$\begin{aligned}S_{Q,Q}(z) &= \frac{C_1}{|z-a^2-d^2|^2 |z-a|^2} + 2\operatorname{Re}\frac{C_2}{|z-a^2-d^2|^2 (z-a)} \\ &+ \frac{C_3}{|z-a^2-d^2|^2} \\ &= \frac{\sigma_3^2 |z-A|^2}{|z-a^2-d^2|^2 |z-a|^2}.\end{aligned}$$

The parameters A and σ_3 can be determined either by the moments or by the equation (3.16). Here we show another way of expressing $S_{Q,Q}$. Suppose that we are going to express the spectrum of linear combination of stationary correlated series $Z_{1,t}$ and $Z_{2,t}$ by the spectra of these processes. It is easy to see that

$$S_{2,\alpha Z_1 + \beta Z_2}(z) = \alpha^2 S_{2,Z_1}(z) + \alpha\beta 2\operatorname{Re} S_{Z_1,Z_2}(z) + \beta^2 S_{2,Z_2}(z).$$

Put

$$Q_t - (a^2 + d^2)Q_{t-1} - (b_1^2 + 2d^2) - 2b_1 d X_{t-1} = \varepsilon_{t-1},$$

where ε_t is defined by equation (3.16), i.e.

$$\begin{aligned}\varepsilon_t &= d^2 Q_{t-1} H_2(\mathrm{w}_{t-1}) + 2ad\mathrm{w}_{t-1} \\ &+ 2X_{t-1}\left[d^2 H_3(\mathrm{w}_{t-1}) + d(a+b_1)H_2(\mathrm{w}_{t-1}) + (2d^2 + ab_1)\mathrm{w}_{t-1}\right] \\ &+ d^2 H_4(\mathrm{w}_{t-1}) + 2b_1 d H_3(\mathrm{w}_{t-1}) + \left[4d^2 + b_1^2\right] H_2(\mathrm{w}_{t-1}) + 4b_1 d\mathrm{w}_{t-1}.\end{aligned}$$

It is an uncorrelated series with expectation $\mathsf{E}\varepsilon_t = 0$ and variance $\sigma_\varepsilon^2 = \mathsf{E}\varepsilon_t^2$. Therefore

$$S_{Q,Q}(z) = \frac{\sigma_\varepsilon^2 + 4b_1 d\mathrm{Re}\left\{[z - (a^2 + d^2)]S_{Q,X}(z)\right\} - 4b_1^2 d^2 S_{XX}(z)}{|1 - (a^2 + d^2)z|^2},$$

since $S_{\alpha(z)Z_1,\beta(z)Z_2}(z) = \alpha(z)\overline{\beta(z)}S_{Z_1,Z_2}(z)$. Now $\sigma_\varepsilon^2 = \mathsf{E}\varepsilon_t^2$ is calculated directly, i.e.

$$\begin{aligned}\sigma_\varepsilon^2 &= \mathsf{E}X_t^4\left[2d^4 + 4a^2d^2\right] \\ &+ 4\mathsf{E}X_t^3\left[2d^3(a+b_1) + 2ad(2d^2 + ab_1)\right] \\ &+ 2\mathsf{E}X_t^2\left[2d^2(4d^2 + b_1^2) + 8ab_1d^2\right] \\ &+ 4\mathsf{E}X_t^2\left[6d^4 + 2d^2(a+b_1)^2 + (2d^2 + ab_1)^2\right] \\ &+ 56d^4 + 56b_1^2d^2 + 2b_1^4.\end{aligned}$$

To get the fourth order moment $\mathsf{E}Q_t^2 = \mathsf{E}X_t^4$ we take the expected value of the square of the equation (3.16) for the quadratic process.

$$\begin{aligned}\mathsf{E}X_t^4 &= \mathsf{E}X_{t-1}^4\left\{2d^4 + 4a^2d^2 + (a^2 + d^2)^2\right\} \\ &+ 4\mathsf{E}X_{t-1}^3\left\{2d^3(a+b_1) + 2ad(2d^2 + ab_1) + (a^2 + d^2)b_1 d\right\} \\ &+ 4\mathsf{E}X_{t-1}^2\left\{6d^4 + 2(d(a+b_1))^2 + (2d^2 + ab_1)^2 + b_1^2 d^2\right\} \\ &+ 2\mathsf{E}X_{t-1}^2\left\{2d^2(4d^2 + b_1^2) + 8ab_1d^2 + (a^2 + d^2)(b_1^2 + 2d^2)\right\} \\ &+ 24d^4 + 6(2b_1d)^2 + 2(4d^2 + b_1^2)^2 + (4b_1d)^2 + (b_1^2 + 2d^2)^2.\end{aligned}$$

Therefore

$$\begin{aligned}\mathsf{E}X_t^4 = &\frac{1}{1 - (2d^4 + 4a^2d^2 + (a^2 + d^2)^2)}\left[60d^4 + 60b_1^2d^2 + 3b_1^4\right. \\ &+ \left\{58d^4 + 12a^2d^2 + 18b_1^2d^2 + 48ab_1d^2 + 6a^2b_1^2\right\}\mathsf{E}X_{t-1}^2 \\ &\left. + 12d\left\{2ad^2 + b_1d^2 + a^2b_1\right\}\mathsf{E}X_{t-1}^3\right].\end{aligned}$$

Bispectrum for the state process

It does not seem too difficult to derive the bispectrum for the state process X_t mainly because of the recursive connection between the cross-moments and the third order cumulants of it. The process X_t is centered, therefore its third order cumulants are

$$\underset{XXX}{\mathrm{Cum}}(k,l) = \mathsf{E}X_{t+k}X_{t+l}X_t,\ k,l = 0,1,2,\ldots.$$

The easy consequence of this fact is that the following properties are fulfilled

$$\begin{aligned}\underset{XXX}{\mathrm{Cum}}(k,l) &= \underset{XXX}{\mathrm{Cum}}(l,k),\\ \underset{XXX}{\mathrm{Cum}}(-k,l) &= \underset{XXX}{\mathrm{Cum}}(k+l,k),\\ \underset{XXX}{\mathrm{Cum}}(-k-l,-k) &= \underset{XXX}{\mathrm{Cum}}(k+l,l).\end{aligned}$$

The bispectrum according to the cumulants $\underset{XXX}{\mathrm{Cum}}(k,l)$ is defined by

$$S_{X,X,X}(z_1,z_2) = \sum_{k=-\infty}^{\infty}\sum_{l=-\infty}^{\infty}\underset{XXX}{\mathrm{Cum}}(k,l)z_1^{-k}z_2^{-l}.$$

The bispectrum can be separated into six parts and we see that according to the above properties of the cumulants it is a symmetrization of some functions, i.e.,

$$\begin{aligned}&S_{X,X,X}(z_1,z_2) =\\ &\sum_{k=1}^{\infty}\sum_{l=1}^{\infty}\underset{XXX}{\mathrm{Cum}}(k+l,l)z_1^{-(k+l)}z_2^{-l} + \sum_{k=1}^{\infty}\sum_{l=1}^{\infty}\underset{XXX}{\mathrm{Cum}}(k+l,l)z_1^{-l}z_2^{-(k+l)}\\ &+\sum_{k=1}^{\infty}\sum_{l=1}^{\infty}\underset{XXX}{\mathrm{Cum}}(-k,l)z_1^{k}z_2^{-l} + \sum_{k=1}^{\infty}\sum_{l=1}^{\infty}\underset{XXX}{\mathrm{Cum}}(-k,l)z_1^{-l}z_2^{k}\\ &+\sum_{k=1}^{\infty}\sum_{l=1}^{\infty}\underset{XXX}{\mathrm{Cum}}(-k-l,-l)z_1^{k+l}z_2^{l} + \sum_{k=1}^{\infty}\sum_{l=1}^{\infty}\underset{XXX}{\mathrm{Cum}}(-k-l,-l)z_1^{l}z_2^{k+l}\\ &+\sum_{l=-\infty}^{\infty}\underset{XXX}{\mathrm{Cum}}(l,l)z_1^{-l}z_2^{-l} + \sum_{l=-\infty}^{\infty}\underset{XXX}{\mathrm{Cum}}(0,l)z_1^{-l} + \sum_{l=-\infty}^{\infty}\underset{XXX}{\mathrm{Cum}}(0,l)z_2^{-l}\\ &-2\underset{XXX}{\mathrm{Cum}}(0,0)\\ &= \mathsf{sym}_{z_{(1:3)}}\left[6\sum_{k=1}^{\infty}\sum_{l=1}^{\infty}\underset{XXX}{\mathrm{Cum}}(k+l,l)z_1^{-(k+l)}z_2^{-l}\right.\\ &\qquad\left.+3\sum_{l=-\infty}^{\infty}\underset{XXX}{\mathrm{Cum}}(0,l)z_1^{-l} - 2\underset{XXX}{\mathrm{Cum}}(0,0)\right]_{z_3=(z_1z_2)^{-1}}.\end{aligned}$$

Let us now turn to our particular case. If we use the equation

$$X_t = aX_{t-1} + dX_{t-1}\mathrm{w}_{t-1} + b_1\mathrm{w}_{t-1} + dH_2(\mathrm{w}_{t-1}),$$

at $t+k+l$, then we arrive at the formula

$$\underset{XXX}{\mathrm{Cum}}(k+l,l) = a^k \underset{XXX}{\mathrm{Cum}}(l,l), \; k \geq 0, l \geq 1.$$

That is why

$$\sum_{k=1}^{\infty}\sum_{l=1}^{\infty} \underset{XXX}{\mathrm{Cum}}(k+l,l) z_1^{-(k+l)} z_2^{-l} = \frac{a}{z_1 - a} S^{+}_{Q,X}(z_3^{-1}),$$

where

$$\begin{aligned} S^{+}_{Q,X}(z) &= \sum_{k=1}^{\infty} \mathrm{Cum}(Q_{t+k}, X_t) z^{-k} \\ &= S_{Q,X}(z) - \frac{1}{1-az} \underset{XXX}{\mathrm{Cum}}(0,0) \\ &= \frac{a^2+d^2}{z-a^2-d^2} \underset{XXX}{\mathrm{Cum}}(0,0) + \frac{2b_1 d}{(1-az^{-1})(z-a^2-d^2)} \underset{XX}{\mathrm{Cum}}(0). \end{aligned}$$

The value of the bispectrum according to the axes of cumulants equals $S_{Q,X}$, i.e.,

$$\sum_{l=-\infty}^{\infty} \underset{XXX}{\mathrm{Cum}}(l,l) z_1^{-l} z_2^{-l} + \sum_{l=-\infty}^{\infty} \underset{XXX}{\mathrm{Cum}}(0,l) z_1^{-l} + \sum_{l=-\infty}^{\infty} \underset{XXX}{\mathrm{Cum}}(0,l) z_2^{-l}$$
$$= \mathrm{sym}_{z_{(1:3)}} \left[3S_{Q,X}(z_1^{-1})\right]_{z_3=(z_1 z_2)^{-1}},$$

so consider first the symmetrized version

$$S^{\mathfrak{P}}_{X,X,X}(z_1,z_2,z_3) = \mathrm{sym}_{z_{(1:3)}} \left\{6\left[\frac{a}{z_1-a}\right] S^{+}_{Q,X}(z_3^{-1}) + 3S_{Q,X}(z_1^{-1})\right\}$$
$$- 2\underset{XXX}{\mathrm{Cum}}(0,0).$$

The **bispectrum** of X_t is given by the function

$$S_{X,X,X}(z_1,z_2) = S^{\mathfrak{P}}_{X,X,X}(z_1,z_2,(z_1 z_2)^{-1}).$$

Consequences for the observation

We assume that the observation Y_t has mean zero, in other words the sample mean has been taken out before fitting a bilinear model. We turn back to the model (3.3). The equations (3.11) for the state process and

(3.11) are the same but for the meaning of the constant b_1. Equation (3.11) shows that

$$Y_t = X_t + \mathrm{w}_t,$$

where X_t and w_t are independent at time t; nevertheless there is a nonlinear relationship between the two processes. The cumulants for the process Y_t and $Y_t + const.$ are equal, so based on the spectra of X_t, we derive the spectra for Y_t independently of its mean being zero or not.

Spectrum of the observation

The autocovariance function of Y_t can be expressed in the following way:

$$\begin{aligned}\underset{YY}{\mathrm{Cum}}(k) &= \underset{XX}{\mathrm{Cum}}(k) + \mathsf{E}X_{t+k}\mathrm{w}_t + \mathsf{E}X_t\mathrm{w}_{t+k} + \mathsf{E}\mathrm{w}_{t+k}\mathrm{w}_t \\ &= \underset{XX}{\mathrm{Cum}}(k) + \delta_{k\geq 1} b_1 a^{k-1} + \delta_{k\leq -1} b_1 a^{-k-1} + \delta_{k=0}. \end{aligned} \tag{3.19}$$

The last equation (3.19) implies that the **spectral density** for Y_t is

$$\begin{aligned} S_{Y,Y}(z) &= S_{XX}(z) + f_1(z) + f_1(z^{-1}) + 1 \\ &= \frac{(1-a^2)(1+\dfrac{d^2}{a(1-a)} + \mathsf{E}X_t^2)}{|z-a|^2} - \frac{d^2}{a(1-a)} \\ &= \sigma_2^2 \frac{|z-b|^2}{|z-a|^2}, \end{aligned} \tag{3.20}$$

where obviously b is some constant. Unfortunately as (3.20) is an $ARMA(1,1)$ spectrum, one should pay attention to estimating the linear coefficient a of the bilinear model.

Cross-spectrum of the observation

The third order moments are calculated in a similar fashion as it has been done before and we leave the details to the reader.

$$\begin{aligned}\mathsf{E}Y_{t+k}^2 Y_t &= \mathsf{E}X_{t+k}^2 X_t + 2\mathsf{E}X_{t+k}\mathrm{w}_{t+k}X_t + \mathsf{E}\mathrm{w}_{t+k}^2 X_t + \mathsf{E}X_{t+k}^2\mathrm{w}_t \\ &= \underset{X^2X}{\mathrm{Cum}}(k) + 2\delta_{k\leq -1} d \underset{XX}{\mathrm{Cum}}(0) a^{-k-1} \\ &\quad + 2\int_0^1\int_0^1 (z_1 z_2)^k \mathrm{sym}_{z_{(2)}} f_2(z_{(2)}{}^{-1}) d\omega_1 d\omega_2 + \int_0^1 z^k q_1(z) d\omega, \end{aligned}$$

where the transfer functions f_2, q_1 are given by (3.12) and (3.17). Now the cross-spectrum can be calculated to be

$$S_{Y^2Y}(z) = S_{X^2X}(z) + 2d\frac{1+\underset{XX}{\mathrm{Cum}}(0)}{z^{-1}-a} + q_1(z).$$

Spectrum for Y_t^2

The spectrum for Y_t^2 will be needed for the estimation according to the cross-spectrum. There are no further theoretical difficulties to derive the formula

$$S_{Y^2Y^2}(z) = S_{Q,Q}(z) + 4\mathsf{E}X_t^2 + 3 + \mathrm{Re}\left\{\frac{C_1}{z-a^2-d^2} + \frac{C_2}{(z-a)(z-a^2-d^2)}\right\},$$

where

$$C_1 = 2[2ad\mathsf{E}X_t^3 + (5d^2+2ab_1)\mathsf{E}X_t^2 + 4d^2 + b_1^2],$$

and

$$C_2 = 4b_1d^2(\mathsf{E}X_t^2+1).$$

Note that Y_t is zero mean process, which fulfils equation (3.3), i.e., $Y_t = X_t + \mathrm{w}_t$, therefore

$$\underset{Y^2Y^2}{\mathrm{Cum}}(0) = \underset{QQ}{\mathrm{Cum}}(0) + 4\underset{XX}{\mathrm{Cum}}(0) + 3.$$

Example 33 *Figure 3.1 contains the theoretical and estimated spectrum of Y_t, the cross-spectrum between Y_t and Y_t^2 and the spectrum of Y_t^2. Process Y_t is a bilinear process with a nonzero mean*

$$Y_t = aY_{t-1} + dY_{t-1}\mathrm{w}_{t-1} + \mathrm{w}_t,$$

with parameters $a = 0.5$, $d = -0.4$ and $\sigma_{\mathrm{w}}^2 = 1$. The sample size of simulation and the number of frequencies were $2^{10} = 1024$ and $2^8 = 256$, respectively.

Bispectrum of the observation

Start with the third order cumulants of Y_t by the domain $k, l \geq 0$. Note here again the assumption that the mean of Y_t has been taken out therefore $\underset{YYY}{\mathrm{Cum}}(k+l,k)$ is

$$\begin{aligned}\mathsf{E}Y_{t+k+l}Y_{t+k}Y_t &= \mathsf{E}X_{t+k+l}X_{t+k}X_t + \mathsf{E}\mathrm{w}_{t+k+l}\mathrm{w}_{t+k}\mathrm{w}_t\\ &+ \mathsf{E}X_{t+k+l}\mathrm{w}_{t+k}X_t + \mathsf{E}\mathrm{w}_{t+k+l}X_{t+k}X_t + \mathsf{E}X_{t+k+l}\mathrm{w}_{t+k}\mathrm{w}_t\\ &+ \mathsf{E}\mathrm{w}_{t+k+l}X_{t+k}\mathrm{w}_t + \mathsf{E}\mathrm{w}_{t+k+l}\mathrm{w}_{t+k}X_t + \mathsf{E}X_{t+k+l}X_{t+k}\mathrm{w}_t\\ &= \underset{XXX}{\mathrm{Cum}}(k+l,l) + \delta_{k=0,l>0}a^{l-1}d\,\underset{XX}{\mathrm{Cum}}(0)\\ &+ \delta_{k>0}a^l\mathsf{E}X_{t+k}^2\mathrm{w}_t + \delta_{l\geq 1}a^{l-1}d\mathsf{E}X_{t+k}X_t\\ &+ \delta_{l\geq 1}a^{l-1}d\mathsf{E}X_{t+k}\mathrm{w}_t + \delta_{k=0,l\geq 1}2da^{l-1}.\end{aligned}$$

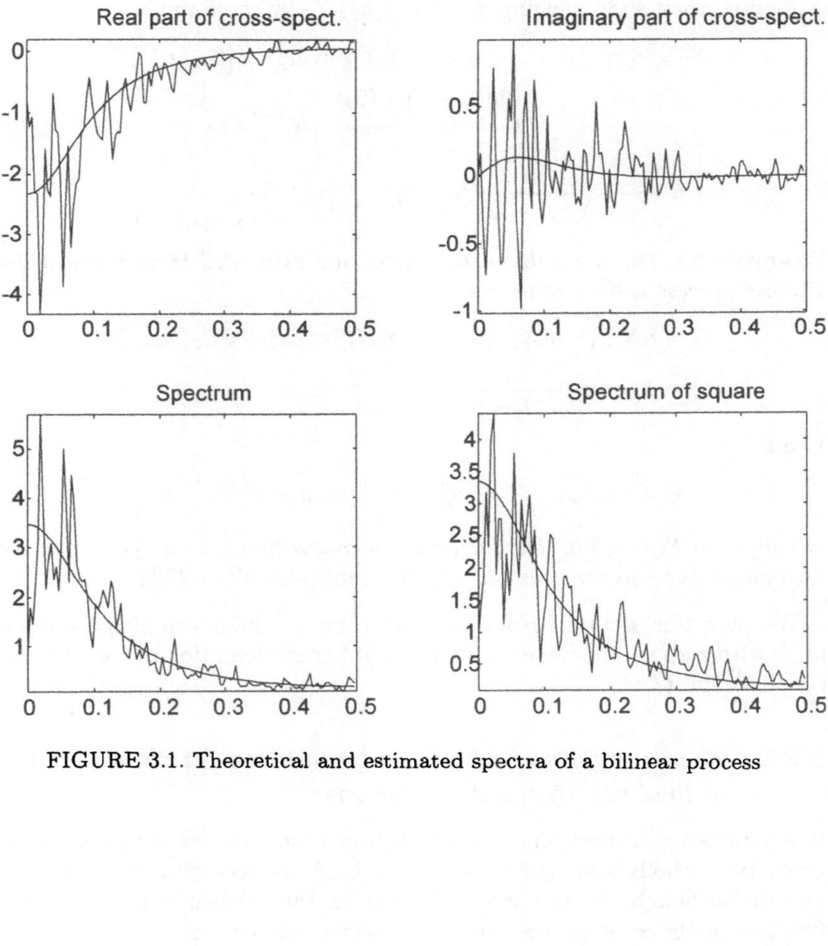

FIGURE 3.1. Theoretical and estimated spectra of a bilinear process

Equation (3.11) has been used to get the various third order moments, for example

$$\begin{aligned} \mathsf{E}X_{t+k+l}\mathrm{w}_{t+k}\mathrm{w}_t &= a\mathsf{E}X_{t+k+l-1}\mathrm{w}_{t+k}\mathrm{w}_t + d\mathsf{E}X_{t+k+l-1}\mathrm{w}_{t+k+l-1}\mathrm{w}_{t+k}\mathrm{w}_t \\ &+ b_1\mathsf{E}\mathrm{w}_{t+k+l-1}\mathrm{w}_{t+k}\mathrm{w}_t + d\mathsf{E}H_2(\mathrm{w}_{t+k+l-1})\mathrm{w}_{t+k}\mathrm{w}_t \\ &= \delta_{l>0}a^{l-1}d\mathsf{E}X_{t+k}\mathrm{w}_t + \delta_{k=0,l\geq 1}2da^{l-1}. \end{aligned}$$

Carrying out the calculation again as in Subsection 3.2.2 by using the cross-spectrum $S_{Y^2Y}(z)$, one can see that the **bispectrum** is given, after symmetrization and changing z_3 into $(z_1z_2)^{-1}$, by the formula

$$\begin{aligned} S_{Y,Y,Y}(z_1,z_2) &= S_{X,X,X}(z_1,z_2) + \operatorname{sym}_{z_{(1:3)}}\Big[3q_1(z_2^{-1}) \\ &+ 3\frac{2d(1+\underset{XX}{\mathrm{Cum}}(0))}{z_1-a} + 6\frac{aq_1(z_2^{-1})}{z_1-a} \\ + 6\frac{d}{z_1-a}&\left(\frac{a}{z_2^{-1}-a}\underset{XX}{\mathrm{Cum}}(0) + f_1(z_2^{-1})\right)\Big]_{z_3=(z_1z_2)^{-1}}. \end{aligned}$$

Example 34 *The normalized theoretical and estimated bispectrum of the bilinear process with nonzero mean*

$$\begin{aligned} X_t &= aX_{t-1} + dX_{t-1}\mathrm{w}_{t-1} + b_1H_1(\mathrm{w}_{t-1}) + dH_2(\mathrm{w}_{t-1}), \\ Y_t &= X_t + \mathrm{w}_t + \frac{d}{1-a}, \end{aligned}$$

where

$$a = -0.7,\ d = 0.3,\ \sigma_{\mathrm{w}}^2 = 1,\ b_1 = a + d^2/(1-a),$$

is plotted in Figure 3.2. The factor of normalization is $(\operatorname{var} Y_t)^{3/2}$ *and the estimation is based on a simulation of sample size* $2^{12} = 4096$.

We close this section by noting again that the difference of the spectra of Y_t with zero and nonzero means is the different definition of constant b_1, see (3.3) and (3.11).

3.2.3 The necessary and sufficient condition for the existence of the 2nth moment, scalar case

It is interesting to note that the assumptions we need are similar to those given by Nicholls and Quinn [91] for ARMA models with random coefficients, although the random coefficient in the bilinear case is a linear function of the noise process instead of being independent.

Let $\{Y_t,\ t \in \mathbf{Z}\}$ fulfil the bilinear equation (3.2) with state space equation

$$\begin{aligned} X_t &= aX_{t-1} + dX_{t-1}\mathrm{w}_{t-1} + dH_2(\mathrm{w}_{t-1}) + b_1H_1(\mathrm{w}_{t-1}), \qquad (3.21) \\ Y_t &= X_t + \mathrm{w}_t, \qquad t \in \mathbb{Z}, \end{aligned}$$

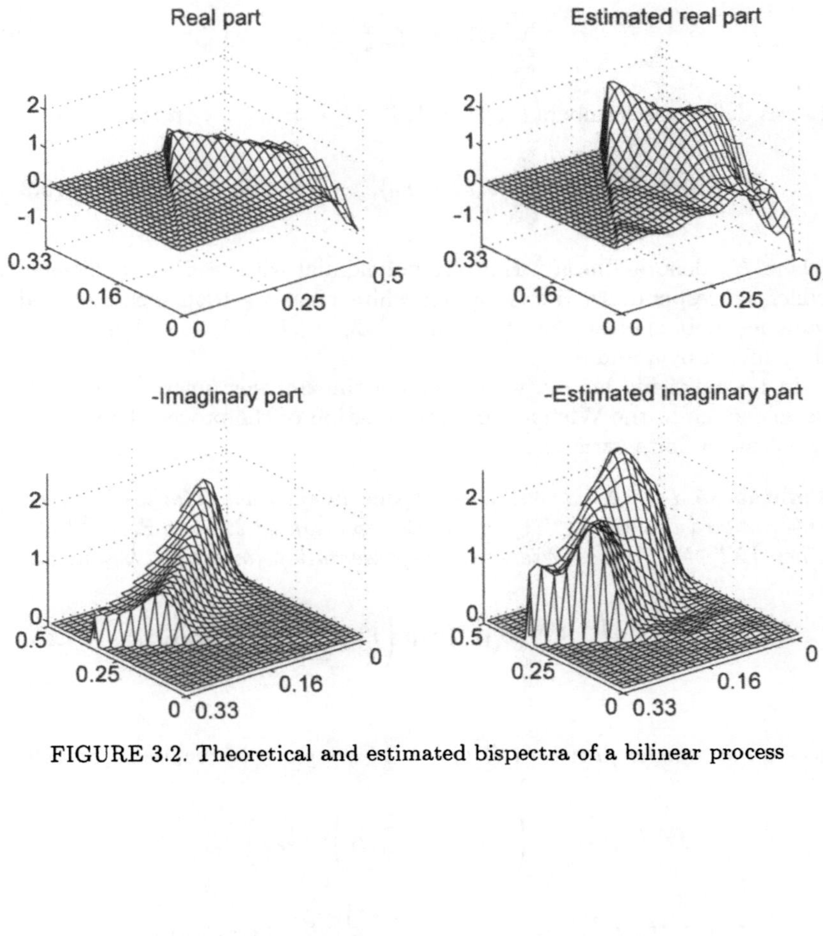

FIGURE 3.2. Theoretical and estimated bispectra of a bilinear process

with Wiener–Itô representation, i.e.,

$$X_t = \sum_{r=1}^{\infty} \int_{D^r} \exp\left(i2\pi t \sum_{j=1}^{r} \omega_j \right) f_r(\omega_{(1:r)}) W(d\omega_{(1:r)}), \qquad t \in \mathbb{Z},$$

where $f_n \in \overline{L_\Phi^n}$, see (3.7), the integrals are r–fold Wiener–Itô stochastic integrals with respect to the Gaussian stochastic measure $W(d\omega)$ and suppose that

$$\sum_{r=0}^{\infty} r! \| \operatorname{sym} f_r \|_\Phi^2 < \infty. \tag{3.22}$$

Let us define the numbers b_{nk}, $n = 1, 2, \ldots$, $k = 0, 1, \ldots, n$ such that

$$(d\mathrm{w}_0 + a)^n = \sum_{k=0}^{n} b_{nk} H_k(\mathrm{w}_0), \qquad \forall x \in \mathbf{R}, n \in \mathbb{Z}_+, \tag{3.23}$$

where H_n denotes the nth Hermite polynomial with leading coefficient 1, which corresponds to the Gaussian white noise w_t with mean zero and variance 1. It is clear that $\{b_{nk},\ n = 1, 2, \ldots, k = 0, \ldots, n\}$ is uniquely determined by a and d.

In the following lemma we are giving the transfer functions when r is large enough in the Wiener–Itô representation of the process $\{X_t^n, t \in \mathbb{Z}\}$ if X_t^n has a finite variance.

Lemma 35 *Let $\{X_t, t \in \mathbb{Z}\}$ be physically realizable stochastic process, subordinated to $\{\mathrm{w}_t, t \in \mathbb{Z}\}$, satisfy the equation (3.21) and $\mathsf{E}(X_t)^{2n} < \infty$. Then $\{X_t^k, t \in \mathbb{Z}\}$ has Wiener–Itô representation for all $1 \le k \le n$, i.e.,*

$$X_t^k = f_0^{(k)} + \sum_{r=1}^{\infty} \int_{D^r} f_r^{(k)}(\omega_{(1:r)}) \exp\left(i2\pi t \sum_{j=1}^{r} \omega_j \right) W(d\omega_{(1:r)}), \quad t \in \mathbb{Z}, \tag{3.24}$$

where the transfer functions $f_r^{(k)}$, $r > 2n$ are given by recursive formula

$$f_r^{(k)}(\omega_{(1:r)}) = \left(\exp\left(i2\pi \sum_{j=1}^{r} \omega_j \right) - b_{k0} \right)^{-1}$$
$$\times \left(\sum_{l=1}^{k} b_{kl} f_{r-l}^{(k)}(\omega_{(1:r-l)}) + \sum_{i=1}^{k-1} \sum_{j=0}^{K} c_{ij}^{(k)} f_{r-j}^{(i)}(\omega_{(1:r-j)}) \right), \tag{3.25}$$

where K depends only on n and $c_{ij}^{(k)}$ are appropriate constants to be defined later, see (3.27).

Proof. Let $1 \le k \le n$. Then $\mathsf{E}X_t^{2k} < \infty$. Therefore $\{X_t^k, t \in \mathbb{Z}\}$ is second order stationary, physically realizable and subordinated to $\{\mathrm{w}_t, t \in \mathbb{Z}\}$, i.e., it has a Wiener–Itô representation (3.24) for all $1 \le k \le n$. It remains to prove (3.25). Since X_t is independent of w_t, we can argue similarly to the proof of the formula (3.6) and have

$$\int_{D^r} f_r^{(k)}(\omega_{(1:r)}) \exp\left(i2\pi t \sum_{j=1}^{r} \omega_j\right) W(d\omega_{(1:r)}) H_p(\mathrm{w}_t)$$
$$= \int_{D^{r+p}} f_r^{(k)}(\omega_{(1:r)}) \exp(i2\pi t \sum_{j=1}^{r+p} \omega_j) W(d\omega_{(1:r+p)}), \tag{3.26}$$

for all $r, p \in \mathbb{Z}_+$, i.e., there are no other terms but the first one applying the diagram formula (2.7), page 42, for the right hand side of (3.26). Raising equation (3.21) to the kth power we have

$$X_t^k = (a + d\mathrm{w}_{t-1})^k X_{t-1}^k + P(X_{t-1}, \mathrm{w}_{t-1}),$$

where the polynomial

$$P(X_{t-1}, \mathrm{w}_{t-1}) = \sum_{i=1}^{k-1} X_{t-1}^i \sum_{j=0}^{k_i^{(k)}} c_{ij}^{(k)} H_j(\mathrm{w}_{t-1}) + \sum_{l=0}^{2k} d_l^{(k)} H_l(\mathrm{w}_{t-1}), \tag{3.27}$$

contains X_{t-1} of power less then k, moreover $k_i^{(k)} \in \mathbb{Z}_+$, $k = 1, \dots, n$, $i = 1, \dots, k-1$, $c_{ij}^{(k)}, d_l^{(k)} \in \mathbf{R}$, $k = 1, \dots, n$, $i = 1, \dots, k-1$, $j = 1, \dots, k_i^{(k)}, l = 0, \dots, 2k$. Let

$$K = \max_{\substack{1 \le k \le n \\ 0 \le i \le k}} \{k_i^{(k)}\}.$$

From equation (3.23) we have

$$X_t^k = \sum_{l=0}^{n} b_{nl} H_l(\mathrm{w}_{t-1}) X_{t-1}^k + \sum_{i=1}^{k-1} X_{t-1}^i \sum_{j=0}^{K} c_{ij}^{(k)} H_j(\mathrm{w}_{t-1}) \tag{3.28}$$
$$+ \sum_{j=0}^{2n} d_j^{(k)} H_j(\mathrm{w}_{t-1}),$$

where if $c_{ij}^{(k)}$ or $d_j^{(k)}$ are not defined by (3.27) then let $c_{ij}^{(k)} = d_j^{(k)} = 0$. From equation (3.26) and (3.28) we have the recursive formula (3.25) for the transfer functions $f_r^{(k)}$. ■

Now we state a lemma about the inner product of transfer functions $f_r^{(k)}$.

Lemma 36 *For the transfer functions $f_r^{(k)}$ defined by the recursive formula (3.25) we have*

$$\left(f_{r-l}^{(i)}, f_{r-l'}^{(i')}\right)_r^{\mathfrak{P}} = \begin{cases} l!\sigma^{2l}\left(f_{r-l}^{(i)}, f_{r-l}^{(i')}\right)_{r-l}^{\mathfrak{P}} & \text{if} \quad l = l', \\ 0 & \text{if} \quad l \neq l'. \end{cases}$$

Proof. Set

$$c(\mathfrak{p}_1, \mathfrak{p}_2) = \int_{D^r} f_{r-l}^{(i)}(\omega_{\mathfrak{p}_1(1)}, \dots, \omega_{\mathfrak{p}_1(r-l)}) \times \overline{f_{r-l'}^{(i')}(\omega_{\mathfrak{p}_2(1)}, \dots, \omega_{\mathfrak{p}_2(r-l')})}\Phi(d\omega_{(1:r)}),$$

$$d(\tau_1, \tau_2) = \int_{D^{r-l}} f_{r-l}^{(i)}(\omega_{\tau_1(1)}, \dots, \omega_{\tau_1(r-l)}) \times \overline{f_{r-l}^{(i')}(\omega_{\tau_2(1)}, \dots, \omega_{\tau_2(r-l)})}\Phi(d\omega_{(1:r-l)}),$$

where $\mathfrak{p}_j \in \mathfrak{P}_r$ and $\tau_j \in \mathfrak{P}_{r-l}$, $j = 1, 2$. For $\mathfrak{p} \in \mathfrak{P}_r$ let $\overline{\mathfrak{p}}$ denote those indices which are defined by $\overline{\mathfrak{p}}(j) = \mathfrak{p}(j)$ for $1 \leq j \leq r - l$. One can prove that

$$c(\mathfrak{p}_1, \mathfrak{p}_2) = \begin{cases} \sigma^{2l} d(\overline{\mathfrak{p}}_1, \overline{\mathfrak{p}}_2) & \text{if } l = l' \text{ and } \overline{\mathfrak{p}}_1 = \overline{\mathfrak{p}}_2, \\ 0 & \text{otherwise,} \end{cases}$$

which follows from the independence of X_t and w_t, see formula (3.6). So we have proved our assertion for the case $l \neq l'$. If $l = l'$, we have

$$\begin{aligned} \left(f_{r-l}^{(i)}, f_{r-l'}^{(i')}\right)_r^{\mathfrak{P}} &= \frac{1}{r!} \sum_{\mathfrak{p}_1, \mathfrak{p}_2 \in \mathfrak{P}_r} c(\mathfrak{p}_1, \mathfrak{p}_2) \\ &= \frac{1}{r!} \sum_{\mathfrak{p}_1, \mathfrak{p}_2 \in \mathfrak{P}_r} \sigma^{2l} d(\overline{\mathfrak{p}}_1, \overline{\mathfrak{p}}_2) \\ &= \frac{1}{r!}\binom{r}{r-l}(l!)^2\sigma^{2l} \sum_{\tau_1, \tau_2 \in \mathfrak{P}_{r-l}} d(\tau_1, \tau_2) \\ &= l!\sigma^{2l}\left(f_{r-l}^{(i)}, f_{r-l}^{(i')}\right)_{r-l}^{\mathfrak{P}}, \end{aligned}$$

which was to be proved. ■

Let $\otimes$ denote the Kronecker– or tensor–product and $(\cdot)^{\otimes n}$ the n–times tensor–product , i.e., $\underbrace{(\cdot) \otimes \ldots \otimes (\cdot)}_{n}$. Our main theorem here is

Theorem 37 *Let $\{X_t\,, t \in \mathbb{Z}\}$ be physically realizable, subordinated to $\{\mathrm{w}_t\,, t \in \mathbb{Z}\}$ and satisfy (3.21). Then the following statements are equivalent*

(i) $\mathsf{E}(X_t)^{2n} < \infty$,

(ii) $\sum_{k=0}^{n} k!\sigma^{2k} b_{nk}^2 < 1$,

(iii) $\mathsf{E}(a + dw_0)^{2n} < 1$.

Proof. (i) $\Longrightarrow$ (ii)
Let $f_r^{(k)}$, b_{kl} and $c_{ij}^{(k)}$ denote the same as earlier. Put

$$g_r = \left(f_r^{(n)}, \dots, f_r^{(1)}\right)^T, \qquad r \in \mathbb{Z}_+,$$

$$A_j = \begin{bmatrix} \tilde{b}_{nj} & c_{1j}^{(n)} & c_{2j}^{(n)} & \cdots & c_{n-1j}^{(n)} \\ & \tilde{b}_{n-1j} & c_{1j}^{(n-1)} & \cdots & c_{n-2j}^{(n-1)} \\ & & \tilde{b}_{n-2j} & \cdots & c_{n-3j}^{(n-2)} \\ & & & \ddots & \vdots \\ & & & & \tilde{b}_{1j} \end{bmatrix}, \quad j = 1, \dots, K,$$

where

$$\tilde{b}_{ij} = \begin{cases} b_{ij} & \text{if} \quad i \geq j, \\ 0 & \text{otherwise}, \end{cases}$$

A^T denotes the transpose of matrix A and

$$H(\omega) = \operatorname{diag}(h_n(\omega), \dots, h_1(\omega)), \qquad \omega \in D,$$

where $h_j(\omega) = (\exp(i\omega) - b_{j0})^{-1}$, $j = 1, \dots, n$. Then one can see that

$$g_r(\omega_{(1:r)}) = H\left(\sum_{j=1}^{r} \omega_j\right) \sum_{j=1}^{K} A_j g_{r-j}(\omega_{(1:r-j)}),$$

for all $\omega_{(1:r)} \in D^r$ and $r > K + 2n$ by Lemma 35. So we have

$$\operatorname{sym} g_r(\omega_{(1:r)}) \otimes \overline{\operatorname{sym} g_r(\omega_{(1:r)})} = \left(H(\sum_{j=1}^{r} \omega_j) \otimes \overline{H(\sum_{j=1}^{r} \omega_j)} \right)$$

$$\times \sum_{i,j=1}^{K} (A_i \otimes A_j)(\operatorname*{sym}_{\omega_{(1:r)}} g_{r-i}(\omega_{(1:r-i)}) \otimes \overline{\operatorname*{sym}_{\omega_{(1:r)}} g_{r-j}(\omega_{(1:r-j)})}). \tag{3.29}$$

If we set

$$G_r = \operatorname{Vec}\left(\left(f_r^{(i)}, f_r^{(j)} \right)_r^{\mathfrak{P}} \right)_{i,j=1}^{n}, \tag{3.30}$$

see Proposition 84, page 214, of the Appendix B for the definition of operator Vec, then by integrating both sides of equation (3.29), first by replacing $\sum_{j=1}^{r}\omega_j$ with λ_r and $\lambda_j=\omega_j$, $j=1,2,\ldots,r-1$ then by using Lemma 36, we have the recursive formula for G_r

$$G_r = H_0 \sum_{k=1}^{K} A_k^{\otimes 2} k!\sigma^{2k} G_{r-k},$$

where

$$H_0 = \operatorname{diag}(h_{nn},\ldots,h_{n1},h_{n-1n},\ldots,h_{1n},\ldots,h_{11}), \quad h_{ij}=(1-b_{i0}b_{j0})^{-1}.$$

It follows from (i) and (3.22) that $\sum_{r=0}^{\infty} G_r$ converges, which is equivalent to the fact that every root of the polynomial

$$P(\lambda) = \det(\lambda^K I - H_0 \sum_{k=1}^{K} k!\sigma^{2k}\lambda^{K-k}A_k^{\otimes 2}),$$

must be smaller than 1 in modulus. Since H_0 is diagonal and $A_j^{\otimes 2}$ is an upper triangular matrix we have $P=\prod_{i,j=1}^{n} P_{ij}$ where

$$P_{ij}(\lambda) = \lambda^K - (1-b_{i0}b_{j0})^{-1}\sum_{k=1}^{K} k!\sigma^{2k}\lambda^{K-k}\tilde{b}_{ik}\tilde{b}_{jk}.$$

Therefore all roots of polynomial P_{nn} must be smaller than 1 in modulus, i.e., all roots of the equation

$$\lambda^{K-n}(\lambda^n - (1-b_{n0}^2)^{-1}\sum_{k=1}^{n} k!\sigma^{2k}\lambda^{n-k}b_{nk}^2) = 0,$$

must be smaller than 1 in modulus. Then Corollary 82, page 213, of the Appendix B gives us (ii).

The equivalence of (ii) and (iii) can be proven by elementary algebra.

(iii) $\Longrightarrow$ (i)

By the Lyapunov inequality $(\mathsf{E}(a+dw_0)^{2k})^{1/2k}$ is non decreasing in k so we have $\mathsf{E}(a+dw_0)^{2k}<1$ for all $1\le k\le n$. Since (iii) $\Longrightarrow$ (ii),

$$\sum_{l=0}^{k} l!\sigma^{2l}b_{kl}^2 < 1, \qquad \text{for all } 1\le k\le n.$$

By Corollary 82 of the Appendix B we see that all roots of the equations

$$\lambda^{K-k}(\lambda^k I - (1-b_{k0}^2)^{-1}\sum_{j=1}^{k} j!\sigma^{2j}\lambda^{k-j}b_{kj}^2) = 0, \quad 1\le k\le n$$

are smaller than 1 in modulus, i.e., the polynomials P_{kk} have all roots smaller than 1 in modulus. Therefore by Proposition 83 of the Appendix B polynomials P_{ij} $(1 \leq i, j \leq n)$ have all roots smaller than 1 in modulus so that polynomial P has the same property as well. But this is equivalent to the convergence of $\sum_{r=0}^{\infty} G_r$, i.e., every moment exists up to $2n$ to prove our assertion. ■

Remark 38 *The process $\{X_t\,,\, t \in \mathbb{Z}\}$ can be considered as a random coefficient autoregressive (RCA) model, see [91]. The autoregressive coefficient is the random variable $a + d\mathrm{w}_{t-1}$. Similarly, $\{X_t^n\,,\, t \in \mathbb{Z}\}$ also can be considered as a RCA–model with the autoregressive coefficient $(a + d\mathrm{w}_{t-1})^n$. The necessary and sufficient condition for stationarity of RCA–models is that the second moment of autoregressive coefficient must be smaller than 1 under the assumption of the independence between the autoregressive coefficient and the noise. In Theorem 37 we had a similar result but in our case the condition of independence is not satisfied.*

We can state a similar theorem for process $\{Y_t\,,\, t \in \mathbb{Z}\}$ since Y_t is different from X_t in w_t. An immediate consequence of Theorem 37 is that not all the moments exist for a scalar bilinear process.

Corollary 39 *If $\{Y_t\,,\, t \in \mathbb{Z}\}$ is a scalar bilinear process such that $d \neq 0$ (really nonlinear) then it can have only finite number of moments.*

Proof. We prove it for $d > 0$. Let $z_1 = \dfrac{-2-a}{d}$ and $z_2 = \dfrac{2-a}{d}$ $(-\infty < z_1 < z_2 < \infty)$. If $\mathrm{w} \in N(0, \sigma^2)$, then using the Markov inequality we have

$$\mathsf{E}(a + d\mathrm{w})^{2n} \geq 2^{2n} \mathsf{P}\left(|a + d\mathrm{w}| > 2\right),$$

therefore

$$\mathsf{E}(a + d\mathrm{w})^{2n} \geq 2^{2n} \left(1 - \frac{1}{\sqrt{2\pi}\sigma} \int_{z_1}^{z_2} e^{-\frac{x^2}{2\sigma^2}} dx \right).$$

So if

$$n \geq \frac{1}{2} \log_2 \left(1 - \frac{1}{\sqrt{2\pi}\sigma} \int_{z_1}^{z_2} e^{-\frac{x^2}{2\sigma^2}} dx \right)^{-1},$$

then $\mathsf{E}(a + d\mathrm{w})^{2n} \geq 1$, i.e., $\mathsf{E}X_t^{2n} = \infty$ by Theorem 37. ■

3.3 Identification of bilinear processes, general case

3.3.1 State space form of lower triangular bilinear models

In this section we generalize the scalar bilinear process considered in Section 3.2. We show first by an example that every lower triangular bilinear model

Y_t

$$\sum_{m=0}^{P} a_m Y_{t-m} = \sum_{m=0}^{Q} b_m \mathrm{w}_{t-m} + \sum_{m=1}^{R}\sum_{n=0}^{S} d_{m,m+n} Y_{t-m-n}\mathrm{w}_{t-m} + f_0, \quad (3.31)$$

where $a_0 = b_0 = 1$, can be put into the so called state representation form

$$\mathbf{X}_t = A\mathbf{X}_{t-1} + D\mathbf{X}_{t-1}\mathrm{w}_{t-1} + \mathbf{b}\mathrm{w}_t + \mathbf{f}_0,$$
$$Y_t = \mathbf{c}'\mathbf{X}_t. \quad (3.32)$$

Variable $\mathbf{X}_t$ is called a state variable and Y_t is referred to as observation. Indeed, consider the following particular case

$$\sum_{m=0}^{3} a_m Y_{t-m} = \sum_{m=0}^{2} b_m \mathrm{w}_{t-m} + \sum_{m=1}^{3}\sum_{n=0}^{3-m} d_{m,m+n} Y_{t-m-n}\mathrm{w}_{t-m} + f_0, \quad (3.33)$$

where $a_0 = b_0 = 1$. Now, if we put

$$\begin{bmatrix} X \\ V_1 \\ V_2 \\ Z_1 \\ Z_2 \end{bmatrix}_t = \begin{bmatrix} -a_1 & 1 & 0 & 0 & 0 \\ -a_2 & 0 & 1 & 0 & 0 \\ -a_3 & 0 & 0 & 0 & 0 \\ 1 & 0 & 0 & 0 & 0 \\ 0 & 0 & 0 & 1 & 0 \end{bmatrix} \begin{bmatrix} X \\ V_1 \\ V_2 \\ Z_1 \\ Z_2 \end{bmatrix}_{t-1} + \begin{bmatrix} 1 \\ b_1 \\ b_2 \\ 0 \\ 0 \end{bmatrix} \mathrm{w}_t$$
$$+ \begin{bmatrix} d_{11} & 0 & 0 & d_{12} & d_{13} \\ d_{22} & 0 & 0 & d_{23} & 0 \\ d_{33} & 0 & 0 & 0 & 0 \\ 0 & 0 & 0 & 0 & 0 \\ 0 & 0 & 0 & 0 & 0 \end{bmatrix} \begin{bmatrix} X \\ V_1 \\ V_2 \\ Z_1 \\ Z_2 \end{bmatrix}_{t-1} \mathrm{w}_{t-1} + \begin{bmatrix} f_0 \\ 0 \\ 0 \\ 0 \\ 0 \end{bmatrix},$$

and

$$Y_t = \begin{bmatrix} 1 & 0 & 0 & 0 & 0 \end{bmatrix} \begin{bmatrix} X \\ V_1 \\ V_2 \\ Z_1 \\ Z_2 \end{bmatrix}_t$$
$$= X_t,$$

then we see that X_t satisfies equation (3.33) because

$$\begin{aligned} X_t &= -a_1 X_{t-1} + V_{1,t-1} + \mathrm{w}_t \\ &\quad + d_{11} X_{t-1}\mathrm{w}_{t-1} + d_{12}. Z_{1,t-1}\mathrm{w}_{t-1} + d_{13} Z_{2,t-1}\mathrm{w}_{t-1} + f_0, \\ V_{1,t} &= -a_2 X_{t-1} + V_{2,t-1} + b_1 \mathrm{w}_t + d_{22} X_{t-1}\mathrm{w}_{t-1} + d_{23} Z_{1,t-1}\mathrm{w}_{t-1}, \\ V_{2,t} &= -a_3 X_{t-1} + b_2 \mathrm{w}_t + d_{33} X_{t-1}\mathrm{w}_{t-1}, \\ Z_{1,t} &= X_{t-1}, \\ Z_{2,t} &= Z_{1,t-1}. \end{aligned}$$

Notice that variable Z is nothing else but shifting back the first entry of the state variable. The general treatment follows easily from this example, see also in Mohler [87].

3.3.2 *Vector valued bilinear model with scalar input*

The lower triangular bilinear model (3.31) will be investigated in the state space form (3.32) because all the properties, cumulants, spectra and so on, of Y_t can be obtained from the state space representation. There are some advantages of the tensor product $\mathbf{X}_t \otimes \mathbf{X}_t$ against the usual matrix product $\mathbf{X}_t\mathbf{X}_t'$ in the case of the bilinear equation (3.32). Therefore as a consequence of this we are using vector form for the covariance and the spectrum of a vector valued series instead of the usual matrix form. This does not mean major differences because the vectors have the same components as the corresponding matrices.

A process $\{\mathbf{Y}_t\,,\, t \in \mathbb{Z}\}$ with m–dimensional random variables is called bilinear if it satisfies the bilinear state-space equation similarly to the scalar case (3.2) above, i.e.

$$\mathbf{Y}_t = A\mathbf{Y}_{t-1} + D\mathbf{Y}_{t-1}\mathrm{w}_{t-1} + \mathbf{b}\mathrm{w}_t + \mathbf{f}_0,$$

where $A,\ D \in \mathbf{R}^{m\times m}$, $\mathbf{b}$, $\mathbf{f}_0$, $\mathbf{c} \in \mathbf{R}^m$ and $\mathbf{f}_0 = -\sigma^2 D\mathbf{b}$ to keep $\mathsf{E}\mathbf{Y}_t = 0$. We assume that there is no eigenvalue of A with modulus 1, w_t is a Gaussian white noise scalar process and $\{\mathbf{Y}_t\,,\, t \in \mathbb{Z}\}$ is physically realizable and subordinated to $\{\mathrm{w}_t\,,\, t \in \mathbb{Z}\}$.

Put $\mathbf{X}_t = \mathbf{Y}_t - \mathbf{b}\mathrm{w}_t$, i.e., use similar transformation to the one used in previous section for $\mathbf{Y}_t$. Then $\mathsf{E}\mathbf{X}_t = 0$ and $\{\mathbf{X}_t\,,\, t \in \mathbb{Z}\}$ satisfies the following equations for all $t \in \mathbb{Z}$,

$$\mathbf{X}_t = A\mathbf{X}_{t-1} + D\mathbf{X}_{t-1}\mathrm{w}_{t-1} + \mathbf{b}_1 H_1(\mathrm{w}_{t-1}) + \mathbf{b}_2 H_2(\mathrm{w}_{t-1}), \qquad (3.34)$$

where $\mathbf{b}_2 = D\mathbf{b}$, $\mathbf{b}_1 = A\mathbf{b}$. It is seen that $\mathbf{X}_t$ is independent of w_t.

We have the Wiener–Itô representation of the stochastic process $\{\mathbf{X}_t\,,\, t \in \mathbb{Z}\}$ from the Wiener–Itô representation of its entries $\{(\mathbf{X}_t)_i\,,\, t \in \mathbb{Z}\}$, $i = 1, \ldots, m$, i.e.,

$$\mathbf{X}_t = \sum_{r=1}^{\infty} \int_{\mathcal{D}^r} \exp\left(i2\pi t \sum_{j=1}^{r} \omega_j\right) \mathbf{f}_r(\omega_{(1:r)}) W(d\omega_{(1:r)}), \qquad t \in \mathbb{Z},$$

where $(f_r)_i \in \overline{L_\Phi^r}$, $i = 1, \ldots, m$ are the transfer functions of $\{(\mathbf{X}_t)_i\,,\, t \in \mathbb{Z}\}$, $i = 1, \ldots, m$, respectively. One may repeat the argument of Subsection 3.2.1 to get

$$\mathbf{X}_{t-1} H_k(\mathrm{w}_{t-1}) = \sum_{r=1}^{\infty} \int_{\mathcal{D}^{r+k}} \exp(i2\pi\,(t-1) \sum_{j=1}^{r} \omega_j) \mathbf{f}_r(\omega_{(1:r)}) W(d\omega_{(1:r)}). \qquad (3.35)$$

This fact yields the following recursive formula for the transfer functions:

$$\mathbf{f}_1(\omega_1) = (\exp(i2\pi\omega_1)I - A)^{-1}\mathbf{b}_1,$$
$$\mathbf{f}_2(\omega_{(1:2)}) = (\exp(i2\pi(\omega_1+\omega_2)I - A)^{-1}(D\mathbf{f}_1(\omega_1) + \mathbf{b}_2),$$

and if $r \geq 3$,

$$\mathbf{f}_r(\omega_{(1:r)}) = (\exp\left(i2\pi\sum_{j=1}^{r}\omega_j\right)I - A)^{-1}D\mathbf{f}_{r-1}(\omega_{(1:r-1)}), \tag{3.36}$$

which is similar to (3.7). There is no question of the existence of inverses because of the physical realizability of the process all the eigenvalues of A must be inside the unit circle. In our previous paper Terdik and Ispány [133] we pointed out that the above Wiener–Itô representation exists if and only if all the eigenvalues of the matrix $A^{\otimes 2} + \sigma^2 D^{\otimes 2}$ are less than one in modulus, which is the necessary and sufficient condition for the existence of a second order stationary physically realizable solution of equation (3.34), see also in Liu and Brockwell [76]. In addition it has been shown that if ρ denotes the spectral radius, then from $\rho\left(A^{\otimes 2} + \sigma^2 D^{\otimes 2}\right) < 1$ it follows that $\rho(A) < 1$ as well. This time we shall follow the method used for the scalar valued case.

The transfer function system for the state variables $\mathbf{X}_t$ is given by (3.36). The question is under what condition the components of $\mathsf{E}(\mathbf{X}_t \otimes \mathbf{X}_t)$ will be finite, where $\otimes$ denotes the tensor product. For that purpose let us regard the transfer functions $\mathbf{f}_r$ in the form

$$\mathbf{f}_r(\omega_{(r)}) = \mathbf{g}_r(\omega_{(1:r)}) + \mathbf{h}_r(\omega_{(1:r)}),$$

where

$$\mathbf{g}_1(\omega_1) = \mathbf{f}_1(\omega_1) = (e^{i2\pi\omega_1}I - A)^{-1}\mathbf{b}_1,$$
$$\mathbf{g}_2(\omega_{(1:2)}) = (e^{i2\pi\Sigma\omega_{(1:2)}}I - A)^{-1}D\mathbf{f}_1(\omega_1),$$
$$\mathbf{g}_r(\omega_{(1:r)}) = (\exp\left(i2\pi\sum_{j=1}^{r}\omega_j\right)I - A)^{-1}D\mathbf{g}_{r-1}(\omega_{(1:r-1)}),$$

and

$$\mathbf{h}_1(\omega_1) = \mathbf{0},$$
$$\mathbf{h}_2(\omega_{(1:2)}) = (e^{i2\pi\Sigma\omega_{(1:2)}}I - A)^{-1}\mathbf{b}_2,$$
$$\mathbf{h}_r(\omega_{(1:r)}) = (\exp\left(i2\pi\sum_{j=1}^{r}\omega_j\right)I - A)^{-1}D\mathbf{h}_{r-1}(\omega_{(1:r-1)}).$$

The functions $\mathbf{g}_r$ and $\mathbf{h}_r$ are orthogonal in the sense of

$$\int_{\mathcal{D}^k}\mathbf{h}_k(\omega_{(1:k)}) \otimes \operatorname{sym}\overline{\mathbf{g}_k(\omega_{(1:k)})}d\omega_{(1:k)} = \mathbf{0},$$

the proof of this for $k = 1, 2$ is very simple and one may use Lemma 32 for the induction step. Let us now consider the expectation of the tensor product of the kth term, $k \geq 2$,

$$\begin{aligned}
&\mathsf{E}\left[\int_{\mathcal{D}^k} \mathbf{h}_k(\omega_{(1:k)}) W\left(d\omega_{(1:k)}\right) \otimes \int_{\mathcal{D}^k} \mathbf{h}_k(\omega_{(1:k)}) W\left(d\omega_{(1:k)}\right)\right] \\
&= \sigma^{2k} \int_{\mathcal{D}^k} \mathbf{h}_k(\omega_{(1:k)}) \otimes \operatorname{sym} \overline{\mathbf{h}_k(\omega_{(1:k)})} d\omega_{(1:k)} \\
&= \sigma^{2k} 2 \int_{\mathcal{D}^k} \mathbf{h}_k(\omega_{(1:k)}) \otimes \overline{\mathbf{h}_k(\omega_{(1:k)})} d\omega_{(1:k)} \\
&= 2 \int_{\mathcal{D}} (e^{i2\pi\omega} I - A)^{-1} \otimes (e^{-i2\pi\omega} I - A)^{-1} d\omega \sigma^2 D^{\otimes 2} \\
&\otimes \sigma^{2(k-1)} \int_{\mathcal{D}^{k-1}} \mathbf{h}_{k-1}(\omega_{(1:k-1)}) \otimes \overline{\mathbf{h}_{k-1}(\omega_{(1:k-1)})} d\omega_{(1:k-1)} \\
&= 2\sigma^4 \left[(I - A^{\otimes 2})^{-1} \sigma^2 D^{\otimes 2}\right]^{k-2} (I - A^{\otimes 2})^{-1} \mathbf{b}_2^{\otimes 2},
\end{aligned}$$

where $\operatorname{sym} \mathbf{h}_k(\omega_{(k)})$ denotes the symmetrized version of $\mathbf{h}_k(\omega_{(k)})$ by its variables. We have used the property $(AB) \otimes (CD) = (A \otimes C)(B \otimes D)$ of the tensor product. A similar calculation gives that

$$\begin{aligned}
\mathsf{E}\left(\int_{\mathcal{D}^k} \mathbf{g}_k(\omega_{(1:k)}) W\left(d\omega_{(1:k)}\right) \otimes \int_{\mathcal{D}^k} \mathbf{g}_k(\omega_{(1:k)}) W\left(d\omega_{(1:k)}\right)\right) \\
= \sigma^2 \left[(I - A^{\otimes 2})^{-1} \sigma^2 D^{\otimes 2}\right]^{k-1} (I - A^{\otimes 2})^{-1} \mathbf{b}_1^{\otimes 2}.
\end{aligned}$$

From this it follows that $\mathsf{E}\mathbf{X}_t \otimes \mathbf{X}_t$ is finite if and only if all the eigenvalues of the matrix $\sigma^2 (I - A \otimes A)^{-1} (D \otimes D)$ are less than 1 in absolute value. Summing up these we get the variance of $\mathbf{X}_t$ by

$$\mathsf{E}\left(\mathbf{X}_t \otimes \mathbf{X}_t\right) = \sigma^2 \left[(I - A \otimes A - \sigma^2 D \otimes D)\right]^{-1} (\mathbf{b}_1^{\otimes 2} + 2\sigma^2 \mathbf{b}_2^{\otimes 2}).$$

Theorem 40 *Let us suppose that the m-dimensional state space variables fulfil the bilinear state equation*

$$\mathbf{X}_t = A\mathbf{X}_{t-1} + D\mathbf{X}_{t-1}\mathrm{w}_{t-1} + \mathbf{b}_1 H_1(\mathrm{w}_{t-1}) + \mathbf{b}_2 H_2(\mathrm{w}_{t-1}), \qquad t \in \mathbb{Z},$$

where the noise process w_t is Gaussian i.i.d., $E\mathrm{w}_t = 0$, $E\mathrm{w}_t^2 = \sigma^2$, A and D are $m \times m$ matrices, $\mathbf{b}_i \in R^m$, $i = 1, 2$ and all eigenvalues of A are inside the unit circle. The transfer functions $\mathbf{f}_r(\omega_{(r)})$; $r \in \mathbb{Z}_+$ are different from zero in $\overline{L_\Phi^r}$. Then the necessary and sufficient condition for the stationarity of $\mathbf{X}_t$ (which is physically realizable and subordinated to w_t, $t \in \mathbb{Z}$) is that all the eigenvalues of the matrix $\sigma^2 (I - A \otimes A)^{-1} (D \otimes D)$ should be less then 1 in modulus.

unit circle is sufficient for the stationarity of $\mathbf{X}_t$. A sufficient condition of the strict stationarity for the vector valued bilinear process $\mathbf{X}_t$ defined by (3.32) was given by [126], as the eigenvalues of the matrix $A \otimes A + \sigma^2 D \otimes D$ need to be inside of the unit circle. One can show that this condition is equivalent to the condition given in Theorem 40.

3.3.3 Spectra

Spectrum

We derive the spectrum $S_{2,\mathbf{X}}$ for $\mathbf{X}_t$ as the Fourier transform of the covariances $\mathsf{E}(\mathbf{X}_{t+s} \otimes \mathbf{X}_t)$. There is no difference between the usual spectral density matrix and the one derived from $\mathsf{E}(\mathbf{X}_{t+s} \otimes \mathbf{X}_t)$ because both have the same entries, i.e., $S_{2,\mathbf{X}}$ can be considered as the spectral density matrix but it is spanned into a vector. Now take the right tensor product of the equation

$$\mathbf{X}_{t+s} = A\mathbf{X}_{t+s-1} + D\mathbf{X}_{t+s-1}\mathrm{w}_{t+s-1} + \mathbf{b}_2 H_2(\mathrm{w}_{t+s-1}) + \mathbf{b}_1 H_1(\mathrm{w}_{t+s-1}),$$

by $\mathbf{X}_t$. If $s \geq 0$, then

$$\begin{aligned}\mathsf{E}(\mathbf{X}_{t+s} \otimes \mathbf{X}_t) &= (A \otimes I)\,\mathsf{E}(\mathbf{X}_{t+s-1} \otimes \mathbf{X}_t) \\ &= (A \otimes I)^s\,\mathsf{E}\mathbf{X}_0^{\otimes 2}.\end{aligned}$$

If $s \leq 0$, take the left tensor product of $\mathbf{X}_{t+s}$ and

$$\mathbf{X}_t = A\mathbf{X}_{t-1} + D\mathbf{X}_{t-1}\mathrm{w}_{t-1} + \mathbf{b}_1 H_1(\mathrm{w}_{t-1}) + \mathbf{b}_2 H_2(\mathrm{w}_{t-1}),$$

so

$$\mathsf{E}(\mathbf{X}_{t+s} \otimes \mathbf{X}_t) = (I \otimes A)^{-s}\,\mathsf{E}\mathbf{X}_0^{\otimes 2}.$$

The matrices $A \otimes I$ and $I \otimes A$ commute therefore the Fourier transform of the covariances is written as

$$S_{2,\mathbf{X}}(z) = [zI - A \otimes I]^{-1}\left[z^{-1}I - I \otimes A\right]^{-1}\left[I - A^{\otimes 2}\right]\mathsf{E}\mathbf{X}_0^{\otimes 2},$$

or equivalently

$$S_{2,\mathbf{X}}(z) = [zI - A]^{-1} \otimes \left[z^{-1}I - A\right]^{-1}\left[I - A^{\otimes 2}\right]\mathsf{E}\mathbf{X}_0^{\otimes 2},$$

where $z = e^{i2\pi\omega}$ and I is the identity matrix with an appropriate dimension as usual. Notice that the poles of the spectrum does not depend on the bilinear part of process $\mathbf{X}_t$.

The quadratic process

If the state process $\mathbf{X}_t$ is a second order stationary solution of equation (3.34), then it is strictly stationary and the existence of the fourth order moments allows us to use the Wiener Itô spectral representation for the quadratic process $\mathbf{Q}_t$ defined by $\mathbf{Q}_t = \mathbf{X}_t^{\otimes 2}$. Consider

$$\mathbf{Q}_t = \mathbf{q}_0 + \sum_{r=1}^{\infty} \int_{\mathcal{D}^r} \exp(it \sum_{j=1}^{r} \omega_j) \mathbf{q}_r(\omega_{(1:r)}) W(d\omega_{(1:r)}), \qquad t \in \mathbb{Z},$$

where $\mathbf{q}_0 = \mathsf{E}\mathbf{X}_0^{\otimes 2}$. The transfer functions $\mathbf{q}_r$ for the process $\mathbf{Q}_t$ are determined in the same way as it has been done for the scalar valued case. Take the tensor product of both sides of equation (3.34), thus

$$\begin{aligned}\mathbf{Q}_t &= \left[A^{\otimes 2} + \sigma^2 D^{\otimes 2}\right] \mathbf{Q}_{t-1} + \left[A \otimes D + D \otimes A\right] \mathbf{Q}_{t-1} \mathrm{w}_{t-1} \\ &\quad + D^{\otimes 2} \mathbf{Q}_{t-1} H_2(\mathrm{w}_{t-1}) \\ &\quad + F_0\left(\mathbf{X}_{t-1}\right) + F_1\left(\mathbf{X}_{t-1}\right) \mathrm{w}_{t-1} + F_2\left(\mathbf{X}_{t-1}\right) H_2(\mathrm{w}_{t-1}) \\ &\quad + F_3\left(\mathbf{X}_{t-1}\right) H_3(\mathrm{w}_{t-1}) + P\left(\mathrm{w}_{t-1}\right),\end{aligned} \tag{3.37}$$

where

$$\begin{aligned}F_0\left(\mathbf{X}_{t-1}\right) &= \sigma^2 D \otimes I \mathbf{X}_{t-1} \otimes \mathbf{b}_1 + \sigma^2 I \otimes D \mathbf{b}_1 \otimes \mathbf{X}_{t-1}, \\ F_1\left(\mathbf{X}_{t-1}\right) &= A \otimes I \mathbf{X}_{t-1} \otimes \mathbf{b}_1 + I \otimes A \mathbf{b}_1 \otimes \mathbf{X}_{t-1} + 2\sigma^2 D \otimes I \mathbf{X}_{t-1} \otimes \mathbf{b}_2 \\ &\quad + 2\sigma^2 I \otimes D \mathbf{b}_2 \otimes \mathbf{X}_{t-1}, \\ F_2\left(\mathbf{X}_{t-1}\right) &= A \otimes I \mathbf{X}_{t-1} \otimes \mathbf{b}_2 + I \otimes A \mathbf{b}_2 \otimes \mathbf{X}_{t-1} + D \otimes I \mathbf{X}_{t-1} \otimes \mathbf{b}_1 \\ &\quad + I \otimes D \mathbf{b}_1 \otimes \mathbf{X}_{t-1}, \\ F_3\left(\mathbf{X}_{t-1}\right) &= D \otimes I \mathbf{X}_{t-1} \otimes \mathbf{b}_2 + I \otimes D \mathbf{b}_2 \otimes \mathbf{X}_{t-1}, \\ P\left(\mathrm{w}_{t-1}\right) &= \mathbf{b}_2^{\otimes 2} H_4(\mathrm{w}_{t-1}) + \left[\mathbf{b}_1 \otimes \mathbf{b}_2 + \mathbf{b}_2 \otimes \mathbf{b}_1\right] H_3(\mathrm{w}_{t-1}) \\ &\quad + \left[\mathbf{b}_1^{\otimes 2} + 4\sigma^2 \mathbf{b}_2^{\otimes 2}\right] H_2(\mathrm{w}_{t-1}) \\ &\quad + 2\sigma^2 \left[\mathbf{b}_1 \otimes \mathbf{b}_2 + \mathbf{b}_2 \otimes \mathbf{b}_1\right] \mathrm{w}_{t-1} + \sigma^2 \mathbf{b}_1^{\otimes 2} + 2\sigma^4 \mathbf{b}_2^{\otimes 2}.\end{aligned}$$

The tensor product has a priority over the other matrix operation in the above formulae. Equation (3.37) for the quadratic process $\mathbf{Q}_t$ is no longer bilinear. It is expressed in terms of $\mathbf{Q}_t$, $\mathbf{X}_t$ and w_t. The transfer functions of $\mathbf{X}_t$ are known and we use (3.35) to get the transfer functions of $\mathbf{Q}_t$. The coefficient of $\mathbf{Q}_{t-1}$ in (3.37) has eigenvalues only inside the unit circle by the assumption of second order stationarity. Let us introduce the following notations:

$$\mathcal{A}(z) = \left[zI - A^{\otimes 2} - \sigma^2 D^{\otimes 2}\right]^{-1}, \text{ and } D_Q = A \otimes D + D \otimes A.$$

Theorem 41 *If the fourth order moment of the second order stationary solution $\mathbf{X}_t$ of the state space equation (3.37) exists then the transfer function system for the quadratic process $\mathbf{Q}_t = \mathbf{X}_t^{\otimes 2}$ is given by the formulae*

$$\mathbf{q}_0 = \mathsf{E}\mathbf{X}_0^{\otimes 2},$$
$$\mathbf{q}_1(z_1) = \mathcal{A}(z_1)\left[D_Q\mathbf{q}_0 + F_0(\mathbf{f}_1(z_1)) + 2\sigma^2\left[\mathbf{b}_1\otimes\mathbf{b}_2 + \mathbf{b}_2\otimes\mathbf{b}_1\right]\right],$$
$$\mathbf{q}_2(z_{(1:2)}) = \mathcal{A}(z_1 z_2)\left[D_Q\mathbf{q}_1(z_1) + D^{\otimes 2}\mathbf{q}_0 + F_0\left(\mathbf{f}_2(z_{(1:2)})\right) + F_1(\mathbf{f}_1(z_1)) + \mathbf{b}_1^{\otimes 2} + 4\sigma^2\mathbf{b}_2^{\otimes 2}\right],$$
$$\mathbf{q}_3(z_{(1:3)}) = \mathcal{A}\left(z_{(1:3)}^{\vec{1}}\right)\left[D_Q\mathbf{q}_2(z_{(1:2)}) + D^{\otimes 2}\mathbf{q}_1(z_1) + \sum_{i=0}^{2} F_i\left(\mathbf{f}_{3-i}(z_{(1:3-i)})\right) + \mathbf{b}_1\otimes\mathbf{b}_2 + \mathbf{b}_2\otimes\mathbf{b}_1\right],$$
$$\mathbf{q}_4(z_{(1:4)}) = \mathcal{A}\left(z_{(1:4)}^{\vec{1}}\right)\left[D_Q\mathbf{q}_3(z_{(1:3)}) + D^{\otimes 2}\mathbf{q}_2(z_{(1:2)}) + \sum_{i=0}^{3} F_i\left(\mathbf{f}_{4-i}(z_{(1:4-i)})\right) + \mathbf{b}_2^{\otimes 2}\right],$$

if $r > 4$, then the following recursion is valid:

$$\mathbf{q}_r(z_{(1:r)}) = \mathcal{A}\left(z_{(1:r)}^{\vec{1}}\right)\left[D_Q\mathbf{q}_{r-1}(z_{(1:r-1)}) + D^{\otimes 2}\mathbf{q}_{r-2}(z_{(1:r-2)}) + \sum_{i=0}^{3} F_i\left(\mathbf{f}_{r-i}(z_{(1:r-i)})\right)\right].$$

We shall see later that the necessary and sufficient condition for the fourth order stationarity of $\mathbf{X}_t$ is that all the eigenvalues of the matrix

$$\left[A^{\otimes 2} + \sigma^2 D^{\otimes 2}\right]^{\otimes 2} + \sigma^2 D_Q^{\otimes 2} + 2\sigma^4 D^{\otimes 4},$$

should be less than 1 in modulus.

Cross-spectrum of $\mathbf{Q}_t$ and $\mathbf{X}_t$

It has been shown that for the identification of the bilinear process the spectrum itself is not satisfactory. We consider here the cross-spectrum $S_{Q,X}$ of $\mathbf{Q}_t$ and $\mathbf{X}_t$. To get the cross-covariance $\mathsf{E}(\mathbf{Q}_{t+s}\otimes\mathbf{X}_t)$, take equation (3.37) at time $t+s$ for $s \geq 1$ and take the tensor product by $\mathbf{X}_t$:

$$\mathsf{E}(\mathbf{Q}_{t+s}\otimes\mathbf{X}_t) = \left[A^{\otimes 2} + \sigma^2 D^{\otimes 2}\right]\otimes I\,\mathsf{E}(\mathbf{Q}_{t+s-1}\otimes\mathbf{X}_t)$$
$$+\sigma^2\left[DA^{s-1}\otimes I\right]\mathsf{E}(\mathbf{X}_t\otimes\mathbf{b}_1\otimes\mathbf{X}_t) + \sigma^2\left[I\otimes DA^{s-1}\otimes I\right]\mathbf{b}_1\otimes\mathsf{E}(\mathbf{X}_t\otimes\mathbf{X}_t),$$

and for $s \leq -1$ take the equation (3.34) and the tensor product by $\mathbf{Q}_{t+s}$

$$\mathsf{E}(\mathbf{Q}_{t+s}\otimes\mathbf{X}_t) = [I\otimes A]^{-s}\,\mathsf{E}(\mathbf{Q}_t\otimes\mathbf{X}_t).$$

The cross-spectrum is the Fourier transform of these covariances

$$S_{\mathbf{Q},\mathbf{X}}(z) = \left[I - z^{-1}\left[A^{\otimes 2} + \sigma^2 D^{\otimes 2}\right] \otimes I\right]^{-1}$$
$$\times \left[\mathsf{E}\left(\mathbf{Q}_t \otimes \mathbf{X}_t\right) + \sigma^2 z^{-1}\left[D\left(I - z^{-1}A\right)^{-1} \otimes I\right] \mathsf{E}\left(\mathbf{X}_t \otimes \mathbf{b}_1 \otimes \mathbf{X}_t\right)\right.$$
$$\left. + \sigma^2 z^{-1}\left[I \otimes D\left(I - z^{-1}A\right)^{-1} \otimes I\right] \mathbf{b}_1 \otimes \mathsf{E}\left(\mathbf{X}_t \otimes \mathbf{X}_t\right)\right]$$
$$+ z\left[I - zI \otimes A\right]^{-1} I \otimes A\mathsf{E}\left(\mathbf{Q}_t \otimes \mathbf{X}_t\right).$$

The constants are important for the estimation of the parameters. The expression for $\mathsf{E}\mathbf{Q}_t \otimes \mathbf{X}_t$ comes from the product of equations (3.34) and (3.37) above, so

$$\mathsf{E}\left(\mathbf{Q}_t \otimes \mathbf{X}_t\right) = \left[I - A^{\otimes 3} - \sigma^2\left[D^{\otimes 2} \otimes A + A \otimes D^{\otimes 2} + D \otimes A \otimes D\right]\right]^{-1}$$
$$\times \left[\sigma^2\left[D \otimes I \otimes A + A \otimes I \otimes D\right] \mathsf{E}\left(\mathbf{X}_t \otimes \mathbf{b}_1 \otimes \mathbf{X}_t\right)\right.$$
$$+ \sigma^2\left[I \otimes D \otimes A + I \otimes A \otimes D\right] \mathbf{b}_1 \otimes \mathsf{E}\mathbf{X}_t^{\otimes 2}$$
$$+ \sigma^2\left[D \otimes A \otimes I + A \otimes D \otimes I\right] \mathsf{E}\left(\mathbf{X}_t \otimes \mathbf{X}_t\right) \otimes \mathbf{b}_1$$
$$+ 2\sigma^4\left[D \otimes I \otimes D\mathsf{E}\left(\mathbf{X}_t \otimes \mathbf{b}_2 \otimes \mathbf{X}_t\right) + I \otimes D^{\otimes 2}\mathbf{b}_2 \otimes \mathsf{E}\mathbf{X}_t^{\otimes 2}\right.$$
$$\left. + D^{\otimes 2} \otimes I\mathsf{E}\mathbf{X}_t^{\otimes 2} \otimes \mathbf{b}_2\right]$$
$$\left. + 2\sigma^4\left[\mathbf{b}_1 \otimes \mathbf{b}_2 \otimes \mathbf{b}_1 + \mathbf{b}_1^{\otimes 2} \otimes \mathbf{b}_2 + \mathbf{b}_2 \otimes \mathbf{b}_1^{\otimes 2} + 4\sigma^2 \mathbf{b}_2^{\otimes 3}\right]\right],$$

compare this formula with the scalar valued case (3.18). The only missing quantity is

$$\mathsf{E}\left(\mathbf{X}_t \otimes \mathbf{b}_i \otimes \mathbf{X}_t\right) = \left[I - A \otimes I \otimes A - \sigma^2 D \otimes I \otimes D\right]^{-1}$$
$$\times \left[\sigma^2 \mathbf{b}_1 \otimes \mathbf{b}_i \otimes \mathbf{b}_1 + 2\sigma^4 \mathbf{b}_2 \otimes \mathbf{b}_i \otimes \mathbf{b}_2\right],$$

$i = 1, 2.$

Bispectrum of $\mathbf{c}'\mathbf{X}_t$

This time we go directly for the bispectrum of $\mathbf{c}'\mathbf{X}_t$ instead of the bispectrum of the vector valued state process $\mathbf{X}_t$. The reason is that $\mathsf{E}\left(\mathbf{X}_r \otimes \mathbf{X}_s \otimes \mathbf{X}_t\right)$ is not symmetric in r, s, t but

$$\mathsf{E}\left(\mathbf{c}'\mathbf{X}_r \mathbf{c}'\mathbf{X}_s \mathbf{c}'\mathbf{X}_t\right) = \mathbf{c}'^{\otimes 3}\mathsf{E}\left(\mathbf{X}_r \otimes \mathbf{X}_s \otimes \mathbf{X}_t\right),$$

is. This allows us to apply the technique in Subsection 3.2.2, page 79, i.e., the formula

$$S_{X,X,X}(z_1, z_2) = \operatorname*{sym}_{z_{(1:3)}} \left[6 \sum_{k=1}^{\infty} \sum_{l=1}^{\infty} \underset{XXX}{\mathrm{Cum}}(k+l, l) z_1^{-k} z_2^{-l}\right.$$
$$\left. \left. + 3 \sum_{l=-\infty}^{\infty} \underset{XXX}{\mathrm{Cum}}(0, l) z_1^{-l} - 2\, \underset{XXX}{\mathrm{Cum}}(0,0)\right]\right|_{z_3 = (z_1 z_2)^{-1}}$$

Now in our case

$$\mathsf{E}\left(\mathbf{X}_{k+l}\otimes\mathbf{X}_l\otimes\mathbf{X}_0\right)=\left(A\otimes I\right)^k\mathsf{E}\left(\mathbf{Q}_l\otimes\mathbf{X}_0\right),$$

therefore

$$\sum_{k=1}^{\infty}\sum_{l=1}^{\infty}\mathsf{E}\left(\mathbf{X}_{k+l}\otimes\mathbf{X}_l\otimes\mathbf{X}_0\right)) z_1^{-(k+l)}z_2^{-l}=\left(A\otimes I\right)\left(z_1I-A\otimes I\right)^{-1} \times S^{+}_{\mathbf{Q},\mathbf{X}}(z_3^{-1}),$$

where

$$\begin{aligned}S^{+}_{\mathbf{Q},\mathbf{X}}(z)=&\left[I-z^{-1}\left[A^{\otimes 2}+\sigma^2D^{\otimes 2}\right]\otimes I\right]^{-1}\\&\times\Big[\mathsf{E}\left(\mathbf{Q}_t\otimes\mathbf{X}_t\right)+\sigma^2z^{-1}\left[D\left(I-z^{-1}A\right)^{-1}\otimes I\right]\\&\times\mathsf{E}\left(\mathbf{X}_t\otimes\mathbf{b}_1\otimes\mathbf{X}_t\right)\\&+\sigma^2z^{-1}\left[I\otimes D\left(I-z^{-1}A\right)^{-1}\otimes I\right]\mathbf{b}_1\otimes\mathsf{E}\left(\mathbf{X}_t\otimes\mathbf{X}_t\right)\Big]\\&-\mathsf{E}\left(\mathbf{Q}_t\otimes\mathbf{X}_t\right).\end{aligned}$$

The bispectrum of $\mathbf{c}'\mathbf{X}_t$ is given by

$$S_{3,c'X}(z_1,z_2)=\mathbf{c}'^{\otimes 3}\operatorname*{sym}_{z_{(1:3)}}\left[6\left(A\otimes I\right)\left(z_1I-A\otimes I\right)^{-1}S^{+}_{\mathbf{Q},\mathbf{X}}(z_3^{-1})\right.\\ \left.+3S_{\mathbf{Q},\mathbf{X}}\left(z_1^{-1}\right)\right]-\mathbf{c}'^{\otimes 3}2\mathsf{E}\left(\mathbf{Q}_t\otimes\mathbf{X}_t\right).$$

remember that$z_3=(z_1z_2)^{-1}$.

Spectra for the observation with zero mean

Similarly to the bilinear model $BL(1,0,1,0)$ one can easily derive the spectra of the scalar observation

$$Y_t=\mathbf{c}'\mathbf{X}_t+w_t.$$

Spectrum

$$S_{2,Y}\left(z\right)=\mathbf{c}'\otimes\mathbf{c}'S_{2,\mathbf{X}}\left(z\right)+2\sigma^2\mathbf{c}'\operatorname{Re}\left(\mathbf{f}_1\left(z\right)\right)+\sigma^2,$$

and

Cross – spectrum

$$S_{Y^2,Y}\left(z\right)=\mathbf{c}'^{\otimes 3}S_{\mathbf{Q},\mathbf{X}}\left(z\right)+2\sigma^2\mathbf{c}'^{\otimes 2}I\otimes\left(D\left(Iz^{-1}-A\right)^{-1}\right)\mathsf{E}\mathbf{X}_t^{\otimes 2}\\+2\sigma^2\mathbf{c}'\left(Iz^{-1}-A\right)^{-1}\mathbf{b}_2+\sigma^2\mathbf{c}'^{\otimes 2}\mathbf{q}_1(z),$$

where the identity $(I\otimes D)\left(I\otimes A\right)^k=I\otimes DA^k$ was applied.

Bispectrum

The calculation of the bispectrum

$$S_{3,Y}(z_1, z_2) = S_{3,\mathbf{c}'\mathbf{X}+\mathrm{w}}(z_1, z_2),$$

is based on the previous formulae and the independence of the state $\mathbf{X}_t$ of the noise w_{t-k} $k \geq 0$. Start with the third order cumulants of Y_t in the domain $k, l \geq 0$. Note here again the assumption that the mean of Y_t has been taken out, therefore the third order cumulants are equal to the moments

$$\begin{aligned}
\mathsf{E}Y_{t+k+l}Y_{t+k}Y_t &= \mathbf{c}'^{\otimes 3}\mathsf{E}\mathbf{X}_{t+k+l} \otimes \mathbf{X}_{t+k} \otimes \mathbf{X}_t + \mathsf{E}\mathrm{w}_{t+k+l}\mathrm{w}_{t+k}\mathrm{w}_t \\
&+ \mathbf{c}'^{\otimes 2}\mathsf{E}\mathbf{X}_{t+k+l} \otimes \mathbf{X}_{t+k}\mathrm{w}_t + \mathbf{c}'^{\otimes 2}\mathsf{E}\mathbf{X}_{t+k+l} \otimes \mathbf{X}_t\mathrm{w}_{t+k} \\
&+ \mathbf{c}'^{\otimes 2}\mathsf{E}\mathrm{w}_{t+k+l}\mathbf{X}_{t+k} \otimes \mathbf{X}_t + \mathbf{c}'\mathsf{E}\mathbf{X}_{t+k+l}\mathrm{w}_{t+k}\mathrm{w}_t \\
&+ \mathbf{c}'\mathsf{E}\mathrm{w}_{t+k+l}\mathbf{X}_{t+k}\mathrm{w}_t + \mathbf{c}'\mathsf{E}\mathrm{w}_{t+k+l}\mathrm{w}_{t+k}\mathbf{X}_t \\
&= \mathbf{c}'^{\otimes 3}\mathsf{E}\mathbf{X}_{t+k+l} \otimes \mathbf{X}_{t+k} \otimes \mathbf{X}_t \\
&+ \delta_{k=0,l>0}\sigma^2\mathbf{c}'^{\otimes 2}\left(A^{l-1}D \otimes I\right)\mathsf{E}\mathbf{X}_t^{\otimes 2} \\
+ \delta_{k>0}\mathbf{c}'^{\otimes 2}\left(A \otimes I\right)^l \mathsf{E}\mathbf{X}_{t+k}^{\otimes 2}\mathrm{w}_t &+ \delta_{l\geq 1}\sigma^2\mathbf{c}'^{\otimes 2}\left(A^{l-1}D \otimes I\right)\mathsf{E}\mathbf{X}_{t+k} \otimes \mathbf{X}_t \\
+ \delta_{k>0,l\geq 1}\sigma^2\mathbf{c}'A^{l-1}D\mathsf{E}\mathbf{X}_{t+k}\mathrm{w}_t &+ \delta_{k=0,l\geq 1}2\sigma^4\mathbf{c}'A^{l-1}\mathbf{b}_2.
\end{aligned}$$

Equation (3.34) has been used to get the various third order moments, for example

$$\begin{aligned}
\mathsf{E}\mathbf{X}_{t+k+l}\mathrm{w}_{t+k}\mathrm{w}_t &= A\mathsf{E}\mathbf{X}_{t+k+l-1}\mathrm{w}_{t+k}\mathrm{w}_t + D\mathsf{E}\mathbf{X}_{t+k+l-1}\mathrm{w}_{t+k+l-1}\mathrm{w}_{t+k}\mathrm{w}_t \\
&+ \mathbf{b}_1\mathsf{E}\mathrm{w}_{t+k+l-1}\mathrm{w}_{t+k}\mathrm{w}_t + \mathbf{b}_2\mathsf{E}H_2(\mathrm{w}_{t+k+l-1})\mathrm{w}_{t+k}\mathrm{w}_t \\
&= \delta_{l>0}\sigma^2 DA^{l-1}\mathsf{E}\mathbf{X}_{t+k}\mathrm{w}_t + \delta_{k=0,l\geq 1}2\sigma^4 A^{l-1}\mathbf{b}_2.
\end{aligned}$$

Carrying out the calculation similar to the one given in Subsection 3.2.2 and 3.2.2, one can see that the **bispectrum** is given, after the symmetrization and changing z_3 into $(z_1z_2)^{-1}$, by the formula

$$\begin{aligned}
S_{Y,Y,Y}(z_1, z_2) &= S_{3,\mathbf{c}'\mathbf{X}}(z_1, z_2) \\
&+ \sigma^2\mathrm{sym}_{z_{(1:3)}}\Big[6\mathbf{c}'^{\otimes 2}\left(A\left(z_1I - A\right)^{-1} \otimes I\right)\mathbf{q}_1(z_3^{-1}) \\
&+ 6\mathbf{c}'^{\otimes 2}\left(\left(z_1I - A\right)^{-1}D\right) \otimes I\left(A\left(z_3^{-1}I - A\right)^{-1} \otimes I\right)\mathsf{E}\mathbf{X}_t^{\otimes 2} \\
&+ 6\sigma^2\mathbf{c}'\left(z_1I - A\right)^{-1}D\mathbf{f}_1(z_3^{-1}) \\
&+ 6\mathbf{c}'^{\otimes 2}\left(\left(Iz_1 - A\right)^{-1}D\right) \otimes I\mathsf{E}\mathbf{X}_t^{\otimes 2} \\
&+ 6\mathbf{c}'\left(Iz_1 - A\right)^{-1}\mathbf{b}_2 + 3\mathbf{c}'^{\otimes 2}\mathbf{q}_1(z_1^{-1})\Big]_{z_3=(z_1z_2)^{-1}}.
\end{aligned}$$

Example 42 *We show the utilization of the above results for the process*

$$Y_t = a_1 Y_{t-1} + a_2 Y_{t-2} + d_{11} Y_{t-1} \mathrm{w}_{t-1} + d_{12} Y_{t-2} \mathrm{w}_{t-1} + d_{22} Y_{t-2} \mathrm{w}_{t-2} + \mathrm{w}_t + f_0,$$

it is transformed into

$$\mathbf{X}_t = A\mathbf{X}_{t-1} + D\mathbf{X}_{t-1}\mathrm{w}_{t-1} + \mathbf{b}_1 \mathrm{w}_{t-1} + \mathbf{b}_2 H_2(\mathrm{w}_{t-1}),$$
$$Y_t = \mathbf{c}'\mathbf{X}_t + \mathrm{w}_t,$$

where

$$\mathbf{X}_t = \begin{bmatrix} U \\ V \\ Z \end{bmatrix}_t, \quad A = \begin{bmatrix} a_1 & 1 & 0 \\ a_2 & 0 & 0 \\ 1 & 0 & 0 \end{bmatrix}, \quad D = \begin{bmatrix} d_{11} & 0 & d_{12} \\ d_{22} & 0 & 0 \\ 0 & 0 & 0 \end{bmatrix},$$
$$\mathbf{b} = \mathbf{c} = \begin{bmatrix} 1 \\ 0 \\ 0 \end{bmatrix}, \quad \mathbf{b}_1 = A\mathbf{b}, \ \mathbf{b}_2 = D\mathbf{b}.$$

The parameters were chosen as $a_1 = -.2$, $a_2 = .15$, $d_{11} = -.5$, $d_{12} = .1$, $d_{22} = .1$, $\sigma^2 = 1$. *The sample size of simulated series was* $n = 2^{11}$ *and the number of frequencies for the spectra was* 2^6.

	$\mathsf{E}Y_t^2$	$\mathsf{E}Y_t^3$
Theoretical	1.4285	−0.3473
Estimated	1.3319	−0.2788

The spectra are plotted in Figure 3.3.

3.3.4 Necessary and sufficient condition for the existence of $2n^{th}$ order moments of the state process

In this section we are giving a similar condition for the existence of even order moments of the vector valued process $\{\mathbf{X}_t, t \in \mathbb{Z}\}$ as in the previous section where we investigated the scalar valued process $\{X_t^n, t \in \mathbb{Z}\}$. Now our attention is focussed on the process $\{\mathbf{X}_t^{\otimes n}, t \in \mathbb{Z}\}$. We omit the proofs where the difference between this and the previous section is only a simple matrix operation.

Suppose $\{\mathbf{X}_t, t \in \mathbb{Z}\}$ with $\mathsf{E}\mathbf{X}_t = 0$ satisfying the following equations

$$\mathbf{X}_t = A\mathbf{X}_{t-1} + D\mathbf{X}_{t-1}\mathrm{w}_{t-1} + \mathbf{b}_1 H_1(\mathrm{w}_{t-1}) + \mathbf{b}_2 H_2(\mathrm{w}_{t-1}), \qquad t \in \mathbb{Z}, \tag{3.38}$$

where $A,\ D \in \mathbf{R}^{m\times m}$, $\mathbf{b},\ \mathbf{c} \in \mathbf{R}^m$ $\mathbf{b}_2 = D\mathbf{b}$, $\mathbf{b}_1 = A\mathbf{b}$. We can see that $\mathbf{X}_t$ is independent of w_t.

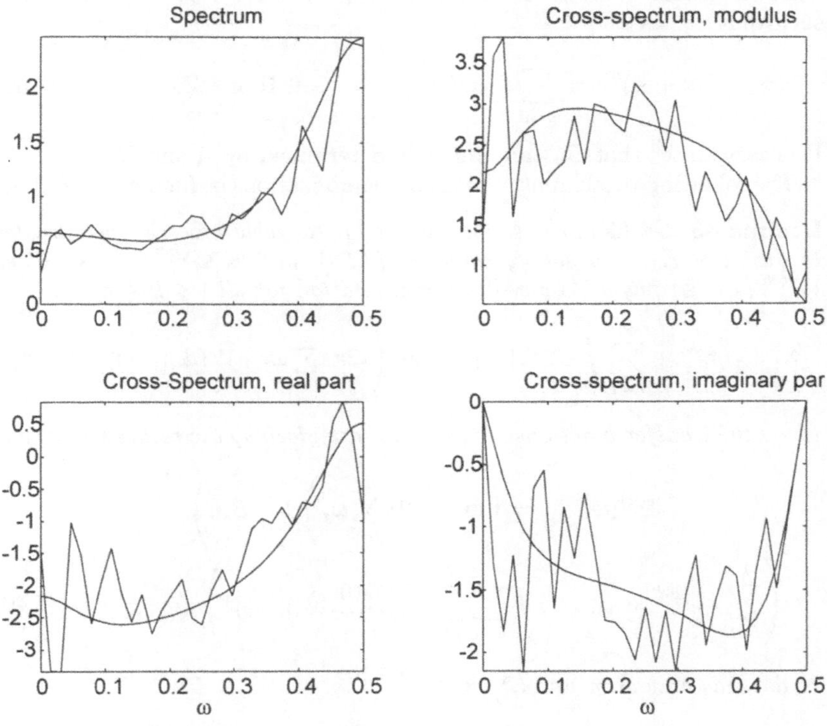

FIGURE 3.3. Theoretical and estimated spectra of a bilinear process, general case

We have the Wiener–Itô chaotic representation of the stochastic process $\{\mathbf{X}_t\,,\,t\in\mathbb{Z}\}$, see subsection 3.3.2 for details,

$$\mathbf{X}_t=\sum_{r=1}^{\infty}\int_{D^r}\exp\left(i2\pi t\sum_{j=1}^{r}\omega_j\right)\mathbf{f}_r(\omega_{(1:r)})W(d\omega_{(1:r)}),\qquad t\in\mathbb{Z},$$

where $(f_r)_i\in\overline{L^r_\Phi}$, $i=1,\ldots,m$ are the transfer functions of $\{(\mathbf{X}_t)_i\,,\,t\in\mathbb{Z}\}$, $i=1,\ldots,m$, respectively.

Let us define the following matrices B_{nk} $(n=1,2,\ldots,\ k=0,1,\ldots,n)$ such that

$$(Dx+A)^{\otimes n}=\sum_{k=0}^{n}B_{nk}H_k(x),\qquad \forall x\in\mathbf{R},n\in\mathbb{Z}_+,\tag{3.39}$$

It is easy to see that B_{nk} are uniquely determined by A and D.

The following two lemmas are clear generalizations of Lemma 35 and 36.

Lemma 43 *Let $\{\mathbf{X}_t\,,\,t\in\mathbb{Z}\}$ be physically realizable process, subordinated to $\{w_t\,,\,t\in\mathbb{Z}\}$, and satisfy equation (3.38), and $\mathsf{E}(\mathbf{X}_t)^{\otimes 2n}<\infty$. Then $\{\mathbf{X}_t^{\otimes k}\,,\,t\in\mathbb{Z}\}$ has a Wiener–Itô representation for all $1\le k\le n$, i.e.*

$$\mathbf{X}_t^{\otimes k}=\mathbf{f}_0^{(k)}+\sum_{r=1}^{\infty}\int_{D^r}\mathbf{f}_r^{(k)}(\omega_{(1:r)})\exp\left(i2\pi t\sum_{j=1}^{r}\omega_j\right)W(d\omega_{(1:r)}),\quad t\in\mathbb{Z},$$

where the transfer functions $\mathbf{f}_r^{(k)}$, $r>2n$, are given by the recursive formula

$$\mathbf{f}_r^{(k)}(\omega_{(1:r)})=\left(\exp\left(i2\pi\sum_{j=1}^{r}\omega_j\right)I-B_{k0}\right)^{-1}$$

$$\times\left(\sum_{l=1}^{k}B_{kl}\mathbf{f}_{r-l}^{(k)}(\omega_{(1:r-l)})+\sum_{i=1}^{k-1}\sum_{j=0}^{K}C_{ij}^{(k)}\mathbf{f}_{r-j}^{(i)}(\omega_{(1:r-j)})\right),\tag{3.40}$$

K depends only on n and $C_{ij}^{(k)}\in\mathbf{R}^{m^k\times m^i}$.

Lemma 44 *For the transfer functions $\mathbf{f}_r^{(k)}$ which are defined by the recursive formula (3.40)*

$$\left((\mathbf{f}_{r-l}^{(i)})_{j},(\mathbf{f}_{r-l'}^{(i')})_{j'}\right)_r^{\mathfrak{P}}=\begin{cases}l!\sigma^{2l}\left((\mathbf{f}_{r-l}^{(i)})_j,\mathbf{f}_{r-l}^{(i')})_{j'}\right)_{r-l}^{\mathfrak{P}} & \text{if}\ \ l=l',\\ 0 & \text{if}\ \ l\neq l',\end{cases}$$

where $j=1,\ldots,m^i$, $j'=1.\ldots,m^{i'}$.

Now we are in a position to prove our main theorem on the existence of the $2n^{\text{th}}$ moment of the state–process, which is a generalization of the one given by Terdik [130] for the case $n=2$. Denote by $\rho(A)$ the spectral radius of the matrix A.

Theorem 45 *Let $\{\mathbf{X}_t\,,\,t\in\mathbb{Z}\}$ be physically realizable, subordinated to $\{w_t\,,\,t\in\mathbb{Z}\}$ and a solution of (3.38). Then the following statements are equivalent*
(i) $\mathsf{E}(\mathbf{X}_t)^{\otimes 2n}<\infty$,
(ii) $\rho\left(\sum\limits_{k=0}^{n}k!\sigma^{2k}B_{nk}^{\otimes 2}\right)<1$,
(iii) $\rho\left(\mathsf{E}(A+Dw_0)^{\otimes 2n}\right)<1$.

Proof. (i) $\Longrightarrow$ (ii)
Denote by $\mathbf{f}_r^{(k)}$, B_{kl} and $C_{ij}^{(k)}$ the same as earlier, see (3.39) and (3.40). Put

$$\mathbf{g}_r=\left((\mathbf{f}_r^{(n)})^T,\dots,(\mathbf{f}_r^{(1)})^T\right)^T,\qquad r\in\mathbb{Z}_+,$$

$$A_j=\begin{bmatrix}\tilde{B}_{nj} & C_{1j}^{(n)} & C_{2j}^{(n)} & \cdots & C_{n-1j}^{(n)}\\ 0 & \tilde{B}_{n-1j} & C_{1j}^{(n-1)} & \cdots & C_{n-2j}^{(n-1)}\\ 0 & 0 & \tilde{B}_{n-2j} & \cdots & C_{n-3j}^{(n-2)}\\ \vdots & \vdots & \vdots & \ddots & \vdots\\ 0 & 0 & 0 & \cdots & \tilde{B}_{1j}\end{bmatrix},\quad j=1,\dots,K,$$

where

$$\tilde{B}_{ij}=\begin{cases}B_{ij} & \text{if}\quad i\geq j,\\ 0 & \text{otherwise,}\end{cases}$$

and H_0 equals to

$$\begin{bmatrix}M_n & 0 & \cdots & 0 & 0\\ 0 & (I-B_{n0}\otimes B_{n-1})^{-1} & \cdots & 0 & 0\\ \vdots & \vdots & \ddots & \vdots & \vdots\\ 0 & 0 & \cdots & (I-B_{n0}\otimes B_{10})^{-1} & 0\\ \vdots & \vdots & \cdots & \vdots & \vdots\\ 0 & 0 & \cdots & 0 & M_1\end{bmatrix},$$

where $M_n=(I-B_{n0}^{\otimes 2})^{-1}$ and $M_1=(I-B_{10}^{\otimes 2})^{-1}$, then $\mathbf{g}_r\in R^s$, $A_j\in\mathbf{R}^{s\times s}$ and $H_0\in\mathbf{R}^{s^2\times s^2}$ where $s=m+m^2+\ldots+m^n$. If we set

$$G_r=\operatorname{Vec}\left(((\mathbf{g}_r)_i,(\mathbf{g}_r)_j)_r^{\mathfrak{P}}\right)_{i,j=1}^{s},$$

then by an argument similar to that in Theorem 37 G_r, $r\geq K+2n$, satisfies the recursive formula

$$G_r=H_0\sum_{k=1}^{K}k!\sigma^{2k}A_k^{\otimes 2}G_{r-k}.$$

From (i) it follows that $\sum_{r=0}^{\infty} G_r$ is convergent, which is equivalent to the fact that every root of the polynomial

$$P(\lambda) = \det(\lambda^K I - H_0 \sum_{k=1}^{K} k!\sigma^{2k}\lambda^{K-k} A_k^{\otimes 2}),$$

must be smaller than 1 in modulus. By Lemma 94, page 221, of the Appendix B we have $P = \prod_{i,j=1}^{n} P_{ij}$, where

$$P_{ij}(\lambda) = \det(\lambda^K I - (I - B_{i0} \otimes B_{j0})^{-1} \sum_{k=1}^{K} k!\sigma^{2k}\lambda^{K-k} \tilde{B}_{ik} \otimes \tilde{B}_{jk}).$$

So all roots of polynomial P_{nn} must be smaller than 1 in modulus, i.e., all roots of the equation

$$\lambda^{(K-n)m^n} \det(\lambda^n I - (I - B_{n0}^{\otimes 2})^{-1} \sum_{k=1}^{n} k!\sigma^{2k}\lambda^{n-k} B_{nk}^{\otimes 2}) = 0,$$

must be smaller than 1 in modulus. Applying Corollary 91, page 220, in the Appendix B we have proved our assertion.

The equivalence of (ii) and (iii) can be proven by elementary algebra.

(iii) $\Longrightarrow$ (i)

It can be supposed that $\mathsf{E}(\mathbf{X}_t)^{\otimes k} < \infty$ for all $1 \leq k < n$. Since we have already proved (i) $\Longrightarrow$ (ii) we have

$$\rho(\sum_{l=0}^{k} l!\sigma^{2l} B_{kl}^{\otimes 2}) < 1 \qquad \text{for all } 1 \leq k \leq n,$$

where the case $k = n$ is given by (iii). Applying Corollary 91, page 220, in the Appendix B we find that all roots of the equations

$$\lambda^{(K-k)m^k} \det(\lambda^k I - (I - B_{k0}^{\otimes 2})^{-1} \sum_{j=1}^{k} j!\sigma^{2j}\lambda^{k-j} B_{kj}^{\otimes 2}) = 0, \quad 1 \leq k \leq n,$$

are smaller than 1 in modulus, i.e., the polynomials P_{kk}, $k = 1, \ldots, n$ have all roots smaller than 1 in modulus. Applying Corollary 93, page 221, in the Appendix B we have that the polynomials P_{ij}, $1 \leq i, j \leq n$ have no roots outside the open unit disc, i.e., polynomial P has all roots smaller than 1 in modulus. This fact shows the convergence of $\sum_{r=0}^{\infty} G_r$, which is equivalent to the existence of all moments up to $2n$. ■

It is easy to see that if we consider the state–process as an RCA–model, we can give an interpretation of Theorem 45 similar to that in the remark to Theorem 37. Corollary 39 can also be generalized as follows.

Corollary 46 *If* $\{\mathbf{X}_t, t \in \mathbb{Z}\}$ *is a state–process such that* $\rho(D) \neq 0$, *then it can only have a finite number of moments.*

Proof. Since $B_{nn} = D^{\otimes n}$ we have that

$$\rho\left(\sum_{k=0}^{n} k!\sigma^{2k} B_{nk}^{\otimes 2}\right) \geq \rho\left(n!\sigma^{2n} B_{nn}^{\otimes 2}\right) = \rho\left(n!\sigma^{2n} D^{\otimes 2n}\right)$$
$$= n!\left(\sigma^2 \rho^2(D)\right)^n \to \infty \quad \text{as } n \to \infty.$$

This fact and Theorem 45 show the truth of our assertion. ■

3.4 Identification of multiple-bilinear models

3.4.1 Chaotic representation and stationarity

Multivariate Bilinear Model

Consider the following state space model of the p_1-dimensional output Y_t with the d-dimensional input w_t,

$$\mathbf{X}_t = A\mathbf{X}_{t-1} + \sum_{j=1}^{d} \mathbf{b}_j \mathrm{w}_{j,t} + \sum_{j=1}^{d} D_j \mathbf{X}_{t-1} \mathrm{w}_{j,t-1} + \mathbf{f}_0,$$
$$\mathbf{Y}_t = C\mathbf{X}_t, \tag{3.41}$$

where $(\mathrm{w}_{1,t}, \mathrm{w}_{2,t}, \ldots, \mathrm{w}_{d,t})' = \mathrm{w}_t$ is a vector valued Gaussian white noise with $\mathsf{E}\mathrm{w}_t = 0$ and the diagonal covariance matrix $\mathsf{E}\mathrm{w}_t\mathrm{w}_s' = \delta_{t=s}\mathrm{diag}(\sigma_1^2, \sigma_2^2, \ldots, \sigma_d^2)$, $\sigma_j^2 > 0$, $j = 1, 2, \ldots, d$. Furthermore, A, C and D_j, $j = 1, 2, \ldots, d$, are appropriate matrices, $\mathbf{b}_j \in \mathbf{R}^{p_2}$, and $\mathbf{f}_0$ is a constant vector chosen as

$$\mathbf{f}_0 = -\sum_{j=1}^{d} \sigma_j^2 D_j \mathbf{b}_j,$$

to keep the expectation of the p_2-dimensional state vectors $\mathbf{X}_t$ at zero. The diagonality of the noise covariance matrix $\mathsf{E}\mathrm{w}_t\mathrm{w}_t'$ is not really a restriction because one can orthogonalize the entries of w_t by a linear transformation, say, as

$$\mathrm{w}_{j,t} = \sum_{k=1}^{d} \alpha_{j,k} \mathrm{w}_{k,t}^0,$$

and get the same type of state space equation as (3.41).

The state equation also has the form

$$\mathbf{X}_t = A\mathbf{X}_{t-1} + B\mathrm{w}_t + D(\mathrm{w}_{t-1} \otimes I)\mathbf{X}_{t-1} + \mathbf{f}_0, \tag{3.42}$$

where the new matrices are defined by

$$B = [\mathbf{b}_1, \mathbf{b}_2, \dots, \mathbf{b}_d],$$
$$D = [D_1, D_2, \dots, D_d],$$

and I is the matrix of unity. Moreover, the symbol $\otimes$ denotes the usual tensor product for matrices and vectors.

The most general lower triangular multiple bilinear time series model

$$\sum_{m=0}^{P} A_m Y_{t-m} = \sum_{m=0}^{Q} B_m \mathrm{w}_{t-m} + \sum_{j=1}^{d}\sum_{m=1}^{R}\sum_{n=0}^{S} D_{j,m,m+n} Y_{t-m-n} \mathrm{w}_{j,t-m} + f_0,$$

where $A_0 = B_0 = I$,discussed by Stensholt and Subba Rao [116] and Stensholt and Tjostheim [117] can be easily transformed to a state space form (3.41). We show this by an example. Put $d = 2$, consider

$$\sum_{m=0}^{3} A_m Y_{t-m} = \sum_{m=0}^{2} B_m \mathrm{w}_{t-m} + \sum_{j=1}^{2}\sum_{m=1}^{3}\sum_{n=0}^{3-n} D_{j,m,m+n} Y_{t-m-n} \mathrm{w}_{j,t-m} + f_0,$$

where $A_0 = B_0 = I$. If we define the matrices

$$A = \begin{bmatrix} -A_1 & I & 0 & 0 & 0 \\ -A_2 & 0 & I & 0 & 0 \\ -A_3 & 0 & 0 & 0 & 0 \\ I & 0 & 0 & 0 & 0 \\ 0 & 0 & 0 & I & 0 \end{bmatrix}, \quad B = \begin{bmatrix} I \\ B_1 \\ B_2 \\ 0 \\ 0 \end{bmatrix},$$

$$D_1 = \begin{bmatrix} D_{111} & 0 & 0 & D_{112} & D_{113} \\ D_{122} & 0 & 0 & D_{123} & 0 \\ D_{133} & 0 & 0 & 0 & 0 \\ 0 & 0 & 0 & 0 & 0 \\ 0 & 0 & 0 & 0 & 0 \end{bmatrix},$$

$$D_2 = \begin{bmatrix} D_{211} & 0 & 0 & D_{212} & D_{213} \\ D_{222} & 0 & 0 & D_{223} & 0 \\ D_{233} & 0 & 0 & 0 & 0 \\ 0 & 0 & 0 & 0 & 0 \\ 0 & 0 & 0 & 0 & 0 \end{bmatrix}, \quad \mathbf{f}_0 = \begin{bmatrix} f_0 \\ 0 \\ 0 \\ 0 \\ 0 \end{bmatrix},$$

then the following state space form will be satisfied

$$\mathbf{X}_t = A\mathbf{X}_{t-1} + B\mathrm{w}_t + \sum_{j=1}^{d} D_j \mathbf{X}_{t-1} \mathrm{w}_{j,t-1} + \mathbf{f}_0,$$
$$Y_t = \begin{bmatrix} C & 0 & 0 & 0 & 0 \end{bmatrix} \mathbf{X}_t.$$

The state process $\mathbf{X}_t$ has the structure

$$\mathbf{X}_t = \begin{bmatrix} Y \\ V_1 \\ V_2 \\ Z_1 \\ Z_2 \end{bmatrix}_t ,$$

where the entries are vectors with the same role as we have seen in Subsection 3.3.1. This method is a slightly generalized version of the transformation given by Mohler [87]. This model is no different from the original one (3.41), but the linear term is put into matrix form.

Wiener–Itô Chaotic Spectral Representation

We need a generalization of the representation of the stationary L_2 functional of a Gaussian stationary process. See Major [78] for details of this representation. In our case we have the independent Gaussian white noise series $\mathrm{w}_{1,t}, \mathrm{w}_{2,t}, \ldots, \mathrm{w}_{d,t}$, $t = 0, \pm 1, \pm 2, \ldots$, and we are interested in a representation for a vector valued L_2 functional $\mathbf{X}_t$ of these series.

The process $\mathbf{X}_t$ is called *strictly physically realizable* by w_t if it is measurable with respect to σ-algebra $\mathfrak{B}_t$ generated by the random variables $\mathrm{w}_{1,t-s}, \mathrm{w}_{2,t-s}, \ldots, \mathrm{w}_{d,t-s}$, $s = 0, 1, 2, \ldots$, and the shift transformation of w_t is the shift transformation for $\mathbf{X}_t$ as well, i.e., it is subordinated with respect to w_t.

First, let us look at the Hilbert space of the L_2 functionals measurable with respect to $\mathrm{w}_{1,t}, \mathrm{w}_{2,t}, \ldots, \mathrm{w}_{d,t}$, $t = 0, \pm 1, \pm 2, \ldots$. Define the set

$$\mathfrak{K}_r = \left\{ \mathbf{k} \mid \mathbf{k} \in \mathbb{Z}_+^d = \{0, 1, 2, \ldots\}^d, \sum_{j=1}^{d} k_j = r \right\},$$

for $r \in \mathbb{Z}_+$ and denote the elements of $\mathfrak{K}_r$ by $k_{(1:d)}^{(r)} = (k_1, k_2, \ldots, k_d)$. Moreover, let the Gaussian stochastic spectral measure for $\mathrm{w}_{j,t}$ be $W_j(d\omega)$, i.e.,

$$\mathrm{w}_{j,t} = \int_0^1 e^{i2\pi t\omega} W_j(d\omega),$$

and the corresponding multiple Wiener spectral measure of order p be $W_{j,p}(d\omega_{(1:p)})$, where $\omega_{(1:p)} = (\omega_1, \omega_2, \ldots, \omega_p) \in [0,1]^p$, see Major [78]. The measures $W_1, W_2, \ldots, W_d$ are orthogonal (independent) and so are their multiple versions $W_{1,k_1}, W_{2,k_2}, \ldots, W_{d,k_d}$. Let the product measure of degree r be given by

$$W_{k_{(1:d)}^{(r)}}(d\omega_{(1:r)}) = \prod_{j=1}^{d} W_{j,k_j}(d\omega_{(1:k_j)}),$$

where $W_{j,0} = 1$.

The Hilbert space $\mathbf{H}$ generated by the random variables having finite second moments and measurable with respect to σ-algebra $\mathfrak{B}$ spanned by the noise processes
$w_{1,t}, w_{2,t}, \ldots, w_{d,t}$, $t = 0, \pm 1, \pm 2, \ldots$, is the direct sum of Hilbert spaces $\mathbf{H}_{k^{(r)}_{(1:d)}}$, $k^{(r)}_{(1:d)} \in \mathfrak{K}_r$, $r = 0, 1, 2, \ldots$. That is,

$$\mathbf{H} = \sum_{r=0}^{\infty} \sum_{k^{(r)}_{(1:d)} \in \mathfrak{K}_r} \mathbf{H}_{k^{(r)}_{(1:d)}},$$

where $\mathbf{H}_0$ is the space of constants and $\mathbf{H}_{k^{(r)}_{(1:d)}}$ is the linear Hilbert space generated by the random variables defined by all possible products of Hermite polynomials with degree $k^{(r)}_{(1:d)}$ of the noise series, i.e., by

$$H_{k_1}(w_{1,t_1^1}, w_{1,t_2^1}, \ldots, w_{1,t_{k_1}^1}) H_{k_2}(w_{2,t_1^2}, w_{2,t_2^2}, \ldots, w_{2,t_{k_2}^2}) \\ \cdots \times H_{k_d}(w_{d,t_1^d}, w_{d,t_2^d}, \ldots, w_{d,t_{k_d}^d}); \qquad t_j^i \in \mathbb{Z},$$

and H_{k_j} denotes the Hermite polynomial of degree k_j. Note that the time points t_j^i are not necessarily different for different indices i, j. $\mathbf{H}_{k^{(r)}_{(1:d)}}$ can be characterized also in the following way. Let $g_r(\omega_{(1:r)}) \in \overline{L^n_{\Phi}}$ and recall that $g_r(-\omega_{(1:r)}) = \overline{g}_r(\omega_{(1:r)})$, then the random variable

$$\eta = \int_{\mathfrak{D}^r} g_r(\omega_{(1:r)}) W_{k^{(r)}_{(1:d)}}(d\omega_{(1:r)}), \qquad \mathfrak{D}^r = [0,1]^r, \tag{3.43}$$

belongs to $\mathbf{H}_{k^{(r)}_{(1:d)}}$, and conversely if $\eta \in \mathbf{H}_{k^{(r)}_{(1:d)}}$, there exists a function $g_r(\omega_{(1:r)}) \in L_2[0,1]^r$ such that (3.43) holds. The function g_r is uniquely determined up to the permutations of its first k_1, then k_2, ..., and so on, up to the last k_d variables. The unique symmetrized version of g_r ($k^{(r)}_{(1:d)}$-symmetrization) is given by

$$\widetilde{g_{k^{(r)}_{(1:d)}}}(\omega_{(1:r)}) = \frac{1}{k^{(r)}_{(1:d)}!} \sum_{j=1}^{d} \sum_{\mathfrak{p}_j \in \mathfrak{P}_{k_j}} g_r(\omega_{\mathfrak{p}_1(1:k_1)}, \omega_{\mathfrak{p}_2(1:k_2)}, \ldots, \omega_{\mathfrak{p}_d(1:k_d)}), \tag{3.44}$$

where $\mathfrak{P}_{k_j}$ denotes all the possible permutations of numbers $1, \ldots, k_j$ and $k^{(r)}_{(1:d)}! = \prod_{j=1}^{d} k_j!$. The scalar product of two elements in $\mathbf{H}_{k^{(r)}_{(1:d)}}$ is

$$\mathsf{E} \int_{\mathfrak{D}^r} g_r(\omega_{(1:r)}) W_{k^{(r)}_{(1:d)}}(d\omega_{(1:r)}) \int_{\mathfrak{D}^r} f_r(\omega_{(1:r)}) W_{k^{(r)}_{(1:d)}}(d\omega_{(1:r)}) \\ = k^{(r)}_{(1:d)}! \int_{\mathfrak{D}^r} \widetilde{g_{k^{(r)}_{(1:d)}}}(\omega_{(1:r)}) \widetilde{f_{k^{(r)}_{(1:d)}}}(-\omega_{(1:r)}) d\omega_{(1:r)} \sigma_{(1:d)}^{2k^{(r)}_{(1:d)}},$$

where $\sigma_{(1:d)}^{2k_{(1:d)}^{(r)}} = \prod_{j=1}^{d} \sigma_j^{2k_j}$.

Now if $\mathbf{X}_t$ is a second-order stationary process in $\mathbf{H}$ and strictly physically realizable with respect to w_t, then the Wiener–Itô spectral representation theorem gives

$$\mathbf{X}_t = \sum_{r=0}^{\infty} \sum_{k_{(1:d)}^{(r)} \in \mathfrak{K}_r} \int_{\mathfrak{D}^r} e^{i2\pi t \Sigma\omega_{(1:r)}} \mathbf{f}_{k_{(1:d)}^{(r)}}(\omega_{(1:r)}) W_{k_{(1:d)}^{(r)}}(d\omega_{(1:r)}), \tag{3.45}$$

where $\sum \omega_{(1:r)} = \sum_{j=1}^{r} \omega_j$ and the transfer functions $\mathbf{f}_{k_{(1:d)}^{(r)}}$ are determined up to the $k_{(1:d)}^{(r)}$-symmetrization.

Transfer function system

We determine the transfer function system for the stationary bilinear process with the help of the Wiener–Itô representation theorem and equation (3.41), assuming that the spectral radius of A is less then 1, i.e., $\rho(A) < 1$. This assumption for the linear case is the necessary and sufficient one for the stationarity, but this is not so when the model is bilinear. To do this, the spectral representation for the product term $\mathbf{X}_t \mathrm{w}_{j,t}$ is needed.

Lemma 47 *If the solution $\mathbf{X}_t$ of equation (3.41) is given by (3.45) then*

$$\mathbf{X}_t w_{j,t} = \sum_{r=1}^{\infty} \sum_{k_{(1:d)}^{(r)} \in \mathfrak{K}_r} \int_{\mathfrak{D}^{r+1}} e^{i2\pi t \Sigma\omega_{(1:r+1)}}$$
$$\times \mathbf{f}_{k_{(1:d)}^{(r)}}(\omega_{(1:r+1)\setminus \Sigma k_{(j)}+1}) W_{k_{(1:d)}^{(r)}+e_j}(d\omega_{(1:r+1)}),$$

where $e_j = (0, 0, \ldots, 0, 1_{(j)}, 0, \ldots, 0) \in \mathbf{R}^d$ is the jth unit vector, and $\omega_{(1:r+1)\setminus \Sigma k_{(j)}+1}$ denotes the r-dimensional vector obtained from vector $(\omega_1, \omega_2, \ldots, \omega_{r+1})$ with variable ω_p and with $p = \sum_{i=1}^{j} k_i + 1$ missing.

The proof of this lemma is as easy as in the scalar case. Use the Diagram Formula, and equation (3.41), see Terdik and Subba Rao [139].

Include both sides of the equation

$$\mathbf{X}_t = A\mathbf{X}_{t-1} + \sum_{j=1}^{d} \mathrm{w}_{j,t} \mathbf{b}_j + \sum_{j=1}^{d} D_j \mathbf{X}_{t-1} \mathrm{w}_{j,t-1} + \mathbf{f}_0,$$

in the spectral representation and compare the transfer functions of the same order. If $r = 1$, we obtain

$$e^{i2\pi t\omega} \mathbf{f}_{e_j}(\omega) = A e^{i2\pi(t-1)\omega} \mathbf{f}_{e_j}(\omega) + \mathbf{b}_j e^{i2\pi t\omega}; \qquad k^{(1)} = e_j, j = 1, 2, \ldots, d;$$

consequently,

$$\mathbf{f}_{e_j}(\omega) = (I - Ae^{-i\omega})^{-1}\mathbf{b}_j; \qquad j = 1, 2, \ldots, d. \tag{3.46}$$

In general, for $r > 1$,

$$e^{i2\pi t \Sigma\omega_{(1:r)}}\mathbf{f}_{k^{(r)}_{(1:d)}}(\omega_{(1:r)}) = Ae^{i2\pi(t-1)\Sigma\omega_{(1:r)}}\mathbf{f}_{k^{(r)}_{(1:d)}}(\omega_{(1:r)})$$
$$+ \sum_{j=1}^{d} \delta_{k_j>0} D_j e^{i2\pi(t-1)\Sigma\omega_{(1:r)}}\mathbf{f}_{k^{(r)}_{(1:d)}-e_j}(\omega_{(1:r)\backslash\Sigma k_{(j)}}).$$

This equation leads to the following recursive formula for the transfer functions

$$\mathbf{f}_{k^{(r)}_{(1:d)}}(\omega_{(1:r)}) = (Ie^{i2\pi\Sigma\omega_{(1:r)}} - A)^{-1}\sum_{j=1}^{d}\delta_{k_j>0}D_j\mathbf{f}_{k^{(r)}_{(1:d)}-e_j}(\omega_{(1:r)\backslash\Sigma k_{(j)}}). \tag{3.47}$$

Note again that here $\mathbf{f}_{k^{(r)}_{(1:d)}-e_j}$ denotes an r-variable function and its value is the same for all $\omega_{\Sigma k_{(j)}}$ and $k^{(r)}_{(1:d)} - e_j \in \mathfrak{K}_{r-1}$ if $k_j > 0$.

Theorem 48 *If the spectral radius of the matrix A is less then 1, i.e., $\rho(A) < 1$, than the strictly physically realizable stationary solution $\mathbf{X}_t$ of Equation. (3.41) is given by*

$$\mathbf{X}_t = \sum_{r=0}^{\infty}\sum_{k^{(r)}_{(1:d)}\in\mathfrak{K}_r}\int_{\mathfrak{D}^r} e^{i2\pi t\Sigma\omega_{(1:r)}}\mathbf{f}_{k^{(r)}_{(1:d)}}(\omega_{(1:r)})W_{k^{(r)}_{(1:d)}}(d\omega_{(1:r)}),$$

where the transfer functions are determined by (3.46) and (3.47).

Example 49 *If $D = 0$, then the transfer function of the (linear) state process is given by (3.46)*

$$\mathbf{f}(\omega) = (I - Ae^{-i2\pi\omega})^{-1}B,$$

where matrix B is as defined by (3.42), i.e., the first-order transfer function determines a linear (Gaussian) process.

Example 50 *The transfer functions $\mathbf{f}_{(r,0,\ldots,0)}$ are the same as the transfer functions of the bilinear process with the single input $\mathrm{w}_{1,t}$, i.e., when $D_2 = D_3 = \cdots = D_d = 0$.*

It is easy to construct a bilinear system (3.41) with

- finite by many nonzero transfer functions

- infinitely many nonzero transfer functions, but they are zero if their first q indices are greater than some fixed constants, say

$$\begin{aligned}\mathbf{f}_{k^{(r)}_{(1:d)}}(\omega_{(1:r)}) &= 0, \quad \text{if} \quad k_1 > K_1, k_2 > K_2, \ldots, k_q > K_q, \quad q < d,\\ \mathbf{f}_{k^{(r)}_{(1:d)}}(\omega_{(1:r)}) &\neq 0, \quad \text{if} \quad k_1 \leq K_1, k_2 \leq K_2, \ldots, k_q \leq K_q. \end{aligned} \tag{3.48}$$

The model (3.41) is not sensitive to the order of components of the noise, therefore we can change the order such that $K_1 \leq K_2 \leq \cdots \leq K_q < \infty$. One can call the bilinear model (3.41) having transfer functions with the properties (3.48) a *bilinear model of Hermite order* $(K_1, K_2, \ldots, K_q, \infty, \ldots, \infty)$. This means that the highest degree of $\mathrm{w}_{j,s}$, $s \leq t$, is K_j in the polynomial representation of the process $\mathbf{Y}_t$. The lower triangular multiple bilinear model has degree $K_1 = \infty$, i.e., $(\infty, \infty, \ldots, \infty)$.

Stationarity

In the preceding section we assumed that the solution $\mathbf{X}_t$ of the bilinear model (3.41) is stationary, i.e., the second-order moments are finite. This means that

$$\begin{aligned}\sum_{r=0}^{\infty} \sum_{k^{(r)}_{(1:d)} \in \mathfrak{K}_r} E \int_{\mathfrak{D}^r} & e^{i2\pi t \Sigma\omega_{(1:r)}} \mathbf{f}_{k^{(r)}_{(1:d)}}(\omega_{(1:r)}) W_{k^{(r)}_{(1:d)}}(d\omega_{(1:r)}) \\ & \otimes \int_{\mathfrak{D}^r} e^{i2\pi t \Sigma\omega_{(1:r)}} \mathbf{f}_{k^{(r)}_{(1:d)}}(\omega_{(1:r)}) W_{k^{(r)}_{(1:d)}}(d\omega_{(1:r)}) \\ = \sum_{r=0}^{\infty} \sum_{k^{(r)}_{(1:d)} \in \mathfrak{K}_r} k^{(r)}_{(1:d)}! \int_{\mathfrak{D}^r} & \widetilde{\mathbf{f}_{k^{(r)}_{(1:d)}}}(\omega_{(1:r)}) \otimes \widetilde{\mathbf{f}_{k^{(r)}_{(1:d)}}}(-\omega_{(1:r)}) d\omega_{(1:r)} \sigma_{(1:d)}^{2k^{(r)}_{(1:d)}} \\ & < \infty.\end{aligned}$$

Conversely, if the above series is convergent for the transfer functions given by (3.46) and (3.47), the process

$$\mathbf{X}_t = \sum_{r=0}^{\infty} \sum_{k^{(r)}_{(1:d)} \in \mathfrak{K}_r} \int_{\mathfrak{D}^r} e^{i2\pi t \Sigma\omega_{(1:r)}} \mathbf{f}_{k^{(r)}_{(1:d)}}(\omega_{(1:r)}) W_{k^{(r)}_{(1:d)}}(d\omega_{(1:r)}), \tag{3.49}$$

is second-order stationary and it is easy to see that it is the solution of the bilinear equation (3.41). It also follows from this that the necessary and sufficient condition for the stationarity of the model (3.41) of finite degree ($K_d < \infty$), is the same as that for the linear model, i.e., $\rho(A) < 1$. The most interesting case is the model with infinite order $K_1 = \infty$.

Theorem 51 *The bilinear model of order $K_j = \infty$, $j = 1, 2, \ldots, d$,*

$$\mathbf{X}_t = A\mathbf{X}_{t-1} + \sum_{j=1}^{d} w_{j,t}\mathbf{b}_j + \sum_{j=1}^{d} D_j \mathbf{X}_{t-1} w_{j,t-1} + \mathbf{f}_0,$$
$$\mathbf{Y}_t = C\mathbf{X}_t,$$

has stationary, physically realizable, subordinated to $\{w_{j,t}, t \in \mathbb{Z}, j = 1, 2, \ldots, d\}$ solution $\mathbf{Y}_t$ for every matrix C if and only if

$$\rho\left(A^{\otimes 2} + \sum_{j=1}^{d} \sigma_j^2 D^{\otimes 2}\right) < 1, \tag{3.50}$$

and, in this case, the variance of $\mathbf{Y}_t$ is $E\mathbf{Y}_t^{\otimes 2} = C^{\otimes 2} E\mathbf{X}_t^{\otimes 2}$, where the variance for state $\mathbf{X}_t$ is given by formula (3.52) below.

First, we need the following Lemma. Note that the transfer function $\mathbf{f}_{k_{(1:d)}^{(r)}}$ can be considered as a function of $r+1$ variables $\omega_{(1:r+1)}$ whose value remains the same for all ω_{r+1}.

Lemma 52 *Let the transfer functions $\mathbf{f}_{k_{(1:d)}^{(r)}}(\omega_{(1:r)})$ be given by the formulae (3.46) and (3.47). Put $\overline{\mathfrak{p}_i} = (\mathfrak{p}_{1i}, \mathfrak{p}_{2i}, \ldots, \mathfrak{p}_{di})$, where $\mathfrak{p}_{ji} \in \mathfrak{P}_{k_j}$, $i = 1, 2$, see (3.44). If $r > 2$ and either $\overline{\mathfrak{p}_1} \neq \overline{\mathfrak{p}_2}$ or $n \neq m$, then*

$$\int_{\mathfrak{D}^r} (Ie^{i2\pi\Sigma\omega_{(1:r)}} - A)^{-1} \mathbf{f}_{k_{(1:d)}^{(r)} - e_n}(\overline{\mathfrak{p}_1}\omega_{(1:r)\setminus\Sigma k_{(1:n)}}) \otimes (Ie^{-i2\pi\Sigma\omega_{(1:r)}} - A)^{-1}$$
$$\times \mathbf{f}_{k_{(1:d)}^{(r)} - e_m}(-\overline{\mathfrak{p}_2}\omega_{(1:r)\setminus\Sigma k_{(1:m)}}) d\omega_{(1:r)} \sigma^{2k_{(1:d)}^{(r)}}$$
$$= 0.$$

Proof. The first consideration is that if either $\overline{\mathfrak{p}_1} \neq \overline{\mathfrak{p}_2}$ or $n \neq m$, the missing variables from the transfer functions $\mathbf{f}_{k_{(1:d)}^{(r)} - e_n}$ and $\mathbf{f}_{k_{(1:d)}^{(r)} - e_m}$ are different. On the other hand, it follows easily from (3.47) that if $r > 1$, then

$$\int_{\mathfrak{D}} \mathbf{f}_{k_{(1:d)}^{(r)}}(\omega_{(1:r)}) d\omega_j = 0; \qquad j = 1, 2, \ldots, r.$$

So the Fourier series expansion of $\mathbf{f}_{k_{(1:d)}^{(r)}}$, $r > 1$, contains only terms having nonzero coefficients of every $\omega_j, j = 1, 2, \ldots, r$, in the exponent. Now put, for simplicity, ω_p and ω_s as the missing variables and take the integral

$$\int_{\mathfrak{D}^2} (Ie^{i2\pi\Sigma\omega_{(1:r)}} - A)^{-1} \mathbf{f}_{k_{(1:d)}^{(r)}-e_p}(\omega_{(1:r)\backslash p}) \otimes (Ie^{-i2\pi\Sigma\omega_{(1:r)}} - A)^{-1}$$
$$\mathbf{f}_{k_{(1:d)}^{(r)}-e_s}(-\omega_{(1:r)\backslash s}) d\omega_p d\omega_s$$
$$= \sum_{m,n=0}^{\infty} \sum_{j,k=1}^{\infty} A^m \otimes A^n g_{1j}(\omega_{(1:r)\backslash\{p,s\}}) \otimes g_{2k}(-\omega_{(1:r)\backslash\{p,s\}})$$
$$\times \int_{\mathfrak{D}^2} e^{2\pi i(m-n)\Sigma\omega_{(1:r)} + i2\pi(j\omega_s - k\omega_p)} d\omega_p d\omega_s,$$

where g_{1j} and g_{2k} denote the Fourier coefficients of the corresponding transfer functions when the variables $\omega_{(1:r)\backslash\{p,s\}}$; i.e., $\omega_1, \omega_2, \ldots, \omega_r$, except ω_p and ω_s, are fixed. To get a nonzero term, equations $m - n + j = 0$ and $m - n + k = 0$ need to be satisfied simultaneously, which cannot happen. ■

Lemma 53 *Let the transfer functions* $\mathbf{f}_{k_{(1:d)}^{(r)}}(\omega_{(1:r)})$ *be given by the formulae (3.46) and (3.47). If* $r > 2$, *then*

$$k_{(1:d)}^{(r)}! \int_{\mathfrak{D}^r} \widetilde{\mathbf{f}_{k_{(1:d)}^{(r)}}}(\omega_{(1:r)}) \otimes \widetilde{\mathbf{f}_{k_{(1:d)}^{(r)}}}(-\omega_{(1:r)}) d\omega_{(1:r)} \sigma^{2k_{(1:d)}^{(r)}}$$
$$= \sum_{j=1}^{d} \delta_{k_j>0} G_j (k_{(1:d)}^{(r)} - e_j)! \int_{\mathfrak{D}^{r-1}} \widetilde{\mathbf{f}_{k_{(1:d)}^{(r)}-e_j}}(\omega_{(1:r-1)})$$
$$\otimes \widetilde{\mathbf{f}_{k_{(1:d)}^{(r)}-e_j}}(-\omega_{(1:r-1)}) d\omega_{(1:r-1)} \sigma_{(1:d)}^{2(k_{(1:d)}^{(r)}-e_j)},$$

where

$$G_j = (I - A^{\otimes 2})^{-1} \sigma_j^2 D_j^{\otimes 2}.$$

Proof. The recursion formula (3.47) and Lemma 52 yield

$$k_{(1:d)}^{(r)}! \int_{\mathfrak{D}^r} \widetilde{\mathbf{f}_{k_{(1:d)}^{(r)}}}(\omega_{(1:r)}) \otimes \widetilde{\mathbf{f}_{k_{(1:d)}^{(r)}}}(-\omega_{(1:r)}) d\omega_{(1:r)} \sigma^{2k^{(r)}}$$
$$= \int_{\mathfrak{D}^r} (Ie^{i2\pi\Sigma\omega_{(1:r)}} - A)^{-1} \otimes (Ie^{-i2\pi\Sigma\omega_{(1:r)}} - A)^{-1} \sum_{j=1}^{d} \delta_{k_j>0} D_j^{\otimes 2} \sigma_j^2$$
$$\times k_{(1:d)}^{(r)}! k_j \frac{(k_{(1:d)}^{(r)} - e_j)!^2}{k_{(1:d)}^{(r)}!^2} \widetilde{\mathbf{f}_{k_{(1:d)}^{(r)}-e_j}}(\omega_{(1:r-1)})$$
$$\otimes \widetilde{\mathbf{f}_{k_{(1:d)}^{(r)}-e_j}}(-\omega_{(1:r-1)}) d\omega_{(1:r)} \sigma^{2(k^{(r)}-e_j)}$$

$$= (I - A^{\otimes 2})^{-1} \int_{\mathfrak{D}^{r-1}} \sum_{j=1}^{d} \delta_{k_j>0} D_j^{\otimes 2} \sigma_j^2 (k^{(r)} - e_j)! \widetilde{\mathbf{f}_{k_{(1:d)}^{(r)}-e_j}}(\omega_{(1:r-1)})$$
$$\otimes \widetilde{\mathbf{f}_{k_{(1:d)}^{(r)}-e_j}}(-\omega_{(1:r-1)}) d\omega_{(1:r-1)} \sigma^{2(k^{(r)}-e_j)}.$$

In the meantime we change the order of the symmetrization from $k_{(1:d)}^{(r)}$ to $k_{(1:d)}^{(r)} - e_j$, so we obtain the term $(k_{(1:d)}^{(r)} - e_j)!^2 / k_{(1:d)}^{(r)}!^2$, and for a fixed j exactly k_j integrals are equal. Therefore, the coefficient becomes

$$k_{(1:d)}^{(r)}! k_j \frac{(k_{(1:d)}^{(r)} - e_j)!^2}{k_{(1:d)}^{(r)}!^2} = (k_{(1:d)}^{(r)} - e_j)!.$$

■

Proof. For the Theorem 51. To show that the convergence of the series (3.49) is equivalent to the assumption (3.50), let us start with the cases $r = 1, 2$. For $r = 1$,

$$\int_{\mathfrak{D}} \mathbf{f}_{e_j}(\omega) \otimes \mathbf{f}_{e_j}(\omega) \frac{d\omega}{2\pi} \sigma_j^2 = (I - A^{\otimes 2})^{-1} \mathbf{b}_j^{\otimes 2} \sigma_j^2 = F_j.$$

If $r = 2$, and $k_{ij}^{(2)} = (0, 0, \ldots, 1_{(i)}, \ldots, 1_{(j)}, \ldots, 0) = e_i + e_j$, $i \neq j$, then

$$\int_{\mathfrak{D}^2} \widetilde{f_{e_i+e_j}}(\omega_1, \omega_2) \widetilde{f_{e_i+e_j}}(-\omega_1, -\omega_2) d\omega_1 d\omega_2 \sigma_i^2 \sigma_j^2$$
$$= (I - A^{\otimes 2})^{-1} \sigma_i^2 \sigma_j^2 (D_i \otimes D_j \mathbf{b}_j \otimes \mathbf{b}_i + D_j \otimes D_i \mathbf{b}_i \otimes \mathbf{b}_j)$$
$$+ G_i F_j + G_j F_i,$$

and for $k_{jj}^{(2)} = 2e_j$, we have

$$2! \int_{\mathfrak{D}^2} \widetilde{f_{2e_j}}(\omega_1, \omega_2) \otimes \widetilde{f_{2e_j}}(-\omega_1, -\omega_2) d\omega_1 d\omega_2 \sigma_j^4 = \sigma_j^2 G_j \mathbf{b}_j^{\otimes 2} + G_j F_j.$$

The case $r = 2$ can be summarized by

$$k_{ij}^{(2)}! \int_{\mathfrak{D}^2} \widetilde{f_{k_{ij}^{(2)}}}(\omega_1, \omega_2) \otimes \widetilde{f_{k_{ij}^{(2)}}}(-\omega_1, -\omega_2) d\omega_1 d\omega_2 \sigma_i^2 \sigma_j^2 = G_i F_j + M_{ij}$$
$$+ \delta_{i\neq j} G_j F_i,$$

where

$$M_{ij} = (I - A^{\otimes 2})^{-1} \sigma_i^2 \sigma_j^2 (D_i \otimes D_j \mathbf{b}_j \otimes \mathbf{b}_i + \delta_{i\neq j} D_j \otimes D_i \mathbf{b}_i \otimes \mathbf{b}_j).$$

Put

$$\sum_{j=1}^{d} G_j = G; \qquad \sum_{j=1}^{d} F_j = F; \qquad \sum_{i,j=1}^{d} M_{i,j} = M,$$

and obtain

$$\sum_{k_{(1:2)}^{(2)} \in \mathfrak{K}_2} k_{(1:2)}^{(2)}! \int_{\mathfrak{D}^2} \widetilde{\mathbf{f}_{k_{(1:2)}^{(2)}}}(\omega_{(1:2)}) \otimes \widetilde{\mathbf{f}_{k_{(1:2)}^{(2)}}}(-\omega_{(1:2)}) d\omega_{(1:2)} \sigma_{(1:2)}^{2k_{(1:2)}^{(2)}} = GF + M.$$

Now, if $r = 3$, use Lemma 53 and obtain

$$\begin{aligned} k_{(1:3)}^{(3)}! &\int_{\mathfrak{D}^3} \widetilde{\mathbf{f}_{k_{(1:3)}^{(3)}}}(\omega_{(1:3)}) \otimes \widetilde{\mathbf{f}_{k_{(1:3)}^{(3)}}}(-\omega_{(1:3)}) d\omega_{(1:3)} \sigma_{(1:3)}^{2k_{(1:3)}^{(3)}} \\ &= \sum_{j=1}^{d} \delta_{k_j > 0} G_j (k_{(1:3)}^{(3)} - e_j)! \int_{\mathfrak{D}^2} \widetilde{\mathbf{f}_{k_{(1:3)}^{(3)} - e_j}}(\omega_2) \\ &\quad \otimes \widetilde{\mathbf{f}_{k_{(1:3)}^{(3)} - e_j}}(-\omega_2) d\omega_{(1:2)} \sigma_{(1:3)}^{2\left(k_{(1:3)}^{(3)} - e_j\right)}. \end{aligned}$$

Therefore,

$$\begin{aligned} &\sum_{k_{(1:3)}^{(3)} \in \mathfrak{K}_3} k_{(1:3)}^{(3)}! \int_{\mathfrak{D}^3} \widetilde{\mathbf{f}_{k_{(1:3)}^{(3)}}}(\omega_{(1:3)}) \otimes \widetilde{\mathbf{f}_{k_{(1:3)}^{(3)}}}(-\omega_{(1:3)}) d\omega_{(1:3)} \sigma_{(1:3)}^{2k_{(1:3)}^{(3)}} \\ &= \sum_{j=1}^{d} G_j \sum_{k_{(1:2)}^{(2)} \in \mathfrak{K}_2} k^{(2)}! \int_{\mathfrak{D}^2} \widetilde{\mathbf{f}_{k_{(1:2)}^{(2)}}}(\omega_{(1:2)}) \otimes \widetilde{\mathbf{f}_{k_{(1:2)}^{(2)}}}(-\omega_{(1:2)}) d\omega_{(1:2)} \sigma_{(1:2)}^{2k_{(1:2)}^{(2)}} \\ &= G^2 F + GM. \end{aligned}$$

We use induction to prove for $r > 3$ that

$$\begin{aligned} \sum_{k_{(1:d)}^{(r)} \in \mathfrak{K}_r} k_{(1:d)}^{(r)}! &\int_{\mathfrak{D}^r} \widetilde{\mathbf{f}_{k_{(1:d)}^{(r)}}}(\omega_{(1:r)}) \otimes \widetilde{\mathbf{f}_{k_{(1:d)}^{(r)}}}(-\omega_{(1:r)}) d\omega_{(1:r)} \sigma_{(1:d)}^{2k_{(1:d)}^{(r)}} \\ &= G^{r-1} F + G^{r-2} M. \end{aligned} \tag{3.51}$$

The left-hand side of this, Lemma 53, equals

$$\begin{aligned} \sum_{k_{(1:d)}^{(r)} \in \mathfrak{K}_r} \sum_{j=1}^{d} \delta_{k_j > 0} G_j (k_{(1:d)}^{(r)} - e_j)! &\int_{\mathfrak{D}^{r-1}} \widetilde{\mathbf{f}_{k_{(1:d)}^{(r)} - e_j}}(\omega_{(1:r-1)}) \\ &\otimes \widetilde{\mathbf{f}_{k_{(1:d)}^{(r)} - e_j}}(-\omega_{(1:r-1)}) d\omega_{(1:r-1)} \sigma^{2(k_{(1:d)}^{(r)} - e_j)} \end{aligned}$$

$$= \sum_{j=1}^{d} G_j \sum_{k_{(1:d)}^{(r-1)} \in \mathfrak{K}_{r-1}} k_{(1:d)}^{(r-1)}! \int_{\mathfrak{D}^{r-1}} \widetilde{\mathbf{f}_{k_{(1:d)}^{(r-1)}}}(\omega_{(1:r-1)})$$

$$\otimes \widetilde{\mathbf{f}_{k_{(1:d)}^{(r-1)}}}(-\omega_{(r-1)}) d\omega_{(1:r-1)} \sigma_{(1:d)}^{2k_{(1:d)}^{(r-1)}}$$

$$= G\left(G^{r-2}F + G^{r-3}M\right) = G^{r-1}F + G^{r-2}M.$$

Equation (3.51) shows that the series (3.49) is convergent if and only if $\rho(G) < 1$. This result and the following lemma, proved in the paper by Terdik and Ispány [133], make the proof complete.

Lemma 54 *Let $A, D \in \mathbf{R}^{m\times m}$ and $\rho(A) < 1$. Then the following statements are equivalent:*

$$\begin{array}{lll} (i) & \rho(A\otimes A + D\otimes D) & < 1, \\ (ii) & \rho((I - A\otimes A)^{-1} D\otimes D) & < 1. \end{array}$$

Summing these results in the variance of the state process $\mathbf{X}_t$; i.e.,

$$\begin{aligned} \mathsf{E}\left(\mathbf{X}_t \otimes \mathbf{X}_t\right) &= F + \sum_{r=2}^{\infty}(G^{r-1}F + G^{r-2}M) = (I-G)^{-1}(F+M) \\ &= \left(I - A^{\otimes 2} - \sum_{j=1}^{d} \sigma_j^2 D_j^{\otimes 2}\right)^{-1} \\ &\times \left(\sum_{j=1}^{d} \sigma_j^2 \mathbf{b}_j^{\otimes 2} + \sum_{i,j=1}^{d} \sigma_i^2\sigma_j^2 D_i \otimes D_j \mathbf{b}_j \otimes \mathbf{b}_i (1+\delta_{i\neq j})\right). \end{aligned} \tag{3.52}$$

■

Corollary 55 *The necessary and sufficient condition for the stationarity of the bilinear model (3.41) of order $(K_1, K_2, \ldots, K_q, \infty, \ldots, \infty)$, where $K_1 \le K_2 \le \cdots \le K_q < \infty$, is*

$$\rho\left(A^{\otimes 2} + \sum_{j=q+1}^{d} \sigma_j^2 D^{\otimes 2}\right) < 1.$$

3.4.2 *Spectra*

Covariance Structure and Spectrum

Now we are in a position to calculate the covariance function for the stationary multiple bilinear model. Observing that

$$k_{(1:d)}^{(r)}! \int_{\mathfrak{D}^r} e^{i2\pi s\Sigma\omega_{(1:r)}} \widetilde{\mathbf{f}_{k_{(1:d)}^{(r)}}}(\omega_{(1:r)}) \otimes \widetilde{\mathbf{f}_{k_{(1:d)}^{(r)}}}(-\omega_{(1:r)}) d\omega_{(1:r)} \sigma_{(1:d)}^{2k_{(1:d)}^{(r)}}$$

$$= \int_{\mathfrak{D}^r} e^{i2\pi s\Sigma\omega_{(1:r)}} (Ie^{i2\pi\Sigma\omega_{(1:r)}} - A)^{-1} \otimes (Ie^{-i2\pi\Sigma\omega_{(1:r)}} - A)^{-1}$$

$$\times \sum_{j=1}^{d} \delta_{k_j>0} D_j^{\otimes 2} \sigma_j^2 \left(k_{(1:d)}^{(r)} - e_j\right)! \widetilde{\mathbf{f}_{k_{(1:d)}^{(r)}-e_j}}(\omega_{(1:r-1)})$$

$$\otimes \widetilde{\mathbf{f}_{k_{(1:d)}^{(r)}-e_j}}(-\omega_{(1:r-1)}) d\omega_{(1:r)} \sigma^{2(k_{(1:d)}^{(r)}-e_j)},$$

we see that the covariances are

$$\begin{aligned} E(\mathbf{X}_{t+s} \otimes \mathbf{X}_t) &= (A \otimes I)^s E\mathbf{X}_0^{\otimes 2}, & s &= 0, 1, 2, \ldots, \\ &= (I \otimes A)^{-s} E\mathbf{X}_0^{\otimes 2}, & s &= 0, -1, -2, \ldots. \end{aligned}$$

Also from the above equation it follows that the spectral density of the observation X_t is

$$S_{2,X}(\omega) = (e^{i2\pi\omega} I - A \otimes I)^{-1} (e^{-i2\pi\omega} I - I \otimes A)^{-1} (I - A^{\otimes 2}) E\mathbf{X}_0^{\otimes 2}.$$

Note that the covariances and the spectrum of the model does not depend on the Hermite order, see (3.48), of the model. An easy consequence of these formulae is that there is no way to discriminate between the linear ($D_j = 0,\ j = 1, 2, \ldots, d$) non-Gaussian and the bilinear processes on the basis of the second-order moments. In the case of Gaussian noise (input), the main difference is that the linear model is also Gaussian while the bilinear one is not. Moreover, the higher order spectra and cumulants vanish for the linear model, and they do not, for the bilinear one, see also Terdik [128].

Third order properties

We modify the state space equation (3.41) by $\sum_{j=1}^{d} \mathbf{b}_j \mathrm{w}_{j,t}$ as usual so that the state process will not depend on w_t but w_{t-1} only, i.e.

$$\mathbf{X}_t = A\mathbf{X}_{t-1} + \sum_{j=1}^{d} \mathbf{b}_{1,j} \mathrm{w}_{j,t-1} + \sum_{j=1}^{d} D_j \mathbf{X}_{t-1} \mathrm{w}_{j,t-1}$$
$$+ \sum_{j,k=1}^{d} \left(\mathbf{b}_{2,j,k} + \delta_{j\neq k} \mathbf{b}_{2,k,j}\right) H_2\left(\mathrm{w}_{j,t-1}, \mathrm{w}_{k,t-1}\right),$$

$$Y_t = C\left(\mathbf{X}_t + \sum_{j=1}^{d} \mathbf{b}_j \mathrm{w}_{j,t}\right), \tag{3.53}$$

where H_2 is Hermite polynomial of second order, and the new coefficients are

$$\mathbf{b}_{1,j} = A\mathbf{b}_j, \ j = 1, 2, ..., d,$$
$$\mathbf{b}_{2,j,k} = D_j \mathbf{b}_k, \ j, k = 1, 2, ..., d.$$

To study the third order properties the first step is to define the quadratic process $\mathbf{Q}_t = \mathbf{X}_t^{\otimes 2}$ by the following equation. It will be seen also that the necessary and sufficient condition for fourth order stationarity in this case is that all the eigenvalues of the matrix

$$\left[A^{\otimes 2} + \sum_{j=1}^{d} \sigma_j^2 D_j^{\otimes 2}\right]^{\otimes 2} + \sum_{j=1}^{d} \sigma_j^2 \left[A \otimes D_j + D_j \otimes A\right]^{\otimes 2}$$
$$+ \sum_{j,k=1}^{d} \left(1 + \delta_{j=k}\right) \sigma_j^2 \sigma_k^2 \left[D_j \otimes D_k\right]^{\otimes 2},$$

are less than 1 in modulus. Take the tensor product of both sides of the state equation, see the method for getting equation (3.37) for the vector valued case, thus

$$
\begin{aligned}
\mathbf{Q}_t &= \left[A^{\otimes 2} + \sum_{j=1}^{d} \sigma_j^2 D_j^{\otimes 2}\right] \mathbf{Q}_{t-1} \\
&+ \left[\sum_{j=1}^{d} \left(A \otimes D_j + D_j \otimes A\right) \mathrm{w}_{j,t-1}\right] \mathbf{Q}_{t-1} \\
&+ \sum_{j,k=1}^{d} D_j \otimes D_k \mathbf{Q}_{t-1} H_2\left(\mathrm{w}_{j,t-1}, \mathrm{w}_{k,t-1}\right) + F_0\left(\mathbf{X}_{t-1}\right) \\
&+ \sum_{j=1}^{d} F_{1,j}\left(\mathbf{X}_{t-1}\right) \mathrm{w}_{j,t-1} + \sum_{j,k=1}^{d} F_{2,j,k}\left(\mathbf{X}_{t-1}\right) H_2\left(\mathrm{w}_{j,t-1}, \mathrm{w}_{k,t-1}\right) \\
&+ F_3\left(\mathbf{X}_{t-1}, \mathrm{w}_{t-1}\right) + P_4\left(\mathrm{w}_{t-1}\right) + P_3\left(\mathrm{w}_{t-1}\right) \\
&+ P_2\left(\mathrm{w}_{t-1}\right) + P_1\left(\mathrm{w}_{t-1}\right) + P_0,
\end{aligned} \tag{3.54}
$$

where the significant terms for third order moments are

$$
F_0\left(\mathbf{X}_{t-1}\right) = \sum_{j=1}^{d} \sigma_j^2 D_j \otimes I \mathbf{X}_{t-1} \otimes \mathbf{b}_{1,j} + I \otimes \sum_{j=1}^{d} \sigma_j^2 D_j \mathbf{b}_{1,j} \otimes \mathbf{X}_{t-1},
$$

$$
\begin{aligned}
F_{1,j}\left(\mathbf{X}_{t-1}\right) &= A \otimes I \mathbf{X}_{t-1} \otimes \mathbf{b}_{1,j} + I \otimes A \mathbf{b}_{1,j} \otimes \mathbf{X}_{t-1} \\
&+ \sigma_j^2 D_j \otimes I \mathbf{X}_{t-1} \otimes \sum_{k=1}^{d} \sigma_k^2 \left(\mathbf{b}_{2,j,k} + \mathbf{b}_{2,k,j}\right) \\
&+ \sigma_j^2 I \otimes D_j \sum_{k=1}^{d} \sigma_k^2 \left(\mathbf{b}_{2,j,k} + \mathbf{b}_{2,k,j}\right) \otimes \mathbf{X}_{t-1},
\end{aligned}
$$

$$
\begin{aligned}
F_{2,j,k}\left(\mathbf{X}_{t-1}\right) &= A \otimes I \mathbf{X}_{t-1} \otimes \left(\mathbf{b}_{2,j,k} + \delta_{j \neq k} \mathbf{b}_{2,k,j}\right) \\
&+ I \otimes A \left(\mathbf{b}_{2,,j,k} + \delta_{j \neq k} \mathbf{b}_{2,k,j}\right) \otimes \mathbf{X}_{t-1} \\
&+ D_j \otimes I \mathbf{X}_{t-1} \otimes \mathbf{b}_{1,k} + I \otimes D_j \mathbf{b}_{1,k} \otimes \mathbf{X}_{t-1} \\
&+ \delta_{j \neq k} \left(D_k \otimes I \mathbf{X}_{t-1} \otimes \mathbf{b}_{1,j} + I \otimes D_k \mathbf{b}_{1,j} \otimes \mathbf{X}_{t-1}\right),
\end{aligned}
$$

$$
\begin{aligned}
P_2\left(\mathrm{w}_{t-1}\right) = \sum_{j,k=1}^{d} H_2\left(\mathrm{w}_{j,t-1}, \mathrm{w}_{k,t-1}\right) \Bigg[&\left(\mathbf{b}_{1,j} \otimes \mathbf{b}_{1,k} + \delta_{j \neq k} \mathbf{b}_{1,k} \otimes \mathbf{b}_{1,j}\right) \\
&+ \sum_{l=1}^{d} \sigma_l^2 \left(\mathbf{b}_{2,j,l} + \mathbf{b}_{2,l,j}\right) \otimes \left(\mathbf{b}_{2,l,k} + \mathbf{b}_{2,k,l}\right) \Bigg],
\end{aligned}
$$

$$P_1\left(\mathrm{w}_{t-1}\right) = \sum_{j,k=1}^{d} \left[\sigma_k^2 \mathbf{b}_{1,k} \otimes \left(\mathbf{b}_{2,j,k} + \mathbf{b}_{2,k,j}\right)\right.$$
$$\left. + \sigma_k^2 \left(\mathbf{b}_{2,j,k} + \mathbf{b}_{2,k,j}\right) \otimes \mathbf{b}_{1,k}\right] \mathrm{w}_{j,t-1},$$

$$P_0 = \sum_{j=1}^{d} \sigma_j^2 \mathbf{b}_{1,j}^{\otimes 2} + \sum_{j,k=1}^{d} \left(1 + \delta_{j=k}\right) \sigma_j^2 \sigma_k^2 \left(\mathbf{b}_{2,j,k} + \delta_{j\neq k} \mathbf{b}_{2,k,j}\right)$$
$$\otimes \left(\mathbf{b}_{2,j,k} + \delta_{j\neq k} \mathbf{b}_{2,k,j}\right).$$

Now the cross-covariances $\mathsf{E}\left(\mathbf{Q}_{t+s} \otimes \mathbf{X}_t\right)$ are obtained similarly to the previous section at time $t+s$ for $s \geq 1$

$$\mathsf{E}\left(\mathbf{Q}_{t+s} \otimes \mathbf{X}_t\right) = \left[A^{\otimes 2} + \sum_{j=1}^{d} \sigma_j^2 D_j^{\otimes 2}\right] \otimes I \mathsf{E}\left(\mathbf{Q}_{t+s-1} \otimes \mathbf{X}_t\right)$$
$$+ \left[\sum_{j=1}^{d} \sigma_j^2 D_j A^{s-1} \otimes I\right] \left[\mathsf{E}\left(\mathbf{X}_t \otimes \mathbf{b}_{1,j} \otimes \mathbf{X}_t\right)\right]$$
$$+ \left[I \otimes \sum_{j=1}^{d} \sigma_j^2 D_j A^{s-1} \otimes I\right] \left[\mathbf{b}_{1,j} \otimes \mathsf{E}\left(\mathbf{X}_t \otimes \mathbf{X}_t\right)\right],$$

and for $s \leq -1$

$$\mathsf{E}\left(\mathbf{Q}_{t+s} \otimes \mathbf{X}_t\right) = \left(I \otimes A\right)^{-s} \mathsf{E}\left(\mathbf{Q}_t \otimes \mathbf{X}_t\right).$$

Therefore the cross-spectrum $S_{Q,X}(z)$ is given by

$$S_{Q,X}(z) = \left[I - z^{-1}\left[A^{\otimes 2} + \sum_{j=1}^{d} \sigma_j^2 D_j^{\otimes 2}\right] \otimes I\right]^{-1}$$
$$\times \left[\mathsf{E}\left(\mathbf{Q}_t \otimes \mathbf{X}_t\right) + z^{-1}\left[\sum_{j=1}^{d} \sigma_j^2 D_j \left(I - z^{-1}A\right)^{-1} \otimes I\right]\right.$$
$$\times \left[\mathsf{E}\left(\mathbf{X}_t \otimes \mathbf{b}_{1,j} \otimes \mathbf{X}_t\right)\right]$$
$$+ z^{-1}\left[I \otimes \sum_{j=1}^{d} \sigma_j^2 D_j \left(I - z^{-1}A\right)^{-1} \otimes I\right]$$
$$\left.\times \left[\mathbf{b}_{1,j} \otimes \mathsf{E}\left(\mathbf{X}_t \otimes \mathbf{X}_t\right)\right]\right] + z\left[I - zI \otimes A\right]^{-1} I \otimes A \mathsf{E}\left(\mathbf{Q}_t \otimes \mathbf{X}_t\right).$$

Some constants are also necessary for the calculations. They are

$$\begin{aligned}\mathsf{E}\left(\mathbf{X}_t \otimes \mathbf{b}_r \otimes \mathbf{X}_t\right) = &\left[I - A \otimes I \otimes A - \sum_{j=1}^{d} \sigma_j^2 D_j \otimes I \otimes D_j\right]^{-1} \\ &\times \left[\sum_{k=1}^{d} \sigma_k^2 \mathbf{b}_{1,k} \otimes \mathbf{b}_r \otimes \mathbf{b}_{1,k}\right. \\ &\left. + \sum_{m,k=1}^{d} \sigma_k^2 \sigma_m^2 \left(\mathbf{b}_{2,m,k} + \mathbf{b}_{2,k,m}\right) \otimes \mathbf{b}_r \otimes \left(\mathbf{b}_{2,m,k} + \mathbf{b}_{2,k,m}\right)\right],\end{aligned}$$

where $\mathbf{b}_r$ is constant vector which can be chosen as $\mathbf{b}_{1,j}$ and $\mathbf{b}_{2,j,k}$ as well. The third order moment $\mathsf{E}\mathbf{Q}_t \otimes \mathbf{X}_t$ is calculated as

$$\begin{aligned}\mathsf{E}\left(\mathbf{Q}_t \otimes \mathbf{X}_t\right) = &\left[I - A^{\otimes 3}\right. \\ &\left. - \left[\sum_{j=1}^{d} \sigma_j^2 D_j^{\otimes 2} \otimes A + A \otimes \sum_{j=1}^{d} \sigma_j^2 D_j^{\otimes 2} + \sum_{j=1}^{d} \sigma_j^2 D_j \otimes A \otimes D_j\right]\right]^{-1} \\ &\times \left[\sum_{j=1}^{d} \sigma_j^2 \left[D_j \otimes I \otimes A + A \otimes I \otimes D_j\right] \mathsf{E}\left(\mathbf{X}_t \otimes \mathbf{b}_{1,j} \otimes \mathbf{X}_t\right)\right. \\ &+ \sum_{j=1}^{d} \sigma_j^2 \left[I \otimes D_j \otimes A + I \otimes A \otimes D_j\right] \mathbf{b}_{1,j} \otimes \mathsf{E}\mathbf{X}_t^{\otimes 2} \\ &+ \sum_{j=1}^{d} \sigma_j^2 \left[D_j \otimes A \otimes I + A \otimes D_j \otimes I\right] \mathsf{E}\left(\mathbf{X}_t \otimes \mathbf{X}_t\right) \otimes \mathbf{b}_{1,j} \\ &+ \sum_{j,k=1}^{d} \sigma_j^2 \sigma_k^2 \left[D_j \otimes I \otimes D_j \mathsf{E}\left(\mathbf{X}_t \otimes \left(\mathbf{b}_{2,j,k} + \mathbf{b}_{2,k,j}\right) \otimes \mathbf{X}_t\right)\right. \\ &+ I \otimes D_j^{\otimes 2} \left(\mathbf{b}_{2,j,k} + \mathbf{b}_{2,k,j}\right) \otimes \mathsf{E}\mathbf{X}_t^{\otimes 2} \\ &\left. + D_j \otimes D_k \otimes I \mathsf{E}\mathbf{X}_t^{\otimes 2} \otimes \left(\mathbf{b}_{2,j,k} + \mathbf{b}_{2,k,j}\right)\right] \\ &+ \sum_{j,k=1}^{d} \sigma_j^2 \sigma_k^2 \left[\mathbf{b}_{1,k} \otimes \left(\mathbf{b}_{2,j,k} + \mathbf{b}_{2,k,j}\right) \otimes \mathbf{b}_{1,j}\right. \\ &+ \left(\mathbf{b}_{2,j,k} + \mathbf{b}_{2,k,j}\right) \otimes \mathbf{b}_{1,k} \otimes \mathbf{b}_{1,j} \\ &\left. + \mathbf{b}_{1,j} \otimes \mathbf{b}_{1,k} \otimes \left(\mathbf{b}_{2,j,k} + \mathbf{b}_{2,k,j}\right)\right] \\ &\left. + \sum_{j,k,l=1}^{d} \sigma_j^2 \sigma_k^2 \sigma_l^2 \left(\mathbf{b}_{2,j,l} + \mathbf{b}_{2,l,j}\right) \otimes \left(\mathbf{b}_{2,l,k} + \mathbf{b}_{2,k,l}\right) \otimes \left(\mathbf{b}_{2,j,k} + \mathbf{b}_{2,k,j}\right)\right].\end{aligned}$$

3.5 State space realization

The question of realizability, i.e., of constructing a state space model for a system given in terms of a set of transfer functions, is a central problem in many cases, for example in engineering. There are several results in this direction for systems with deterministic inputs see Kailath [66], Rugh[109]. The realization of linear systems with a stochastic Gaussian input was treated by Akaike[1] and studied intensively afterwards, see Lindquist and Pavon[74]. The bilinear system seems to be the next step leading from the linear towards non-linear ones. The realization problem of bilinear systems appeared in the papers by Gabr and Subba Rao [40], Pham [95], Priestley[99] mainly from the time series side.

In this section we follow the early Akaike work and the results for the deterministic systems Rugh[109], and also Brillinger[17], who investigated the identification of finite degree functional series by spectral analysis. First the bilinear realization problem is considered together with the simplest non-linear system, i.e., the second Hermite degree bilinear process. Then we show that some basic properties of the transfer function system are necessary and sufficient for the transfer functions to be bilinearly realizable. In section four the stationarity of the bilinear state space model is investigated. We construct the abstract bilinear minimal realization for a time invariant stationary degree-N non-linear system driven by Gaussian white noise.

3.5.1 The bilinear realization problem

We consider a second order stationary stochastic series Y_t, $t \in \mathbb{Z}$, where $\mathbb{Z} = \{0, \pm 1, \pm 2, \dots\}$, which is measurable with respect to the σ -algebra $\mathcal{B}_t$ generated by a Gaussian white noise series $\mathrm{w}_s, s \leq t$ with $E\mathrm{w}_s = 0, E\mathrm{w}_s^2 = \sigma^2$, that is, Y_t is *physically realizable.* We assume that if T_s is the shift transformation for w_t, i.e., $T_s \mathrm{w}_t = \mathrm{w}_{t+s}$, $t, s \in \mathbb{Z}$ then it is also the shift transformation for Y_t. In that case Y_t, $t \in \mathbb{Z}$, is referred as *subordinated* to w_t, $t \in \mathbb{Z}$. The series Y_t is called *bilinear realizable* if there exist $m \times m$ matrices A, D and m-dimensional vectors $\mathbf{b}, \mathbf{c}$ such that Y_t is connected to w_t by the state space equations

$$\begin{aligned} \mathbf{X}_t &= A\mathbf{X}_{t-1} + D\mathbf{X}_{t-1}\mathrm{w}_{t-1} + \mathbf{b}\mathrm{w}_t + \mathbf{f}_0 \,, \\ Y_t &= \mathbf{c}'\mathbf{X}_t, \end{aligned} \tag{3.55}$$

where $\mathbf{c}'$ denotes the transpose of $\mathbf{c}$ and $\mathbf{f}_0 = -\sigma^2 D\mathbf{b}$, keeping $E\mathbf{X}_t = \mathbf{0}$. We should mention at once that the realization (3.55) is not unique at all. We discuss that problem later. It is easy to see that a lower triangular bilinear model Y_t, i.e.,

$$\sum_{m=0}^{P} a_m Y_{t-m} = \sum_{m=0}^{Q} b_m \mathrm{w}_{t-m} + \sum_{m=1}^{R} \sum_{n=0}^{S} c_{m,m+n} Y_{t-m-n} \mathrm{w}_{t-m},$$

is bilinear realizable ($a_0 = b_0 = 1$).

Model (3.55) is more general in that sense because it can produce finite degree nonlinear processes as the following simple example shows.

Example 56 *Let*

$$A = \begin{pmatrix} a_1 & 0 \\ 0 & a_2 \end{pmatrix}; \quad \mathbf{b} = \begin{pmatrix} 1 \\ 0 \end{pmatrix};$$
$$D = \begin{pmatrix} 0 & 0 \\ d_{2,1} & 0 \end{pmatrix}; \quad \mathbf{c} = \begin{pmatrix} 0 \\ c_2 \end{pmatrix}; \quad |a_1|,\ |a_2| < 1 ,$$

then the state space equations (3.55) are

$$\begin{aligned} x_t^{(1)} &= a_1 x_{t-1}^{(1)} + \mathrm{w}_t , \\ x_t^{(2)} &= a_2 x_{t-1}^{(2)} + d_{2,1} x_t^{(1)} \mathrm{w}_{t-1} - \sigma^2 d_{2,1} , \\ Y_t &= c_2 x_t^{(2)} . \end{aligned}$$

Direct calculation leads to the Wiener–Itô spectral representation of Y_t which is

$$Y_t = c_2 d_{2,1} \int_{\mathcal{D}^2} \frac{e^{i2\pi(t-1)\Sigma\omega_{(1:2)}}}{(1 - a_2 e^{-i2\pi\Sigma\omega_{(1:2)}})(1 - a_1 e^{-i2\pi\omega_1})} W(d\omega_{(1:2)}). \tag{3.56}$$

The process Y_t given by (3.56) is physically realizable and stationary if and only if

$$|a_1|,\ |a_2| < 1.$$

Wiener–Itô chaotic representation for the bilinear stationary states

Let us suppose that the state vector process $\mathbf{X}_t$ in (3.55) is also physically realizable and subordinated to w_t. Moreover, the eigenvalues of the matrix A (of linearity) are inside of the unit circle. Then the stationary solution for equation

$$\mathbf{X}_t = A\mathbf{X}_{t-1} + D\mathbf{X}_{t-1}\mathrm{w}_{t-1} + \mathbf{b}\mathrm{w}_t + \mathbf{f}_0, \tag{3.57}$$

can be given as follows. Under the above assumptions $\mathbf{X}_t$ has the Wiener–Itô representation

$$\mathbf{X}_t = \sum_{r=1}^{\infty} \int_{\mathcal{D}^r} e^{i2\pi t\Sigma\omega_{(1:r)}} \mathbf{f}_r(\omega_{(1:r)}) W(d\omega_{(1:r)}); \quad \mathbf{f}_0 = -\sigma^2 D\mathbf{b} ,$$

where the vector functions $\mathbf{f}_r(\omega_{(1:r)})$ are uniquely determined up to permutations.

Now from the uniqueness of the r-dimensional transfer function we have

$$\mathbf{f}_1(\omega_{(1)}) = (I - Ae^{-i2\pi\omega_1})^{-1}\mathbf{b} ,$$
$$\mathbf{f}_r(\omega_{(1:r)}) = (I - Ae^{-i2\pi\Sigma\omega_{(1:r)}})^{-1}e^{-i2\pi\Sigma\omega_{(1:r)}}D\mathbf{f}_{r-1}(\omega_{(1:r-1)}); \ r \geq 2.$$

Clearly these transfer functions have the following properties.

Property 57 $\mathbf{f}_r(\omega_{(1:r)})$ *is a function of variables*

$$e^{-i2\pi\omega_1}, e^{-i2\pi(\omega_1+\omega_2)}, \ldots e^{-i2\pi\Sigma\omega_{(1:r)}}.$$

Put $\mathbf{z}_k = e^{-i2\pi\Sigma\omega_{(1:k)}}$ *and call the function*

$$\mathbf{f}_r^R(\mathbf{z}_1, \mathbf{z}_2, \ldots, \mathbf{z}_r) = \mathbf{f}_r^R(\mathbf{z}_{(1:r)}) = \prod_{k=2}^{r}[(I - Ae^{-i2\pi\Sigma\omega_{(1:r-k+2)}})^{-1}D]\mathbf{f}_1(\omega_{(1)}); \tag{3.58}$$

a ***regular*** *version of the transfer function* $\mathbf{f}_r(\omega_{(1:r)})$.

Property 58 *The regular transfer function* $\mathbf{f}_r^R$ *is* ***strictly proper****, i.e., the components of* $\mathbf{f}_r^R$ *are rational functions and the degree of the numerator polynomial in* $\mathbf{z}_k$ *is less than the degree of the denominator polynomial for each* k.

Property 59 *The regular transfer function* $\mathbf{f}_r^R$ *is* ***recognizable****, i.e., the denominators of its components can be expressed as a product of real-coefficient single variable polynomials.*

These terminologies come from the deterministic (non)linear system theory, see for example [109].

The regular transfer functions of the observation series Y_t given by (3.55) are

$$\begin{aligned} g_1^R(\mathbf{z}_1) &= \mathbf{c}'(I - A\mathbf{z}_1)^{-1}\mathbf{b}, \\ g_2^R(\mathbf{z}_{(2)}) &= \mathbf{c}'(I - A\mathbf{z}_2)^{-1}D(I - A\mathbf{z}_1)^{-1}\mathbf{b}, \\ &\vdots \\ g_r^R(\mathbf{z}_{(1:r)}) &= \mathbf{c}'(I - A\mathbf{z}_r)^{-1}D\cdots(I - A\mathbf{z}_2)^{-1}D(I - A\mathbf{z}_1)^{-1}\mathbf{b}, \\ &\vdots \end{aligned} \tag{3.59}$$

so they are strictly proper recognizable functions of the variables $\mathbf{z}_k = e^{-i2\pi\Sigma\omega_{(k)}}$. The question arises as to whether the process Y_t having transfer functions with Properties 57-59 is bilinear realizable. Property 57 needs some explanation because the transfer functions are determined only up to the permutation. Property 57 means that one can choose a regular transfer function among those transfer functions which have the same symmetrized version.

3.5.2 Realization of the Hermite degree-N homogeneous polynomial model

One sees from Example 1 that the series Y_t in the model (3.55) can have only finitely many nonzero transfer functions. In case it has only one with N variables, i.e.,

$$Y_t = \int_{\mathcal{D}^N} e^{i2\pi t\Sigma\omega_{(N)}} g_N(\omega_{(1:N)}) W(d\omega_{(1:N)}),$$

it will be referred to as *homogeneous of Hermite degree-N* or simply *homogeneous of degree-N*. If all the transfer functions with degree greater then N are identically zero but $g_N(\omega_{(N)}) \not\equiv 0$ then we call it a *degree-N polynomial model*, see [17] in the continuous time case. If we are given a degree-N homogeneous model by the transfer function with Properties 57-59 it is easy to construct its bilinear realization, see [86] for the deterministic case. Rather than making the method explicit we show it by an example.

Example 60 *Let the second degree regular transfer function be given by*

$$g_2^R(\mathbf{z}_{(1:2)}) = \frac{a_{1,1}\mathbf{z}_1\mathbf{z}_2 + a_{1,0}\mathbf{z}_1 + a_{0,1}\mathbf{z}_2 + a_{0,0}}{(1-\alpha_1\mathbf{z}_1)(1-\alpha_2\mathbf{z}_1)(1-\beta_1\mathbf{z}_2)(1-\beta_2\mathbf{z}_2)}, \tag{3.60}$$

where $\alpha_1 \neq \alpha_2,\ \ \beta_1 \neq \beta_2,\ \ |\alpha_i|,\ |\beta_i| < 1,\ \ i = 1,2$. *Now one can put*

$$g_2^R(\mathbf{z}_{(1:2)}) = \frac{[1,\mathbf{z}_2]\begin{bmatrix} a_{1,0}\mathbf{z}_1 + a_{0,0} \\ a_{1,1}\mathbf{z}_1 + a_{0,1}\end{bmatrix}}{(1-\alpha_1\mathbf{z}_1)(1-\alpha_2\mathbf{z}_1)(1-\beta_1\mathbf{z}_2)(1-\beta_2\mathbf{z}_2)}$$

$$= \frac{[1,\mathbf{z}_2]\begin{bmatrix} 1 & \mathbf{z}_1 & 0 & 0 \\ 0 & 0 & 1 & \mathbf{z}_1\end{bmatrix}\begin{bmatrix} a_{0,0} \\ a_{1,0} \\ a_{0,1} \\ a_{1,1}\end{bmatrix}}{(1-\alpha_1\mathbf{z}_1)(1-\alpha_2\mathbf{z}_1)(1-\beta_1\mathbf{z}_2)(1-\beta_2\mathbf{z}_2)},$$

and construct a linear realization for

$$g_{2,1}^R(\mathbf{z}_1) = \frac{\begin{bmatrix} 1 & \mathbf{z}_1 & 0 & 0 \\ 0 & 0 & 1 & \mathbf{z}_1\end{bmatrix}}{(1-\alpha_1\mathbf{z}_1)(1-\alpha_2\mathbf{z}_1)},$$

and

$$g_{2,2}^R(\mathbf{z}_2) = \frac{[1,\mathbf{z}_2]}{(1-\beta_1\mathbf{z}_2)(1-\beta_2\mathbf{z}_2)}.$$

Taking Gilbert's method, see [66] for example, we have

$$g_{2,i}^R(\mathbf{z}_i) = C_i(I - A_i\mathbf{z}_i)^{-1}B_i,$$

where

$$C_1 = \begin{bmatrix} 1 & 0 & 1 & 0 \\ 0 & 1 & 0 & 1 \end{bmatrix}, \quad A_1 = diag(\alpha_1, \alpha_1, \alpha_2, \alpha_2) \ ,$$

$$B_1 = \frac{1}{(\alpha_1 - \alpha_2)} \begin{bmatrix} \alpha_1 & 1 & 0 & 0 \\ 0 & 0 & \alpha_1 & 1 \\ -\alpha_2 & -1 & 0 & 0 \\ 0 & 0 & -\alpha_2 & -1 \end{bmatrix} \begin{bmatrix} a_{0,0} \\ a_{1,0} \\ a_{0,1} \\ a_{1,1} \end{bmatrix} ,$$

$$C_2 = [1, 1] , \quad A_2 = diag(\beta_1, \beta_2), \quad B_2 = \frac{1}{\beta_1 - \beta_2} \begin{bmatrix} \beta_1 & 1 \\ -\beta_2 & -1 \end{bmatrix} .$$

Now we can define the matrices

$$A_{2,2} = \begin{bmatrix} A_1 & 0 \\ 0 & A_2 \end{bmatrix}, \quad D_{2,2} = \begin{bmatrix} 0 & 0 \\ B_2 C_1 & 0 \end{bmatrix},$$

$$\mathbf{b}_{2,2} = \begin{bmatrix} B_1 \\ 0 \end{bmatrix}, \quad \mathbf{c}_{2,2}' = [0, C_2] ,$$

where the 0-s are appropriate zero vectors or matrices. A bilinear realization for g_2^R is

$$\mathbf{X}_t = A_{2,2}\mathbf{X}_{t-1} + D_{2,2}\mathbf{X}_{t-1}\mathrm{w}_{t-1} + \mathbf{b}_{2,2}\mathrm{w}_t + \mathbf{f}_0, \quad \mathbf{f}_0 = -\sigma^2 D\mathbf{b} \ ,$$

$$Y_t = \mathbf{c}_{2,2}'\mathbf{X}_t .$$

The construction is simple for a degree-N polynomial model as well. Let g_2^R be defined by (3.60), and let the linear part be

$$g_1^R(\mathbf{z}_1) = c_{1,1}(I - A_{1,1}\mathbf{z}_1)^{-1}\mathbf{b}_{1,1},$$

then a bilinear realization for the second degree polynomial model Y_t defined by the transfer functions g_1^R, g_2^R is given by the matrices

$$A = \begin{bmatrix} A_{1,1} & 0 \\ 0 & A_{2,2} \end{bmatrix}, \quad D = \begin{bmatrix} D_{1,1} & 0 \\ 0 & D_{2,2} \end{bmatrix},$$

$$\mathbf{b} = \begin{bmatrix} b_{1,1} \\ b_{2,2} \end{bmatrix}, \quad \mathbf{c} = \begin{bmatrix} c_{1,1} \\ c_{2,2} \end{bmatrix} .$$

Theorem 61 *A stationary polynomial model of degree-N has the bilinear realization (3.55) if and only if Properties 57-59 are satisfied by its transfer functions.*

If we truncate a bilinear model Y_t given by its transfer function system $\{g_1, g_2, \dots\}$ given in (3.59), i.e., we take the model Y_t^N defined by the finitely many transfer functions $\{g_1, g_2, \dots, g_N\}$ of Y_t, then Y_t^N is bilinear realizable again but not necessarily with a lower dimensional state space. This principle of truncation is different from the truncation in the summations used by Priestley in [99] for a more general nonlinear case.

The construction of a bilinear realization becomes more complicated if one desires the minimal possible dimension of state X_t. It needs a particular factorization procedure, see [63]. However, an abstract bilinear realization is given by [109], see also [38], in terms of the space of regular transfer functions. This method has given us an idea which leads to a natural generalization of linear stochastic realization theory by Akaike for bilinear models.

3.5.3 Minimal realizations

Linear model

Let us consider briefly the well known linear realization only to show our method we are going to use for the bilinear case. The problem is to find a quadruple $(m, A, \mathbf{b}, \mathbf{c})$ for the linearly connected scalar input w_t and output Y_t such that

$$\begin{aligned} \mathbf{X}_t &= A\mathbf{X}_{t-1} + \mathbf{b}\mathrm{w}_t, \\ Y_t &= \mathbf{c}'\mathbf{X}_t, \end{aligned} \tag{3.61}$$

where m is the dimension of the realization, A is an $m \times m$ matrix and $\mathbf{b}, \mathbf{c} \in \mathbf{R}^m$. The main property of $(m, A, \mathbf{b}, \mathbf{c})$ is that if

$$Y_t = \sum_{k=0}^{\infty} g_k \mathrm{w}_{t-k}, \tag{3.62}$$

then

$$\mathbf{c}' A^k \mathbf{b} = g_k \quad . \tag{3.63}$$

Akaike's idea for the construction of $(m, A, \mathbf{b}, \mathbf{c})$ is the following, see Akaike [1]. Denote by H_{t-}, the Hilbert space spanned by the Gaussian white noise $\mathrm{w}_t, \mathrm{w}_{t-1}, \dots$ and take $Y_{t+k/t-}$ as the orthogonal projection of $Y_{t+k}, k = 0, 1, \dots$ onto H_{t-}. The predictor space $\mathcal{X}_t$ at time t is spanned by the variables $Y_{t+k/t-}, k = 0, 1, \dots$ and provide the state space with a minimal dimension.

It is well known that the statements

1. the existence of the realization (3.61),
2. the finite dimensionality of $\mathcal{X}_t$ and

3. the transfer function of Y_t is rational

are equivalent. The standard method for giving a particular minimal realization called canonical is based on the factorization of the Hankel matrix of the system. For a recent treatment realization of linear systems, see [84], linear Gaussian systems are considred with statistical methods in [5].

Now we are giving a particular abstract realization with a minimal dimension for the linear process (3.62), see [109] in the deterministic case, i.e., we are giving linear operators $L_A, L_{\mathbf{b}}, L_{\mathbf{c}'}$ defined as $L_A : \mathcal{X}_t \to \mathcal{X}_t$, $L_{\mathbf{b}} : R \to \mathcal{X}_t$ and $L_{\mathbf{c}'} : \mathcal{X}_t \to R$, having a matrix and vector structure when the state space is finite dimensional, moreover equation (3.63) is satisfied.

Consider the spectral representation of Y_t

$$Y_t = \int_0^1 e^{i2\pi t\lambda} g(e^{-i2\pi\lambda}) W(d\lambda),$$

where

$$g(e^{-i2\pi\lambda}) = \sum_{k=0}^{\infty} g_k e^{-i2\pi\lambda k},$$

and $W(d\lambda)$ is the Gaussian stochastic spectral measure of w_t. It is easy to see that

$$\begin{aligned} Y_{t+1/t-} &= \sum_{k=0}^{\infty} g_{k+1} \mathrm{w}_{t-k} \\ &= \int_0^1 e^{i2\pi\lambda t} e^{i2\pi\lambda} \left[g(e^{-i2\pi\lambda}) - \hat{L}_{\mathbf{c}'} g \right] W(d\lambda), \end{aligned}$$

where

$$\hat{L}_{\mathbf{c}'} g = \frac{1}{2\pi} \int_0^1 g(e^{-i2\pi\lambda}) d\lambda.$$

The operation

$$e^{i2\pi\lambda} [g(e^{-i2\pi\lambda}) - \hat{L}_{\mathbf{c}'} g],$$

is clearly linear and well defined for every $g \in L^2[0,1]$. Now we define the operator L_A as

$$L_A Y_{t+k/t-} = Y_{t+k+1/t-}, \quad k = 0, 1, 2, \dots,$$

and similarly

$$\hat{L}_A g = e^{i2\pi\lambda} [g(e^{-i2\pi\lambda}) - \hat{L}_{\mathbf{c}'} g],$$

As

$$L_A^k Y_t = Y_{t+k/t-} = \int_0^1 e^{i2\pi\lambda t} \hat{L}_A^k g(e^{-i2\pi\lambda}) W(d\lambda),$$

the state space $\mathcal{X}_t$ is isomorphic to the subspace of $L^2[0,1]$ generated by

$$g(e^{-i2\pi\lambda}), \quad \hat{L}_A g(e^{-i2\pi\lambda}), \quad \ldots \hat{L}_A^k g(e^{-i2\pi\lambda}), \ldots,$$

say $\mathcal{G}$. It is seen that the spaces $\mathcal{X}_t$ and $\mathcal{G}$ are finite dimensional at the same time. The operators

$$L_{\mathbf{b}} a = aY_t, \quad \hat{L}_{\mathbf{b}} a = ag(e^{-i2\pi\lambda}), \quad a \in R,$$

as well as

$$L_{\mathbf{c}'} x = Ex\mathrm{w}_t, \quad x \in \mathcal{X}_t,$$

and $\hat{L}_{\mathbf{c}'}$ defined above correspond to each other. Moreover,

$$L_{\mathbf{c}'} L_A^k L_{\mathbf{b}} 1 = \hat{L}_{\mathbf{c}'} \hat{L}_A^k \hat{L}_{\mathbf{b}} 1 = g_k, \quad k = 0, 1, 2, \ldots,$$

which solves the realization problem. If $\mathcal{X}_t$, i.e., $\mathcal{G}$ is finite dimensional, say m, and we choose the basis as $Y_{t+m-1/t-}, Y_{t+m-2/t-}, \ldots Y_{t/t-}$ then

$$Y_{t+m/t-} = \sum_{k=1}^{m} a_k Y_{t+m-k/t-},$$

and the observability form of the minimal realization can be given by the matrix representation

$$L_A = A = \begin{bmatrix} a_1 & 1 & 0 & \cdots & 0 \\ a_2 & 0 & 1 & \cdots & 0 \\ \vdots & \vdots & \vdots & \ddots & \vdots \\ a_{m-1} & 0 & 0 & \cdots & 1 \\ a_m & 0 & 0 & \cdots & 0 \end{bmatrix};$$

$$L_{\mathbf{b}} = \mathbf{b} = \begin{bmatrix} g_0 \\ g_1 \\ \vdots \\ g_{m-1} \end{bmatrix}; \quad L_{\mathbf{c}'} = \mathbf{c}' = [1, 0, \ldots, 0].$$

Hermite degree-N homogeneous bilinear model

Let us suppose that we are given a degree-N homogeneous stationary model Y_t by its transfer function $g_N(\omega_{(1:N)})$ according to the Gaussian white noise input w_t, $E\mathrm{w}_t = 0$, $E\mathrm{w}_t^2 = \sigma^2$, i.e.,

$$Y_t = \int_{\mathcal{D}^N} e^{i2\pi t \Sigma\omega_{(1:N)}} g_N(\omega_{(1:N)}) W(d\omega_{(1:N)}). \tag{3.64}$$

We have seen earlier that the necessary and sufficient condition for the bilinear realizability is that the regular transfer function $g_N^R(\mathbf{z}_{(1:N)})$ should be strictly proper and recognizable, see Properties 58, 59. Now, in case it is bilinear realizable, we shall to construct the minimal (abstract) realization for (3.64). The method used here, similarly to the linear case, is based on the construction of the minimal realization for a deterministic bilinear model. The only difficulty is to construct the state space and to show that the operators defined in the stochastic case correspond to their deterministic versions. Let us start by taking the Fourier expansion of $g_N(\omega_{(1:N)})$, put $\mathbf{z}_k = e^{-i2\pi\Sigma\omega_{(1:k)}}$ and get

$$\begin{aligned} g_N(\mathbf{z}_{(1:N)}) &= \sum_{k_{(1:N)}\geq 0} g_{k_{(1:N)}} \mathbf{z}_{(1:N)}^{-k_{(1:N)}} \\ &= \sum_{k_{(1:N)}\geq 0} g_{k_{(1:N)}} \prod_{j=1}^{N} e^{-i2\pi k_j \Sigma\omega_{(1:j)}} \\ &= \sum_{k_{(1:N)}\geq 0} g_{k_{(1:N)}} e^{-i2\pi \sum_{j=1}^{N}(\Sigma k_{(j:N)})\omega_j}, \end{aligned} \tag{3.65}$$

where $k_N = (k_1, k_2, \ldots, k_N)$ and $\mathbf{z}_{(1:N)}^{-k_{(1:N)}} = \prod_{j=1}^{N} \mathbf{z}_j^{-k_j}$.

As we are going to use the Hermite polynomial description of the process, consider the spectral representation of the Nth order Hermite polynomial $H_N(\mathrm{w}_{t-h_1}, \mathrm{w}_{t-h_2}, \ldots, \mathrm{w}_{t-h_N})$ of the noises $\mathrm{w}_{t-h_1}, \mathrm{w}_{t-h_2}, \ldots, \mathrm{w}_{t-h_N}$; $h_1 \geq h_2 \geq \ldots \geq h_N$, i.e.,

$$H_N(\mathrm{w}_{t-h_1}, \mathrm{w}_{t-h_2}, \ldots, \mathrm{w}_{t-h_N}) = \int_{\mathcal{D}^N} e^{i2\pi\sum_{j=1}^{N}(t-h_j)\omega_j} W(d\omega_{(1:N)}).$$

Now, it is easy follows from the state space equation (3.57) and also from the transfer functions (3.59) that the degree of the dependence of Y_t on w_{t-k} is not higher than 2, i.e., the product term in the series expansion of Y_t can not contain more than two of w_{t-h} with the same delay h. In fact

$$\Sigma k_{(1:N)} \geq \Sigma k_{(1:N-1)} > \Sigma k_{(1:N-2)} > \ldots > \Sigma k_{(1:1)} = k_1 > 0.$$

The consequence of this is that the coefficients of ω_j in (3.65) are different unless $j = N-1$ or N, see (3.58), i.e., $k_1, k_2, \ldots, k_{N-1} \geq 1$ and only k_N

can be 0. This implies

$$Y_t = \sum_{k_{(1:N-1)}\geq 1; k_N \geq 0} g_{k_{(1:N)}} \int_{\mathcal{D}^N} e^{i2\pi \sum_{j=1}^{N}(t-\Sigma k_{(j:N)})\omega_j} W(d\omega_{(1:N)}).$$

We define the state space $\mathcal{X}_t$ as the direct sum of the following orthogonal subspaces:

1. $\mathcal{X}_t^1$ is spanned by $E(Y_{t+k}|\mathcal{B}_t)$, $k = 1, 2, \ldots$ where $\mathcal{B}_t$ is the σ-algebra generated by $\mathrm{w}_t, \mathrm{w}_{t-1}, \ldots$
2. $\mathcal{X}_t^p$, $p = 2, 3, \ldots$ is spanned by
   (a) the orthogonal projection of $\mathrm{w}_{t-1}\mathcal{X}_t^{p-1}$ onto the subspace $\mathcal{H}_t^{N-p+1}$ generated by the variables
   $$H_{N-p+1}(\mathrm{w}_{t-h_1}, \mathrm{w}_{t-h_2}, \ldots, \mathrm{w}_{t-h_{N-p+1}}) = \int_{\mathcal{D}^{N-p+1}} e^{i2\pi \sum_{j=1}^{N-p+1}(t-h_j)\omega_j} W(d\omega_{(1:N-p+1)}),$$
   $(h_{(1:N)} \geq 0)$ and
   (b) the conditional expectation with respect to $\mathcal{B}_t$ of the elements given previously by 2a and their forward shifts.

The state space $\mathcal{X}_t$ has the form

$$\mathcal{X}_t = \sum_{p=1}^{N} \mathcal{X}_t^p.$$

The spaces $\mathcal{X}_t^p$ can also be given by the help of the following linear operators. If $u \in \mathcal{H}_t^p$ then put $L_D u = \alpha \mathrm{w}_{t-1}u/\mathcal{H}_t^{p-1}$ where $\mathrm{w}_{t-1}u/\mathcal{H}_t^{p-1}$ denotes the orthogonal projection of $\mathrm{w}_{t-1}u$ onto the space $\mathcal{H}_t^{p-1}$ and α is a constant to be defined later. It is enough to define the operator L_A on the generating systems of the subspaces, i.e., put

$$\begin{aligned} (i) \quad & L_A \mathsf{E}(Y_{t+k}|\mathcal{B}_t) = \mathsf{E}(Y_{t+k+1}|\mathcal{B}_t) = L_A^{k+1} Y_t, \\ (ii) \quad & L_A^k (L_D[L_A^l Y_t])_t = \mathsf{E}(L_D[L_A^l Y_t]_{t+k}|\mathcal{B}_t), \end{aligned} \tag{3.66}$$

and so on.

By these definitions

$$\begin{aligned} \mathcal{X}_t^1 &= Span\{L_A^k Y_t, \ k = 0, 1, \ldots\}, \\ \mathcal{X}_t^2 &= Span\{L_A^{k_2} L_D L_A^{k_1} Y_t, \ k_1, k_2 = 0, 1, \ldots\}, \\ \vdots \ & \ \vdots \\ \mathcal{X}_t^N &= Span\{L_A^{k_N} L_D \cdots L_A^{k_2} L_D L_A^{k_1} Y_t, \ k_{(1:N)} \geq 0\}. \end{aligned}$$

Let us regard the action of L_A :

$$L_A Y_t = \mathsf{E}(Y_{t+1}|\mathcal{B}_t) = \sum_{r=1}^{\infty} \int_{\mathcal{D}^r} e^{i2\pi t\ \Sigma\omega_{(1:r)}}$$

$$g_r(\mathbf{z}_{(1:r)})^A_{\mathbf{z}_k = e^{-i2\pi\Sigma\omega_{(1:k)}}} W(d\omega_{(1:r)}),$$

To get $g_r(\mathbf{z}_{(1:r)})^A$, take the expectation

$$\mathsf{E}Y_{t+1} \int_{\mathcal{D}^N} e^{i2\pi \sum_{j=1}^{r}(t-\Sigma_{(1:r-j+1)})\omega_j}$$

$$= \delta_{r=N} \sum_{k_{(1:N-1)}\geq 1; k_N \geq 0} g_{k_{(1:N)}} N! \times \int_{\mathcal{D}^N} \operatorname{sym} e^{i2\pi \sum_{j=1}^{N}(t+1-\Sigma k_{(j:N)})\omega_j}$$

$$\times \operatorname{sym} e^{-i2\pi \sum_{j=1}^{N}(t-\Sigma h_{(j:N)})\omega_j} \prod_{1}^{N} \sigma^2 d\omega_j$$

$$= \delta_{r=N} g_{h_N+1, h_{(1:N-1)}} \sigma^{2N}(1+\delta_{h_N=0}).$$

It follows easily from this that

$$L_A Y_t = \int_{\mathcal{D}^N} e^{i2\pi t\ \Sigma\omega_{(1:N)}}\ g_N(\mathbf{z}_{(1:N)})^A_{\mathbf{z}_k = e^{-i2\pi\Sigma\omega_{(1:k)}}} W(d\omega_{(1:N)}),$$

where

$$g_N(\mathbf{z}_{(1:N)})^A = \sum_{k_{(1:N-1)}\geq 1; k_N\geq 0} g_{k_1+1, k_{(1:2,N)}} \mathbf{z}_{(1:N)}^{-k_{(1:N)}}. \tag{3.67}$$

In general operator L_A can be defined as the stochastic version of an operator on $L^2(\mathcal{D}^N)$, say $\hat{L}_A$, defined by

$$\hat{L}_A g_N = g_N^A,$$

where g_N^A is defined by (3.67). Moreover,

$$L_A^k Y_t = \int_{\mathcal{D}^N} e^{it2\pi\Sigma\omega_{(1:N)}} \hat{L}_A^k\ g_N(\mathbf{z}_{(1:N)})^A_{\mathbf{z}_j = e^{-i2\pi\Sigma\omega_{(1:j)}}} W(d\omega_{(1:N)}).$$

Consider now the operator L_D. Let $u_t \in \mathcal{X}_t^1$ be given by

$$u_t = \sum_{k_{(1:N-1)}\geq 1; k_N\geq 0} d_{k_1+s, k_{(2,N)}} \int_{\mathcal{D}^N} e^{i2\pi \sum_{j=1}^{N}(t-\Sigma k_{(j:N)})\omega_j} W(d\omega_{(1:N)}),$$

where $s = 0, 1, \ldots$ is fixed. Hence

$$L_D u_t = \alpha \mathrm{w}_{t-1} u_t / \mathcal{H}_t^{N-1}$$

$$= \sum_{k_{(2,N-1)}\geq 1; k_N\geq 0} d_{1+s, k_{(2,N)}}$$

$$\times \int_{\mathcal{D}^{N-1}} e^{i2\pi \sum_{j=1}^{N-1}(t-\Sigma k_{(2,N-j+1)})\omega_j} W(d\omega_{(1:N-1)}),$$

because

$$\mathrm{w}_{t-1}u_t = \sum_{k_{(1:N-1)}\geq 1; k_N\geq 0} d_{k_1+s,k_{(2,N)}}$$
$$\times \left\{ \int_{D^{N+1}} e^{i2\pi\sum_{j=1}^{N}(t-\Sigma k_{(j:N)})\omega_j + i2\pi(t-1)\omega_{N+1}} W(d\omega_{(1:N+1)}) \right.$$
$$+ \delta_{k_1=1}\sigma^2(1+\delta_{N=2}\delta_{k_2=0})$$
$$\left. \times \int_{D^{N-1}} e^{i2\pi\sum_{j=1}^{N-1}(t-\Sigma k_{(2,N-j+1)})\omega_j} W(d\omega_{(1:N-1)}) \right\}.$$

Clearly $\alpha = [\sigma^2(1+\delta_{N=2}\delta_{k_2=0})]^{-1}$. The operator $\hat{L}_D$ corresponding to L_D is defined by

$$\hat{L}_D d_N(\mathbf{z}_{(1:N)}) = \sum_{k_{(2,N-1)}\geq 1; k_N\geq 0} d_{1+s,k_{(2,N)}} \mathbf{z}_{(1:N-1)}^{-k_{(2,N)}} = \hat{d}_{N-1}(\mathbf{z}_{(1:N-1)}), \tag{3.68}$$

where

$$d_N(\mathbf{z}_{(1:N)}) = \sum_{k_{(1:N-1)}\geq 1; k_N\geq 0} d_{k_1+s,k_{(2,N)}} \mathbf{z}_{(1:N)}^{-k_{(1:N)}}.$$

Finally the operators $L_{\mathbf{b}}$ and $L_{\mathbf{c}'}$ are defined, similarly to the linear case, by

$$L_{\mathbf{b}}a = aY_t, \quad \hat{L}_{\mathbf{b}}a = ag_N(\mathbf{z}_{(1:N)}),$$

$$L_{\mathbf{c}'}x = \delta_{x\in\mathcal{X}_t^N}\frac{1}{\sigma^2}Ex\mathrm{w}_t, \quad \hat{L}_{\mathbf{c}'}d_r(\mathbf{z}_{(1:r)}) = \delta_{r=1}\int_0^1 e^{i2\pi\lambda}d_1(e^{-i2\pi\lambda})d\lambda.$$

Easy algebra leads to the following theorem, see [109] p. 153 in the deterministic case.

Theorem 62 *The homogeneous model Y_t of Hermite degree-N is bilinear realizable if and only if its state space $\mathcal{X}_t$ is finite dimensional and then $(L_A, L_D, L_{\mathbf{b}}, L_{\mathbf{c}'})$ is a minimal dimensional abstract realization having the form*

$$L_A = A = \begin{bmatrix} A_{1,1} & 0 & \cdots & 0 \\ 0 & A_{2,2} & \cdots & 0 \\ \vdots & \vdots & \ddots & \vdots \\ 0 & 0 & \cdots & A_{N,N} \end{bmatrix},$$

$$L_D = D = \begin{bmatrix} 0 & \cdots & 0 & 0 \\ D_{2,1} & \cdots & 0 & 0 \\ \vdots & \ddots & \vdots & \vdots \\ 0 & \cdots & D_{N,N-1} & 0 \end{bmatrix}, L_{\mathbf{b}} = \mathbf{b} = \begin{bmatrix} \mathbf{b}_1 \\ 0 \\ \vdots \\ 0 \end{bmatrix},$$

$$L_c = \mathbf{c}' = \begin{bmatrix} 0 & 0 & \cdots & \mathbf{c}'_N \end{bmatrix},$$

where $A_{j,j}$, $D_{j,j-1}$, $\mathbf{b}_1$, $\mathbf{c}'_N$ *are appropriate matrices and vectors.*

Hermite degree-N polynomial and the general bilinear model

We are now in a position to state the same proposition as Theorem 62 (of course with the operators there having particular matrix forms) for the degree-N polynomial model Y_t as well as for the general infinite degree stationary model given by its transfer function system. The most important part of it is the construction of the state space. The operator L_A defined by (3.66) is well defined also in this situation, i.e., L_A^k means shifting the time forward by k steps and take the conditional expectation with respect to $\mathcal{B}_t$. The operator L_D acts by components, i.e., if

$$X = \sum_{r=1}^{\infty} \int_{\mathcal{D}^r} e^{i2\pi s \Sigma\omega_{(1:r)}} d_r(\mathbf{z}_{(1:r)})_{\mathbf{z}_k = e^{-i2\pi\Sigma\omega_{(1:k)}}} W(d\omega_{(1:r)}),$$

then

$$\begin{aligned} L_D X &= \sum_{r=1}^{\infty} L_D \int_{\mathcal{D}^r} e^{i2\pi s \Sigma\omega_{(1:r)}} d_r(\mathbf{z}_{(1:r)})_{\mathbf{z}_k = e^{-i2\pi\Sigma\omega_{(1:k)}}} W(d\omega_{(1:r)}) \\ &= \sum_{r=1}^{\infty} \mathrm{w}_{t-1} \int_{\mathcal{D}^r} e^{i2\pi s \Sigma\omega_{(1:r)}} d_r(\mathbf{z}_{(1:r)})_{\mathbf{z}_k = e^{-i2\pi\Sigma\omega_{(1:k)}}} W(d\omega_{(1:r)}) \Big| \mathcal{H}_t^{r-1} \\ &= \sum_{r=1}^{\infty} \int_{\mathcal{D}^{r-1}} e^{i2\pi s \Sigma\omega_{(1:r-1)}} \hat{L}_D(d_r(\mathbf{z}_{(1:r)})_{\mathbf{z}_k = e^{-i2\pi\Sigma\omega_{(1:k)}}} W(d\omega_{(1:r-1)}), \end{aligned}$$

see (3.68). The state space $\mathcal{X}_t$ is defined as the space generated by the union of the following subspaces

$$\begin{aligned} \mathcal{X}_t^1 &= Span\{L_A^k Y_t,\ k = 0, 1, \ldots\}, \\ \mathcal{X}_t^2 &= Span\{L_A^{k_2} L_D L_A^{k_1} Y_t,\ k_1, k_2 = 0, 1, \ldots\}, \\ \vdots\ & \quad \vdots \\ \mathcal{X}_t^N &= Span\{L_A^{k_N} L_D \ldots L_D L_A^{k_1} Y_t,\ k_{(1:N)} \geq 0 \ldots\}, \\ \vdots\ & \quad \vdots \end{aligned}$$

These subspaces will not be orthogonal if the model Y_t is not homogeneous. In the polynomial case of degree-N the $\mathcal{X}_t^{N+k} = 0, \quad k \geq 1$. Moreover, this construction contains the linear case in a natural way because then only $\mathcal{X}_t^1$ is different from 0, so $\mathcal{X}_t = \mathcal{X}_t^1$. The state vector in our case is a basis in $\mathcal{X}_t$ which is very probably different from

$$(\mathrm{w}_{t-k+1}, \mathrm{w}_{t-k}, \mathrm{w}_{t-k-1}, \dots \mathrm{w}_t, Y_{t-k+1}, Y_{t-k}, Y_{t-k-1}, \dots Y_t),$$

called by Priestley [99] a state vector for nonlinear models.

The time series approach to the construction of the state space is simple for the linear model because one can regard the space

$$H_t = Span\{\mathrm{w}_t, \mathrm{w}_{t-1}, \dots\},$$

as the space of the observations $Span\{Y_t, Y_{t-1}, \dots\}$, i.e., the state space is the projection of the future to the past by the observations. To follow this for the bilinear case it would be necessary to give the subspaces in terms of the past of Y_t. The invertibility of the model would be the key for the nonlinear time series realization theory. Our contribution is also different from the one given by Pham [95].

3.6 Some bilinear models of interest

3.6.1 Simple bilinear model

One of the simplest bilinear models is given by

$$X_t = dX_{t-k}\mathrm{w}_{t-l} + \mathrm{w}_t, \; k, l > 0, \tag{3.69}$$

where w_t is Gaussian white noise series with $\mathsf{E}\mathrm{w}_t = 0$, and $\mathsf{E}\mathrm{w}_t^2 = \sigma^2$. If $k = l$, it is symmetric in the noise w_t and the observation X_t. If $k > l$, it is easy to see that X_t is an uncorrelated non-Gaussian time series. The necessary and sufficient assumption of second order stationarity of X_t (physically realizable, subordinated to $\{\mathrm{w}_t\,, t \in \mathbb{Z}\}$) is equivalent of the existence of the variance $\mathrm{Var}\,(X_t)$, it is $\sigma^2 d^2 < 1$. The necessary and sufficient assumption of existence of $2n^{th}$ moment of X_t is given by Theorem 37, page 88. It is

$$\mathsf{E}(d\mathrm{w}_0)^{2n} = (2n-1)!!\sigma^{2n}d^{2n} < 1,$$

in particular if $2n = 6$ then this condition is that $\sigma^2 d^2 < 0.258$. This has been mentioned by Granger and Andersen [43] as well.

Example 63 *Consider the equation (3.69) if $k = 2, l = 1$ and suppose that $\sigma^2 d^2 < 1$. Then X_t is uncorrelated, therefore spectrum S_2 is constant*

$$S_2\,(\omega) = \frac{\sigma^2}{1 - \sigma^2 d^2}.$$

To derive the bispectrum, consider the cumulant $\operatorname{Cum}(X_r, X_s, X_t)$. *The solution of (3.69) in our particular case is*

$$X_t = \mathrm{w}_t + \sum_{k=1}^{\infty} d^k \mathrm{w}_{t-2k} \prod_{j=0}^{k-1} \mathrm{w}_{t-2j-1},$$

in L_2 *sense. The index set according to the term*

$$\mathrm{w}_{t-2k} \prod_{j=0}^{k-1} \mathrm{w}_{t-2j-1} = H_{k+1}\left(\mathrm{w}_{t-1}, \mathrm{w}_{t-3}, \dots, \mathrm{w}_{t-2k+1}, \mathrm{w}_{t-2k}\right),$$

contains a decreasing series with difference 2 *but the last one with difference* 1. *Therefore the cumulant*

$$\operatorname{Cum}\left(H_{k+1}\left(\mathrm{w}_{r-1}, \dots, \mathrm{w}_{r-2k+1}, \mathrm{w}_{r-2k}\right),\right.$$
$$\left. H_{m+1}\left(\mathrm{w}_{s-1}, \dots, \mathrm{w}_{s-2m+1}, \mathrm{w}_{s-2m}\right), H_{n+1}\left(\mathrm{w}_{t-1}, \dots, \mathrm{w}_{t-2n+1}, \mathrm{w}_{t-2n}\right)\right),$$

might be nonzero only if either k *or* m *or* n *is zero. Indeed, supposing* $k \geq m \geq n$ *the cumulant can be different from zero if* $k = n + m$, *because the indices of* w *are monotone decreasing, and if* $n = 0$ *since the differences of the indices are different. Therefore*

$$\begin{aligned}\operatorname{Cum}(X_r, X_s, X_t) &= \operatorname{Cum}(\mathrm{w}_r, X_s, X_t) + \operatorname{Cum}(X_r, \mathrm{w}_s, X_t) \\ &\quad + \operatorname{Cum}(X_r, X_s, \mathrm{w}_t) \\ &= \frac{d\sigma^4}{1 - \sigma^2 d^2}\left[\operatorname*{sym}_{r,s,t} \delta_{r+1=s=t+2}\right],\end{aligned}$$

from which the bispectrum

$$S_3(z_1, z_2) = \frac{d\sigma^4}{1 - d^2\sigma^2}\left(z_1^{-1} z_2 + z_1 z_2^{-1} + z_1^{-2} z_2^{-1} + z_1^{-1} z_2^{-2} + z_1^2 z_2 + z_1 z_2^2\right),$$

follows.

It is interesting to remark that if X_t is weakly stationary, it is strictly stationary as well since the distribution of the Wiener–Itô representation of X_t is independent of the time shift. Now the necessary and sufficient condition for strict stationarity of X_t (X_t is not necessarily weakly stationary) is that the Lyapunov exponent $\ln|d| + \mathsf{E}\ln|\mathrm{w}_0| < 0$, i.e.,

$$\sigma^2 d^2 < 2e^{\gamma},$$

where γ is Euler's constant, see [102]. Therefore in the range

$$1 \leq \sigma^2 d^2 < 2e^{\gamma} \simeq 3.5621,$$

there exists a strictly stationary process X_t for which equation (3.69) holds and it is not weakly stationary because the second order moment of X_t does not exist.

This model is even more interesting if we wish to consider a bilinear model with the long range dependent input e_t.

Symmetric bilinear model with a long range dependent input

We briefly mention here some of the properties of a fractionally differenced Gaussian noise process, see Beran [11] for more details. Define the fractional Gaussian noise process as

$$e_t = (1 - B)^{-h} \mathrm{w}_t,$$

where B is the back shift operator, i.e., $B\mathrm{w}_t = \mathrm{w}_{t-1}$ and w_t is a Gaussian white noise process as before. Now, if $|h| < 1/2$, then e_t is Gaussian stationary process having both a moving average and an infinite order AR representation. If $0 < h < 1/2$, then e_t is long range dependent process in the sense that its autocorrelation function decreases hyperbolically instead of exponentially as usual in the ARMA theory; in fact, for large k

$$\mathrm{Cov}(e_t, e_{t+k}) = \sigma_e^2 \,\mathrm{corr}(e_t, e_{t+k}) \simeq k^{2h-1}. \tag{3.70}$$

The property (3.70) implies and is implied by the fact that the spectral density of e_t is hyperbolic for small frequencies, i.e.,

$$f_e(z) \simeq \omega^{-2h}, \quad \omega \to 0.$$

The exact expressions for the correlation and spectral density are

$$\mathrm{corr}(e_t, e_{t+k}) = \frac{(-1)^k \Gamma^2(1-h)}{\Gamma(1-h+k)\Gamma(1-h-k)},$$
$$\sigma_e^2 = \sigma^2 \frac{\Gamma(1-2h)}{\Gamma^2(1-h)},$$
$$f_e(z) = \sigma^2 |1 - z^{-1}|^{-2h}.$$

Consider the bilinear model (3.69) with $k = l = 1$, i.e.,

$$x_t = d x_{t-1} e_{t-1} + e_t. \tag{3.71}$$

It is easy to see that if the solution of (3.71) exists it must be

$$\tilde{x}_t = e_t + \sum_{k=1}^{\infty} d^k e_{t-k}^2 \prod_{j=1}^{k-1} e_{t-j},$$

in L_2 sense. The proof of the following theorem is found in Appendix C, page 223.

Theorem 64 *For every $d \neq 0$*

$$\lim_{n\to\infty} \mathsf{E} \left(e_t + \sum_{k=1}^{n} d^k e_{t-k}^2 \prod_{j=1}^{k-1} e_{t-j} \right) = \infty.$$

The consequence of this theorem is that there is no even first order stationary solution of the bilinear model (3.71) with a long range input. Nevertheless, there exists a strictly stationary solution of (3.71) if and only if

$$\sigma^2 d^2 \frac{\Gamma(1-2h)}{\Gamma^2(1-h)} < 2e^{\gamma}.$$

The following figure shows this region of $(h, \sigma|d|)$ bounded by the curve, Figure 3.4,

$$\sigma^2 d^2 \frac{\Gamma(1-2h)}{\Gamma^2(1-h)} = 2e^{\gamma}.$$

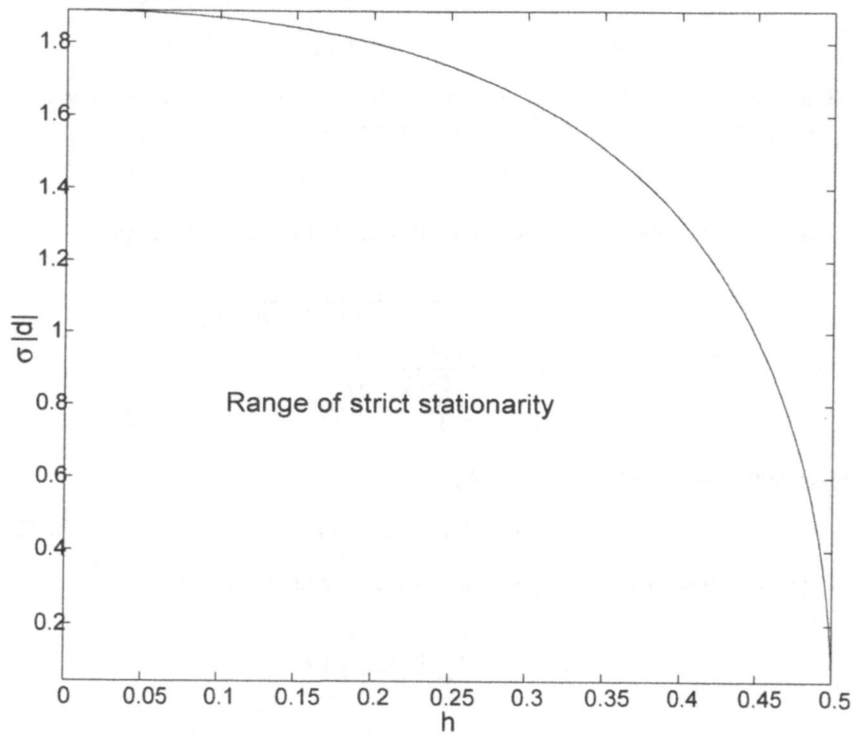

FIGURE 3.4. Range of strict stationarity of long range dependent bilinear models

3.6.2 *Hermite degree-2 bilinear model*

The stochastic stationary series Y_t has Hermite degree-2 according to the Gaussian white noise input w_s, $(Ew_s = 0, Ew_s^2 = \sigma^2)$ if it has the representation

$$Y_t = \sum_{r=1}^{2} \int_{\mathcal{D}^r} g_r(\omega_{(1:r)}) e^{it2\pi \Sigma \omega_{(1:r)}} \mathbf{W}(d\omega_{(1:r)}). \tag{3.72}$$

The necessary and sufficient condition for bilinear realizability of Y_t by Theorem 61 is that the regular version of its transfer functions g_1, g_2 should be strictly proper and recognizable, i.e.,

$$g_1^R(z_1) = \frac{\beta(z_1)}{\alpha_1(z_1)}, \quad g_2^R(z_{(1:2)}) = \frac{\gamma(z_1, z_2)}{\alpha_{21}(z_1)\alpha_{22}(z_2)},$$

where $\alpha_1, \alpha_{21}, \alpha_{22}, \beta$ and γ are polynomials with appropriate degrees. The corresponding transfer functions are

$$g_1(\omega_1) = g_1^R(z_1)_{z_1 = e^{-i2\pi\omega_1}}, \quad g_2(\omega_{(1:2)}) = g_2^R(\mathbf{z}_{(1:2)})_{z_k = e^{-i2\pi\Sigma\omega_k}}.$$

To get the minimal realizations we need to consider operators $\hat{L}_A, \hat{L}_D, \hat{L}_{\mathbf{b}}, \hat{L}_{\mathbf{c}}$ defined in Section 3.5.3. It is easy to see that

$$\hat{L}_A g_1 = \frac{\beta^A(z_1)}{\alpha_1(z_1)}, \quad \hat{L}_A g_2 = \frac{\gamma^A(z_1, z_2)}{\alpha_{21}(z_1)\alpha_{22}(z_2)},$$

$$\hat{L}_D g_1 = const., \quad \hat{L}_D g_2 = \frac{\gamma^D(z_1)}{\alpha_{22}(z_1)}.$$

Now in every particular case one can easily build up a minimal realization for the model based on the matrix representation of these linear operators. Instead of doing this we have found the following direct representation useful. Suppose that

$$\alpha_1(\omega) = \sum_{m=0}^{P} a_m e^{-i2\pi m\omega}; \quad \alpha_{21}(\omega) = \sum_{m=0}^{P_1} a_m^{(1)} e^{-i2\pi m\omega};$$

$$\alpha_{22}(\omega) = \sum_{m=0}^{P_2} a_m^{(2)} e^{-i2\pi m\omega}; \qquad \beta(\omega) = \sum_{m=0}^{Q} b_m e^{-i2\pi m\omega};$$

$$\gamma(\omega, \omega + \lambda) = \sum_{m=1}^{R} \sum_{n=0}^{S} c_{m,m+n} e^{-i2\pi(m+n)\omega + m\lambda}.$$

and put

$$\sum_{m=0}^{P} a_m X_{t-m} = \sum_{m=0}^{Q} b_m \mathrm{w}_{t-m},$$

$$\sum_{m=0}^{P_1} a_m^{(1)} X_{t-m}^{(1)} = \mathrm{w}_t,$$

$$\sum_{m=0}^{P_2} a_m^{(2)} X_{t-m}^{(2)} = \sum_{m=1}^{R} \sum_{n=0}^{S} c_{m,m+n} X_{t-m-n}^{(1)} \mathrm{w}_{t-m}.$$

Let us assume that $a_0 = a_0^{(1)} = a_0^{(2)} = b_0 = 1$ since it does not cause any restriction of generality. It is clear that by this definition of state processes $X_t, X_t^{(1)}, X_t^{(2)}$ the output process is given by the equation

$$Y_t = X_t + X_t^{(2)}.$$

The first step now is to set a minimal canonical realization for the vector $(X_t, X_t^{(1)})$, which is a linear realization problem, then find the minimal canonical realization for the homogeneous bilinear model $X_t^{(2)}$ and combine these variables together to get Y_t.

Second order properties, spectrum

We have seen in the previous section that a bilinear realizable model with Hermite degree-2 can be given by

$$\begin{aligned} Y_t = &\int_{D} e^{i2\pi t\omega_1} \frac{\beta(\omega_1)}{\alpha_1(\omega_1)} W(d\omega_1) \\ &+ \int_{D^2} e^{i2\pi t(\omega_1+\omega_2)} \frac{\gamma(\omega_1, \omega_1+\omega_2)}{\alpha_{21}(\omega_1)\alpha_{22}(\omega_1+\omega_2)} \mathbf{W}(d\omega_1, d\omega_2), \end{aligned} \tag{3.73}$$

where the polynomials α_1, α_{21}, α_{22}, β, are given in Section 6 and note that we put γ as

$$\gamma(\omega_1, \omega_1+\omega_2) = \sum_{m=1}^{R} \sum_{n=0}^{S} c_{m,m+n} e^{-i2\pi(n\omega_1+m(\omega_1+\omega_2))}. \tag{3.74}$$

It follows easily from this representation that the necessary and sufficient condition of stationarity for the physically realizable model Y_t is that all the roots of the polynomials α_1, α_{21}, α_{22}, should be inside the unit circle.

Theorem 65 *Let us suppose that the process Y_t defined by (3.73) is stationary. Then the spectrum S_2 of Y_t is given by the formula*

$$S_2(\omega_1) = \sigma^2 \left| \frac{\beta(\omega_1)}{\alpha_1(\omega_1)} \right|^2 + \sigma^4 \left| \frac{\gamma_0(\omega_1)}{\alpha_{22}(\omega_1)} \right|^2 + \sigma^4 \left| \frac{1}{\alpha_{22}(\omega_1)} \right|^2 \int_{\mathcal{D}} \left| \frac{\gamma(\omega_2, \omega_1)}{\alpha_{21}(\omega_2)} \right|^2 d\omega_2,$$

where

$$\gamma_0(\omega_1) = \sum_{m=1}^{R} c_{m,m} e^{-i2\pi m\omega_1}. \tag{3.75}$$

Proof. The first and second order terms in (3.73) are orthogonal, therefore the covariance function of Y_t is

$$\operatorname{Cov}(Y_t, Y_s) = \sigma^2 \int_{\mathcal{D}} e^{i2\pi(t-s)\omega_1} \left| \frac{\beta(\omega_1)}{\alpha_1(\omega_1)} \right|^2 d\omega_1$$
$$+\sigma^4 2! \int_{\mathcal{D}^2} e^{i2\pi(t-s)(\omega_1+\omega_2)} \left| \operatorname*{sym}_{\omega_{(1:2)}} \frac{\gamma(\omega_1, \omega_1+\omega_2)}{\alpha_{21}(\omega_1)\alpha_{22}(\omega_1+\omega_2)} \right|^2 d\omega_1 d\omega_2.$$

The spectrum is shown by changing variables in the second term; put for simplicity

$$\widetilde{g_2}\left(z_{(1:2)}\right) = \operatorname*{sym}_{\omega_{(1:2)}} \frac{\gamma(\omega_1, \omega_1+\omega_2)}{\alpha_{21}(\omega_1)\alpha_{22}(\omega_1+\omega_2)}.$$

Now

$$\int_{\mathcal{D}^2} (z_1 z_2)^{t-s} \left|\widetilde{g_2}\left(z_{(1:2)}\right)\right|^2 d\omega_1 d\omega_2 = \int_{\mathcal{D}} z^{t-s} \left[\int_{\mathcal{D}} \left|\widetilde{g_2}\left(v, zv^{-1}\right)\right|^2 d\lambda \right] d\omega,$$

and we get the spectrum, see (2.13) page 50, for the general case. In our particular case notice that if we define a function γ_1 by the equation

$$\gamma(\omega_1, \omega_1+\omega_2) = \gamma_s(\omega_1, \omega_1+\omega_2) + \gamma_0(\omega_1+\omega_2),$$

where γ_0 is defined by (3.75), the spectrum can be simplified. Indeed

$$2\int_{\mathcal{D}} \left|\widetilde{g_2}\left(v, zv^{-1}\right)\right|^2 d\lambda = \int_{\mathcal{D}} \overline{\left[\frac{\gamma_s(\lambda,\omega)+\gamma_0(\omega)}{\alpha_{21}(\lambda)\alpha_{22}(\omega)} + \frac{\gamma_s(\omega-\lambda,\omega)+\gamma_0(\omega)}{\alpha_{21}(\omega-\lambda)\alpha_{22}(\omega)} \right]}$$
$$\times \left[\frac{\gamma_s(\lambda,\omega)}{\alpha_{21}(\lambda)\alpha_{22}(\omega)} + \frac{\gamma_0(\omega)}{\alpha_{22}(\omega)} \right] d\lambda$$
$$= \left| \frac{\gamma_0(\omega)}{\alpha_{22}(\omega)} \right|^2 + \int_{\mathcal{D}} \left| \frac{\gamma(\lambda,\omega)}{\alpha_{21}(\lambda)\alpha_{22}(\omega)} \right|^2 d\lambda,$$

since

$$\int_{\mathcal{D}} \frac{1}{\alpha_{21}(-\lambda)\alpha_{21}(\omega-\lambda)} d\lambda = 1.$$

■

The spectrum S_2 is a rational function which can only identify a Gaussian ARMA model. To make a difference between a linear (Gaussian) model and a bilinear one is possible only by considering the higher order moments.

Third order properties, bispectrum

The existence of higher order moments for the stationary bilinear process Y_t given by (3.72) is automatic because the product of two processes having finite Hermite degrees is os a finite degree again. Therefore it is enough to assume the second order stationarity of Y_t. The formula (2.14), page 50, is used for this particular case. After performing some algebra, see [138], the following explicit formula of the bispectrum for bilinear realizable stationary models with Hermite degree-2 can be derived .

Theorem 66 *Let the bilinear model with Hermite degree-2 be given by (3.73). Then the bispectrum of* Y_t *is*

$$S_3\left(\omega_{(1:2)}\right) = S(\omega_1, \omega_2, -\omega_1 - \omega_2),$$

where function S *is*

$$S(\omega_{(1:3)}) = 6\sigma^4 \operatorname*{sym}_{\omega_{(1:3)}} \left[\frac{\beta(\omega_1)\beta(\omega_2)\gamma(-\omega_1,\omega_3)}{\alpha_1(\omega_1)\alpha_1(\omega_2)\alpha_{22}(\omega_3)} \right]$$
$$+ 6\sigma^6 \frac{1}{\alpha_{22}(\omega_1)\alpha_{22}(\omega_2)\alpha_{22}(\omega_3)}$$
$$\times \operatorname*{sym}_{\omega_{(1:3)}} \left[\frac{\gamma_0(\omega_1)\gamma_0(\omega_2)\gamma_0(\omega_3)}{3} + \int_{\mathcal{D}} \frac{\gamma(-\lambda,\omega_1)\gamma(\lambda-\omega_1,\omega_2)\gamma(\lambda,\omega_3)}{|\alpha_{21}(-\lambda)|^2\alpha_{21}(\lambda-\omega_1)} d\lambda \right].$$

The easy consequence of this theorem is that the turning point during the identification of the bilinear model with Hermite degree two is the bispectrum and not the spectrum. The rational spectrum describes the Gaussian stationary model totally and the rational spectrum and the bispectrum do the same job for the bilinear model with Hermite degree-2.

Separable bilinear time series with Hermite degree-2

There is a particular case of model (3.73) which will have some importance later in considering the quadratic prediction of time series.

Definition 67 *The bilinear model (3.73) is called* separable *if the polynomial*

$$\gamma(z,v) = \sum_{m=1}^{R} \sum_{n=0}^{S} c_{m,m+n} z^{-n} v^{-m},$$

is some product of two polynomials of a single variable, i.e.,

$$\gamma(z,v) = \gamma_1(z)\gamma_0(v).$$

Both the spectrum and the bispectrum have particular simple forms for these types of processes.

Spectrum

$$S_2(\omega_1) = \sigma^2 \left| \frac{\beta(\omega_1)}{\alpha_1(\omega_1)} \right|^2 + \sigma^4 \left| \frac{\gamma_0(\omega_1)}{\alpha_{22}(\omega_1)} \right|^2 \left[1 + \int_{\mathcal{D}} \left| \frac{\gamma_1(\omega_2)}{\alpha_{21}(\omega_2)} \right|^2 d\omega_2 \right],$$

Bispectrum

$$\begin{aligned} S_3(\omega_{(1:2)}) = {} & 6\sigma^4 \operatorname*{sym}_{\omega_{(1:3)}} \left[\frac{\beta(\omega_1)\beta(\omega_2)\gamma_1(-\omega_1)\gamma_0(\omega_3)}{\alpha_1(\omega_1)\alpha_1(\omega_2)\alpha_{22}(\omega_3)} \right] \\ & + 6\sigma^6 \frac{\gamma_0(\omega_1)\gamma_0(\omega_2)\gamma_0(\omega_3)}{\alpha_{22}(\omega_1)\alpha_{22}(\omega_2)\alpha_{22}(\omega_3)} \\ & \times \operatorname*{sym}_{\omega_{(1:3)}} \left[\frac{1}{3} + \int_{\mathcal{D}} \frac{|\gamma_1(\lambda)|^2 \gamma_1(\lambda - \omega_1)}{|\alpha_{21}(\lambda)|^2 \alpha_{21}(\lambda - \omega_1)} d\lambda \right], \end{aligned}$$

where $\sum \omega_{(1:3)} = 0$.

3.7 Identification of GARCH(1,1) model

The Autoregressive Conditionally Heteroscedastic (ARCH) models are introduced for the handling of nonlinear dynamics of financial time series (see [41] including its references, also [104]). The following model dues to Bollersev [12] is called by the generalized ARCH, i.e., GARCH model of order $(1,1)$. We shall consider it as a particular stationary bilinear model and show the main spectral properties as spectrum and bispectrum, the properties of autocorrelation has been considered recently in [54]. The method applied here can be generalized for the GARCH model of order (p,q) as well. The GARCH model of order $(1,1)$ is defined by the equations

$$\begin{aligned} h_{0,t} &= c_0 + a_0 h_{0,t-1} + d_0 h_{0,t-1} \mathrm{w}_{t-1}^2, \qquad (3.76) \\ X_{0,t} &= \sqrt{h_{0,t}} \mathrm{w}_t, \end{aligned}$$

where w_t is a Gaussian white noise series with mean zero and variance 1, and the coefficients a_0, d_0, and the constant c_0 are positive. We are interested in stationary solutions only, therefore let us assume the stationarity of $h_{0,t}$. The first consideration is that $X_{0,t}$ is an uncorrelated series with $\mathsf{E}X_{0,t} = 0$ and

$$\operatorname{Var} X_{0,t} = \mathsf{E}h_{0,t} = \frac{c_0}{1 - a_0 - d_0} > 0.$$

Therefore one can normalize $X_{0,t}$ by the standard deviation and get the following equations

$$h_t = 1 - a_0 - d_0 + a_0 h_{t-1} + d_0 h_{t-1} \mathrm{w}_{t-1}^2,$$
$$X_t = \sqrt{h_t}\mathrm{w}_t.$$

where $X_t = X_{0,t}/\sqrt{\operatorname{Var} X_{0,t}}$ and $h_t = h_{0,t}/\operatorname{E} h_{0,t}$. The consequence of this transformation is that the only unknown parameters are a_0 and d_0. We define the following coefficients

$$a = a_0 + d_0,$$
$$d = d_0,$$

and rewrite the equations as

$$V_t = aV_{t-1} + dV_{t-1}H_2(\mathrm{w}_{t-1}) + dH_2(\mathrm{w}_{t-1}), \tag{3.77}$$
$$Y_t = V_t + V_t H_2(\mathrm{w}_t) + H_2(\mathrm{w}_t) + 1, \tag{3.78}$$

where $H_2(\mathrm{w}_t) = \mathrm{w}_t^2 - 1$ is the Hermite polynomial of second order. The equation (3.77) will be called the state equation with state variable

$$V_t = h_t - 1,$$

and equation (3.78) will be called equation of the observations with variable of the observations $Y_t = X_t^2$. The original parameters of the model are given by

$$d_0 = d,$$
$$a_0 = a - d,$$

Note that in order to guarantee the positivity of the parameters of equation (3.76), it is supposed that

$$0 < d < a < 1.$$

Now as we have seen earlier we may look for the solution of equation (3.77) in the chaotic form, i.e.,

$$V_t = \sum_{r=1}^{\infty} \int_{\mathcal{D}^r} \exp\left(i2\pi (t-1) \sum_{j=1}^{r} \omega_j \right) f_r(\omega_{(1:r)}) W(d\omega_{(1:r)}), \qquad t \in \mathbb{Z},$$

where the stochastic spectral measure W is defined by the spectral representation of w_t

$$\mathrm{w}_t = \int_{\mathcal{D}} \exp(i2\pi t\omega) W(d\omega), \qquad t \in \mathbb{Z}.$$

The necessary and sufficient condition for existence of stationary solution of equation (3.77) is that the

$$\sum_{r=0}^{\infty} r! \| \operatorname{sym} f_r \|_{\Phi}^2 < \infty, \tag{3.79}$$

where $\|f_r\|_{\Phi}^2$ was defined earlier, page 34, $\Phi(d\omega) = d\omega$ in this case. Therefore we omit Φ from the notation. The transfer functions f_r are determined by the equation (3.77). First let us consider the product

$$V_t H_2(\mathrm{w}_t) = \sum_{r=1}^{\infty} \int_{\mathcal{D}^{r+2}} \exp\left(i2\pi \left[t \sum_{j=1}^{r+2} \omega_j - \sum_{j=1}^{r} \omega_j \right] \right)$$
$$\times f_r(\omega_{(1:r)}) W(d\omega_{(1:r+2)}),$$

as V_t and $H_2(\mathrm{w}_t)$ are independent. The recursion for transfer functions f_r follows

$$f_2(\omega_{(1:2)}) = d\left(1 - a\exp\left(i2\pi(\omega_1+\omega_2)\right)\right)^{-1},$$
$$f_{2r}(\omega_{(1:2r)}) = d\left(1 - a\exp\left(-i2\pi \sum_{j=1}^{2r} \omega_j\right)\right)^{-1} \exp\left(-i2\pi \sum_{j=1}^{2(r-1)} \omega_j\right)$$
$$\times f_{2(r-1)}(\omega_{(1:2(r-1))}), \qquad r \geq 2, \tag{3.80}$$

and $f_{2r+1}(\omega_{(1:2r+1)}) = 0, \quad r = 0, 1, 2, \ldots$.

Lemma 68 *The transfer functions f_{2r} defined by (3.80) fulfil the equation*

$$(2r)! \| \operatorname{sym} f_{2r} \|^2 = \frac{2! d^2}{1-a^2} (2[r-1])! \| \operatorname{sym} f_{2(r-1)} \|^2, \qquad r \geq 2.$$

This Lemma implies that under Assumption 30 the necessary and sufficient condition for existence of stationary solution of equation (3.77) is that

$$a^2 + 2d^2 < 1,$$

i.e.,

$$(a_0 + d_0)^2 + 2d_0^2 < 1.$$

3.7.1 Spectrum of the state process

The spectra of the processes h_t and V_t are the same. We derive the spectrum by the equation

$$h_t - ah_{t-1} = 1 - a + dh_{t-1}H_2(\mathrm{w}_t),$$

because the variables of the right-hand side are uncorrelated. Taking the spectrum for the both sides we get

$$S_{2,h}(z) = \frac{2d^2\mathsf{E}h_t^2}{|1-az|^2}.$$

The spectrum of V_t is

$$S_{2,V}(z) = \frac{(1-a^2)\operatorname{Var}(V_t)}{|1-az|^2},$$

where

$$\operatorname{Var}(V_t) = \frac{2d^2}{1-a^2-2d^2}.$$

3.7.2 Spectrum of the square of the observations

Remember that $Y_t = X_t^2$ and the equation for it is

$$Y_t = V_t + V_tH_2(\mathrm{w}_t) + H_2(\mathrm{w}_t) + 1.$$

Now the spectrum of Y_t is given by

$$S_{2,Y}(z) = S_{2,V}(z) + S_{2,VH_2}(z) + 2 + 2\operatorname{Re} S_{V,VH_2}(z) \\ + 2\operatorname{Re} S_{V,H_2}(z) + 2\operatorname{Re} S_{VH_2,H_2}(z).$$

The Hermite polynomial of second order $H_2(\mathrm{w}_t)$ will be denoted by $H_{2,t}$ in this section. The cumulants and the corresponding spectra that are necessary for the spectrum of Y_t are listed in page 227, Appendix D. The series $V_tH_{2,t}$ is uncorrelated, therefore its spectrum is constant

$$S_{2,VH_2}(z) = 2\operatorname{Var}(V_t).$$

The spectrum of Y_t is

$$S_{2,Y}(z) = S_{2,V}(z) + 2\mathsf{E}h_t^2 + 4d(\operatorname{Var}(V_t)+1)\operatorname{Re}\frac{1}{z-a},$$

where

$$\mathsf{E}h_t^2 = \frac{1-a^2}{1-a^2-2d^2}.$$

Note

$$\operatorname{Var}(Y_t) = 3\operatorname{Var}(V_t) + 2.$$

3.7.3 *Bispectrum of the state process*

Properties of the V_t^2

Le us denote V_t^2 by Q_t. The equation for the Q_t comes from the square of equation (3.77) and is

$$\begin{aligned} Q_t &= \left(a^2+2d^2\right)Q_{t-1} + 2d\,(a+2d)\,Q_{t-1}H_{2,t-1} + d^2 Q_{t-1}H_{4,t-1} \\ &\quad + 4d^2 V_{t-1} + 2d\,(a+4d)\,V_{t-1}H_{2,t-1} + 2d^2 V_{t-1}H_{4,t-1} \\ &\quad + 2d^2 + 4d^2 H_{2,t-1} + d^2 H_{4,t-1}, \end{aligned} \tag{3.81}$$

where $H_{4,t}$ denotes the Hermite polynomial of fourth order $H_4(\mathrm{w}_t)$. The assumption of fourth order stationarity of V_t is that

$$a^4 + 12a^2d^2 + 32ad^3 + 60d^4 < 1.$$

Some cumulants of interest

$$\begin{aligned} \operatorname{Cum}(Q_t, V_t) &= \frac{12d^2\,(a+2d)\operatorname{Var}(V_t) + 8d^3}{1 - a^3 - 6ad^2 - 8d^3}, \\ \operatorname{Cum}(Q_{t+s}, V_t) &= \left(a^2+2d^2\right)\operatorname{Cum}(Q_{t+s-1}, V_t) \\ &\quad + 4d^2 \operatorname{Cum}(V_{t+s-1}, V_t), \quad s \ge 1. \end{aligned} \tag{3.82}$$

If $S^+_{Q,V}$ denotes the following function

$$S^+_{Q,V}(z) = \sum_{s=1}^{\infty} \operatorname{Cum}(Q_{t+s}, V_t)\, z^{-s},$$

then summing up both sides of equation (3.82) yields

$$S^+_{Q,V}(z) = \frac{z^{-1}\left(a^2+2d^2\right)\operatorname{Cum}(Q_t,V_t) + \frac{4d^2\operatorname{Var}(V_t)}{z-a}}{1-\left(a^2+2d^2\right)z^{-1}}.$$

The cross-spectrum $S_{Q,V}$ is given by

$$S_{Q,V}(z) = S^+_{Q,V}(z) + \frac{\operatorname{Cum}(Q_t, V_t)}{1-az},$$

because the cumulants of negative indices is simply

$$\operatorname{Cum}(Q_{t+s}, V_t) = a^{-s}\operatorname{Cum}(Q_t, V_t), \quad s \le 0. \tag{3.83}$$

Bispectrum of the state process

We refer to the technique applied in Subsection 3.2.2, page 79, i.e., to the formula

$$\begin{aligned} S_{V,V,V}(z_1,z_2) = \operatorname*{sym}_{z_{(1:3)}} \Bigg[& 6\sum_{k=1}^{\infty}\sum_{l=1}^{\infty} \operatorname*{Cum}_{VVV}(k+l,l)\, z_1^{-k-l} z_2^{-l} \\ & + 3\sum_{l=-\infty}^{\infty} \operatorname*{Cum}_{VVV}(0,l)\, z_1^{-l} - 2\operatorname*{Cum}_{VVV}(0,0) \Bigg]_{z_3=(z_1z_2)^{-1}} . \end{aligned} \tag{3.84}$$

for the calculation of the bispectrum. Let us start with the first term

$$\sum_{k=1}^{\infty}\sum_{l=1}^{\infty}\underset{VVV}{\operatorname{Cum}}(k+l,l)z_1^{-k-l}z_2^{-l}=\frac{a}{z_1-a}S_{Q,V}^{+}\left(z_3^{-1}\right),$$

as

$$\underset{VVV}{\operatorname{Cum}}(k+l,l)=a^k\operatorname{Cum}\left(Q_{t+l},V_t\right).$$

The next term to consider is

$$\sum_{l=-\infty}^{\infty}\underset{VVV}{\operatorname{Cum}}(0,l)z_1^{-l}=S_{Q,V}\left(z_1^{-1}\right).$$

Therefore the bispectrum of V_t is

$$S_{V,V,V}(z_1,z_2)=\underset{z_{(1:3)}}{\operatorname{sym}}\left[\frac{6a}{z_1-a}S_{Q,V}^{+}\left(z_3^{-1}\right)+3S_{Q,V}\left(z_1^{-1}\right)\right.$$
$$\left.\left.-2\underset{VVV}{\operatorname{Cum}}(0,0)\right]\right|_{z_3=(z_1z_2)^{-1}}.$$

3.7.4 Bispectrum of the process Y_t

The formula (3.84) of the general expression of the bispectrum applies again for the process Y_t. From the equation

$$Y_t=V_t+V_tH_2\left(\mathrm{w}_t\right)+H_2\left(\mathrm{w}_t\right)+1,$$

it is clear that we have to carefully go through all the cross-cumulants as defined by the terms of this equation. The cumulants of type $\operatorname{Cum}_{YYY}(0,l)$ are $\operatorname{Cum}(Y_{t+l},Y_t,Y_t)=\operatorname{Cum}\left(Y_{t+l},(Y_t-1)^2\right)$ where

$$(Y_t-1)^2=V_t^2+V_t^2H_{2,t}^2+H_{2,t}^2+2V_t^2H_{2,t}+2V_tH_{2,t}+2V_tH_{2,t}^2.$$

One can easily get the third order moment, putting $l=0$, it is

$$\operatorname{Cum}\left(Y_t,Y_t,Y_t\right)=15\operatorname{Cum}\left(Q_t,V_t\right)+36\operatorname{Var}\left(V_t\right)+8.$$

We apply the following notation for restricted Fourier transforms of cumulants of jointly stationary time series X_t, Y_t, and V_t

$$S_{X,Y}^{+}(z)=\sum_{s=1}^{\infty}\operatorname{Cum}\left(X_{t+s},Y_t\right)z^{-s},$$
$$S_{X,Y,V}^{+}(z_1,z_2)=\sum_{k=1}^{\infty}\sum_{l=1}^{\infty}\operatorname{Cum}(X_{t+k+l},Y_{t+l},V_t)z_1^{-k-l}z_2^{-l}.$$

By this notation and by the general formula (3.84), the bispectrum is the following

$$
\begin{aligned}
S_{Y,Y,Y}(z_1,z_2) = \operatorname*{sym}_{z_{(1:3)}} \Big[6 \Big\{ & S^{+}_{V,V,V}(z_1,z_2) + S^{+}_{V,V,VH_2}(z_1,z_2) \\
& + S^{+}_{V,V,H_2}(z_1,z_2) + S^{+}_{V,VH_2,V}(z_1,z_2) \\
& + S^{+}_{V,VH_2,VH_2}(z_1,z_2) + 6S^{+}_{V,QH_2}(z_1) \\
& + S^{+}_{V,VH_2,H_2}(z_1,z_2) + S^{+}_{V,H_2,V}(z_1,z_2) \\
& + S^{+}_{V,H_2,VH_2}(z_1,z_2) + S^{+}_{V,H_2,H_2}(z_1,z_2) \Big\} \\
& + 3S_{V,Q}(z_1) + 3S_{V,QH_2^2}(z_1) + 3S_{V,H_2^2}(z_1) \\
& + 6S_{V,VH_2^2}(z_1) + 18S_{V,VH_2}(z_1) + 9S_{VH_2,Q}(z_1) \\
& + 9S^{+}_{H_2,Q}(z_1) + 12S_{V,H_2}(z_1^{-1}) \\
& - 30\,\mathrm{Cum}(Q_t,V_t) - 72\,\mathrm{Var}(V_t) - 16 \Big]\Big|_{z_3=(z_1z_2)^{-1}} .
\end{aligned}
$$

The expressions for the cumulants and corresponding Fourier transforms are found in page 227, Appendix D.

3.7.5 Simulation

The parameters of the equation (3.76) were chosen as $a_0 = 0.3$ and $d_0 = 0.15$ such that $c_0 = 1-a_0-d_0$. The time series $Y_t = X^2_{0,t}$ was simulated with sample size 2^{12} and the spectrum $S_{2,Y}(z)$ at frequencies $k2^{-6}$ of $[0,1/2]$, and the bispectrum $S_{Y,Y,Y}(z_1,z_2)$ at frequencies $(k2^{-6}, l2^{-6})$ of $[0,1/2] \times [0,1/3]$ were estimated (see figures 3.5, and 3.6).

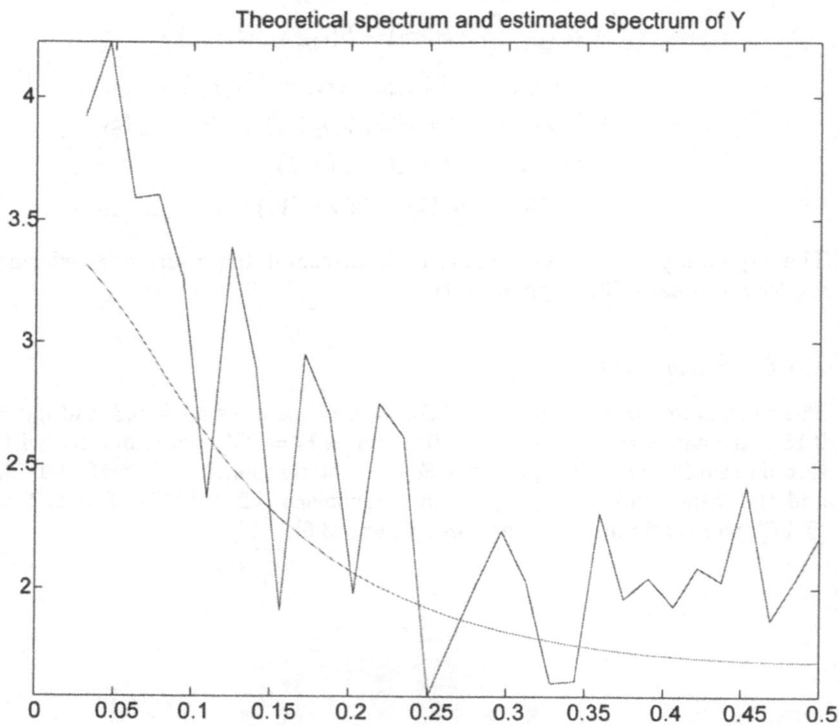

FIGURE 3.5. Theoretical spectrum and estimated spectrum of the square of the GARCH(1,1) model

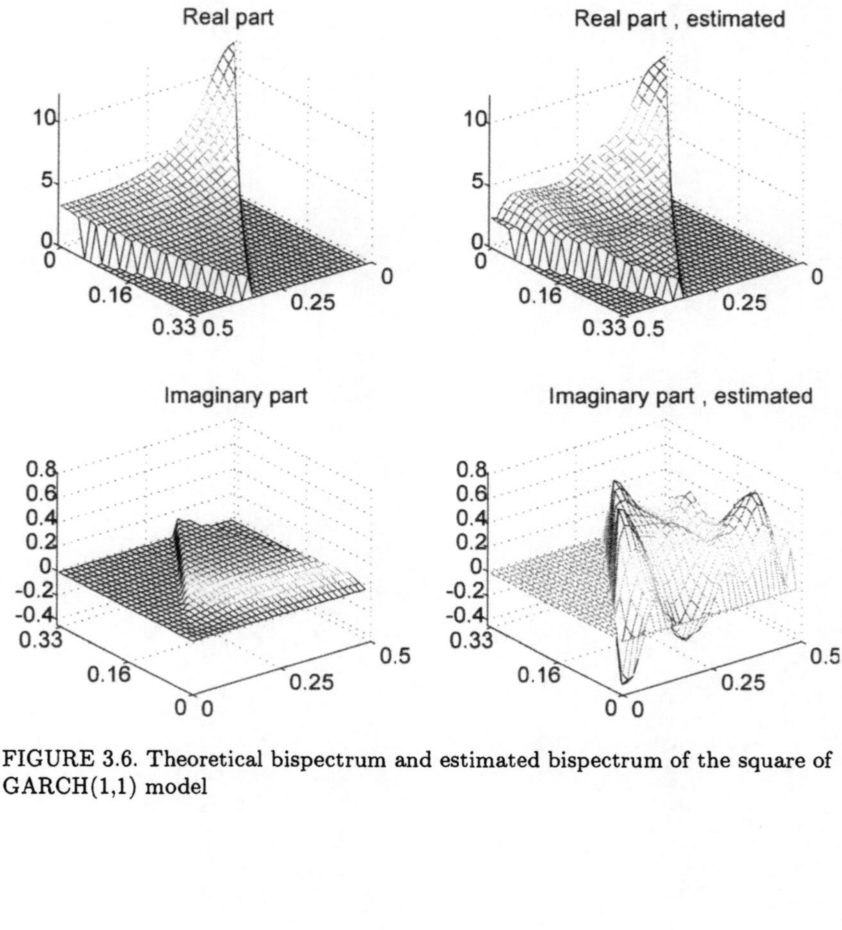

FIGURE 3.6. Theoretical bispectrum and estimated bispectrum of the square of GARCH(1,1) model

4 Non-Gaussian Estimation

Recently, considerable attention has been paid to nonlinear models in time series analysis. The fact is, most of the time series in practice are not Gaussian, and the second order statistics of a bilinear model, for example, does not contain any information about the parameters of nonlinearity, see Terdik and Subba Rao [139]. The methods of parameter estimation are usually based on either the covariances as Yule-Walker equations or the spectrum, see the monographs by Brockwell and Davis [29], Priestley [100], Rosenblatt [107]. These are called Gaussian estimates because they make use of the second order information only. Brillinger [21] started to apply a criterion involving a third order spectrum, i.e., a bispectrum as well as a second order one and found improvement in estimates. The idea is that the theoretical and the estimated spectra are compared by an iteratively reweighted least squares procedure. The idea of Gaussian estimation in the second order case goes back essentially to the method suggested by Whittle [144]. The properties of such types of estimates are studied by Rice [106], using the asymptotic properties of the spectral estimators due to Brillinger and Rosenblatt [27], and discussed by several authors such as Walker [142], [143], [35] and Hannan [50]. The handicap of the nonGaussian parameter estimation is that the exact spectral and bispectral densities are supposed to be known up to some parameters of the model. The models that have been successfully considered are the linear (nonGaussian) Brillinger [21] and the bilinear one Terdik [131].

In this chapter we follow the course set by Brillinger [21] and consider a functional of the spectrum and the bispectrum depending on an unknown single parameter ϑ. We spare some space and formulae by considering one unknown parameter only. It is straightforward to generalize all our results to a higher dimensional parameter. The estimation of the parameter of a non-Gaussian series data is constructed by the minimization of the functional of the theoretical and estimated spectra.. The asymptotic variance of the estimator is calculated for both the cases when the spectra are estimated by the periodogram and by the smoothed periodogram. The simplicity of the variance for the linear model has a particular importance, the variance depending on the skewness and the kurtosis of the error series. The consistency is set down via the general lemma of Brillinger [19]. Asymptotic normality is proved by the results of Brillinger [20] and Rosenblatt [107]. Attention is paid to the fact that in general it can not be assumed that all the moments exist for a nonGaussian time series, see for example the assumptions of the existence of higher order moments of bilinear time series, Theorem 37, page 88, and also Terdik and Ispány [134]. The assumptions of the results published in [72] are extremely strong.

4.1 Estimating a parameter for non-Gaussian data

Let Y_t, $t \in Z$ be a real time series with a zero mean and finite *p-th* order moments ($p \geq 2$). We suppose it to be p-th order weakly stationary, i.e., for all $u \in \mathbb{Z}$ the cumulant function of the p-th order $\operatorname{Cum}(Y_u, Y_{u+t_1}, \ldots, Y_{u+t_{p-1}}) = c_p(t_1, \ldots, t_{p-1})$. Let us take our departure from a few assumptions. They will be supposed bellow for the easier calculation of the formulae.

Assumption 69 *Let us assume that cumulant function of p-th order of the series exists and it depends on a single real unknown parameter. Suppose further that*

$$\sum_{t_1,\ldots,t_{p-1}}^{\infty} (1+|t_j|)\,|c_p(t_1,\ldots,t_{p-1},\vartheta)| < \infty, \qquad j = 1,2,\ldots,p-1.$$

In this case the spectral densities

$$S_r(\omega_1,\ldots,\omega_{r-1},\vartheta) = \sum c_r(t_1,\ldots,t_{r-1},\vartheta) z_1^{t_1}\cdots z_{r-1}^{t_{r-1}},$$

exist up to the p-th order, where $z_j = \exp(2\pi i\omega_j)$, $0 \leq \omega_j < 1$, $j = 1,\ldots,r-1$, $2 \leq r \leq p$. We shall work with $p = 6$ for the next three sections and $p = 12$ later on.

Assumption 70 *The unknown parameter ϑ belongs to a compact set $\Theta \subset \mathbb{R}$. Suppose also that the spectrum $S_2(\omega, \vartheta)$ and the bispectrum $S_3(\omega_1, \omega_2, \vartheta)$ are twice continuously differentiable with respect to $\vartheta \in \Theta$. These derivatives are continuous and bounded above and suppose further that they are bounded away from zero in modulus with respect to the frequencies of the sets Λ_1 and Λ_2 to be defined below.*

If Y_t, $t \in \{0, 1, \ldots, T-1\}$ is an observation of a time series, put

$$d_T(\omega) = \sum_{t=0}^{T-1} Y_t z^{-t}, \quad z = \exp(2\pi i \omega), 0 \le \omega < 1, \tag{4.1}$$

as a discrete Fourier transform. Write

$$I_{2T}(\omega) = \frac{1}{T} d_T(\omega) \overline{d_T(\omega)},$$
$$I_{3T}(\omega_1, \omega_2) = \frac{1}{T} d_T(\omega_1) d_T(\omega_2) \overline{d_T(\omega_1 + \omega_2)},$$

for the second and third order periodograms, respectively.

Note that if $\tilde{y}_t$, $t \in \mathbb{Z}$ is a stationary process with $E|\tilde{y}_t|^3 < \infty$ and $E\tilde{y}_t = m = const \neq 0$, then the periodograms of the second and the third orders may be defined by putting $Y_t = \tilde{y}_t - \frac{1}{T}\sum_{u=0}^{T-1} \tilde{y}_u$ in the formula (4.1). Most of the results that we formulate below are true for such a situation as well.

We are going to apply some common methods for the estimation of the second and the third order spectral densities.

Assumption 71 *The following two assumptions are applied*

- *Let $W_1(\omega)$ be a real valued, even weight function of bounded variation with supp $W_1 \subset [-1, 1]$,*

  $$\int_{-\infty}^{\infty} W_1(\omega) d\omega = 1, \text{ and } \int_{-\infty}^{\infty} W_1^2(\omega) d\omega = \|W_1\|^2 < \infty.$$

  Let B_{1T} be a bandwidth parameter depending on T, satisfying $B_{1T} \to 0$, $TB_{1T}^2 \to \infty$ as $T \to \infty$ and write $W_{1T}(\omega) = W_1(\omega/B_{1T})/B_{1T}$.

- *Let $W_2(\omega_1, \omega_2)$ be a non-negative function of bounded variation with supp $(W_2) \subset [-1, 1]^2$, symmetric in the following sense*

  $$W_2(\omega_1, \omega_2) = W_2(\omega_2, \omega_1) = W_2(\omega_1, -\omega_2 - \omega_1) = W_2(-\omega_1 - \omega_2, \omega_2). \tag{4.2}$$

  Suppose further

  $$\iint_{-\infty}^{+\infty} W_2(\omega_1, \omega_2) d\omega_1 d\omega_2 = 1, \text{ and}$$
  $$\iint_{-\infty}^{+\infty} W_2^2(\omega_1, \omega_2) d\omega_1 d\omega_2 = \|W_2\|^2 < \infty.$$

  Let B_{2T} be a bandwidth parameter satisfying the following conditions: $B_{2T} \to 0$, $TB_{2T}^2 \to \infty$ as $T \to \infty$ and write $W_{2T}(\omega_1, \omega_2) = W_2(\omega_1/B_{2T}, \omega_2/B_{2T})/B_{2T}^2$.

- $B_{1T} \leq B_{2T}$ *and there exists the limit*

$$\lim_{T\to\infty} B_{1T}/B_{2T} = \rho.$$

In second order analysis it is usual assumed only $TB_{1T} \to \infty$ as $T \to \infty$. It will be clear later that we shall need a slower speed of convergence of B_{1T} to *zero*. The bandwidths B_{1T} and B_{2T} can be chosen as $B_{2T} \sim B_{1T}$, i.e., the limit of the ratio B_{1T}/B_{2T} is some constant $\rho \neq 0$ in that case not only $TB_{1T} \to \infty$ but $TB_{1T}^2 \to \infty$ as well. Another choice might be $B_{1T} \sim B_{2T}^2$. In general $B_{1T} \leq B_{2T}$. It should also be noted that the symmetry (4.2) of W_2 is necessary for the smoothed biperiodogram to have the same properties of symmetry as the bispectrum itself at once. This is not so important because the symmetry of the estimator of the bispectrum can be guaranteed after smoothing the biperiodogram, by the window $W_1(\omega_1)W_1(\omega_2)$ as well in the following way. First let us restrict the smoothed biperiodogram on the triangle Δ with vertices $(0,0)$, $(1/2,0)$, $(1/3,1/3)$ then define it on the whole $[0,1]^2$ by the properties of periodicity and symmetry of the bispectrum. This estimator will have the necessary symmetry. The window $W_1(\omega_1)W_1(\omega_2)$ is symmetric on zero which implies that during the smoothing it puts the same weights for the biperiodogram values with the same distances from a central one. Which is reasonable because there is no reason to make a difference between them. The overall effectiveness of these windows are depending on some particular properties of them, [20] p. 52, [122] p. 42.

We shall deal with the discrete Fourier frequencies $f_k = k/T$, $k = 0,\ldots,T-1$. Consider the following smoothed estimate for the spectral density

$$S_{2T}(\omega+m) = \frac{1}{T}\sum_k W_{1T}(\omega - f_k) I_{2T}(f_k),\ \omega \in [0,1),\ m \in \mathbb{Z}.$$

It should be noted that around *zero*, $(\omega \leq B_{1T})$, it is more reasonable consider the convolution because in practice when ω is close to *zero* the weighted ordinates are truncated. For example if $\omega = 0$ only the $W_{1T}(-f_k)I_{2T}(f_k)$ are involved to the estimation. The first and second order moments of $S_{2T}(\omega)$ are derived by Brillinger, see [20], p. 147. The first one is

$$\begin{aligned}\mathsf{E}S_{2T}(\omega) &= \frac{1}{T}\sum_k W_{1T}(\omega - f_k) S_2(f_k) + O(T^{-1}) \\ &= S_2(\omega) + O(B_{1T}) + O(T^{-1}B_{1T}^{-1}),\end{aligned}$$

uniformly in ω, if $\omega \neq 0 \bmod (1)$. Define δ_ω the Kronecker comb,

$$\delta_\omega = \begin{cases} 1 & if \quad \omega = 0 \bmod (1), \\ 0 & otherwise. \end{cases}$$

The covariance is given by

$$\mathsf{Cov}(S_{2T}(\omega_1),\, S_{2T}(\omega_2)) = \frac{1}{T} S_4\left(\omega_1, \omega_2, -\omega_2\right) \tag{4.3}$$
$$+ \frac{\|W_1\|^2}{TB_{1T}}\left(S_2^2(\omega_1) + O(B_{1T})\right)\left(\delta_{\omega_1-\omega_2} + \delta_{\omega_1+\omega_2}\right) + O(T^{-2}B_{1T}^{-2}),$$

uniformly in ω_1, ω_2.

The method of smoothing the biperiodogram is a two-variable analogue of the previous one. A consistent estimate of the bispectrum is

$$S_{3T}(\omega_1 + m, \omega_2 + n) = \frac{1}{T^2}\sum_k \sum_r W_{2T}(\omega_1 - f_k,\, \omega_2 - f_r) I_{3T}(f_k, f_r),$$

where $(f_k, f_r) = (k/T, r/T)$ $k, r = 0, 1, 2, ..., T-1$ are the Fourier frequencies on the plane, $\omega_1, \omega_2 \in [0, 1)$, and $m, n \in \mathbb{Z}$. In practice, for a fixed T and B_{2T}, one should pay attention to the estimation around zero and close to the axis. The following expansion shows that the smoothed estimator is asymptotically unbiased

$$\mathsf{E}S_{3T}(\omega_1,\, \omega_2) = S_3(\omega_1,\, \omega_2) + O(T^{-1}B_{2T}^{-1}) + O(B_{2T}),$$

the error terms O are uniform in ω_1, ω_2. Here and throughout the chapter put $\omega_3 = -\omega_1 - \omega_2$, $\lambda_3 = -\lambda_1 - \lambda_2$ and the indices of the frequencies exceeding 3 will be considered as $\mathsf{mod}(3) + 1$. We exclude the boundary not only of the interval $[0, 1/2]$ but also of the triangle $\triangle$ with vertices $(0, 0)$, $(1/2, 0)$, $(1/3, 1/3)$. The following expression and some other selected ones are proved in the Appendix E, page 231. The cross-covariance between the smoothed periodogram and the biperiodogram is

$$\mathsf{Cov}\left(S_{2T}(\lambda), S_{3T}(\omega_1,\, \omega_2)\right) = \frac{1}{T} S_5\left(\lambda, -\omega_1,\, -\omega_2, -\lambda\right)$$
$$+ \frac{W_{12}\left(\rho\right)}{TB_{2T}} \sum_{k=1}^{3} S_2\left(\omega_k\right) \bar{S}_3\left(\omega_{k+1},\, \omega_{k+2}\right)\left(\delta_{\lambda+\omega_k} + \delta_{\lambda-\omega_k}\right)$$
$$+ \frac{W_{20}}{TB_{2T}} \sum_{k=1}^{3} \left[S_2\left(\omega_k\right) S_3\left(\lambda, 0\right) + O\left(B_{2T}\right)\right] \delta_{\omega_{k+1}} + O(T^{-2}B_{1T}^{-2}), \tag{4.4}$$

where the constants W_{12} and W_{20} are defined by

$$W_{12}\left(\rho\right) = \iint_{-\infty}^{+\infty} W_1\left(\omega_1\right) W_2(\rho\omega_1, \omega_2) d\omega_1 d\omega_2, \tag{4.5}$$
$$W_{20} = \int_{-\infty}^{+\infty} W_2(\omega_1, -\omega_1) d\omega_1,$$

where

$$\rho = \lim_{T\to\infty} \frac{B_{1T}}{B_{2T}},$$

and $W_{12}(\rho) = 0$ if $\rho = 0$. The covariance according to the smoothed biperiodogram is

$$\begin{aligned}
&\mathsf{Cov}\left(S_{3T}(\lambda_1, \lambda_2), S_{3T}(\omega_1, \omega_2)\right) = \frac{1}{T} S_6\left(\lambda_1, \lambda_2, -\omega_1, -\omega_2, \lambda_3\right) \\
&+ \frac{W_{23}}{TB_{2T}} \left[\sum_{m,n=1}^{3} \left(S_2\left(\lambda_m\right) S_4\left(\lambda_{m+1}, \lambda_{m+2}, -\omega_{n+1}\right) + O\left(B_{2T}\right)\right) \delta_{\lambda_m - \omega_n} \right. \\
&\left. + \sum_{m,n=1}^{3} \left(S_3\left(\lambda_m, \lambda_{m+1}\right) S_3\left(\lambda_{m+2}, -\omega_{n+1}\right) + O\left(B_{2T}\right)\right) \delta_{\lambda_m + \lambda_{m+1} - \omega_n} \right] \\
&+ \frac{W_{20}}{TB_{2T}} S_4\left(\omega_1, \omega_2, \omega_3\right) \sum_{m=1}^{3} \left(S_2\left(\lambda_m\right) + O\left(B_{2T}\right)\right) \delta_{\lambda_{m+1}} \\
&+ \frac{W_{20}}{TB_{2T}} S_4\left(\lambda_1, \lambda_2, \lambda_3\right) \sum_{m=1}^{3} \left(S_2\left(\omega_m\right) + O\left(B_{2T}\right)\right) \delta_{\omega_{m+1}} \\
&+ \frac{W_{20}^2}{TB_{2T}^2} \left[S_2(0) \left(\sum_{m=1}^{3} S_2\left(\lambda_m\right) \delta_{\lambda_{m+1}} \right) \left(\sum_{m=1}^{3} S_2\left(\omega_m\right) \delta_{\omega_{m+1}} \right) \right. \\
&\left. + O\left(B_{2T}\right) \sum_{m,n=1}^{3} \delta_{\lambda_m} \delta_{\omega_n} \right] \\
&+ \frac{\|W_2\|^2}{TB_{2T}^2} \left(S_2\left(\omega_1\right) S_2(\omega_2) S_2(\omega_1 + \omega_2) + O\left(B_{2T}\right)\right) \\
&\times \sum_{m=1}^{3} \delta_{\lambda_1 - \omega_m} \left(\delta_{\lambda_2 - \omega_{m+1}} + \delta_{\lambda_2 - \omega_{m+2}}\right) \\
&+ O(T^{-2} B_{2T}^{-2}),
\end{aligned} \tag{4.6}$$

with the $O(T^{-2}B_{2T}^{-2})$ term uniform in the frequencies (λ_1, λ_2), (ω_1, ω_2). Further

$$W_{23} = \iiint_{-\infty}^{+\infty} W_2(\lambda_1, \lambda_2) W_2(\lambda_1 + \lambda_2, \omega) d\lambda_1 d\lambda_2 d\omega. \tag{4.7}$$

The equation (4.6) is based on the expression for $\mathsf{Cov}\left(I_{3T}(\lambda_1, \lambda_2), I_{3T}(\omega_1, \omega_2)\right)$. The latter one is calculated by the formula for expressing the cumulants of products by cumulants,(see formula (1.11), page 15, Brillinger [20], Leonov and Shiryaev [73]). From now on we shall use that $B_{1T} \leq B_{2T}$.

Most of the results listed above are in the monographs by Brillinger [20] and Rosenblatt [107], further investigations can be found in Van Ness [90] and Alekseev [3].

The symmetry relations of $S_3(\omega_1,\omega_2)$, i.e.,

$$S_3(\omega_1,\omega_2)=S_3(\omega_2,\omega_1)=S_3(\omega_1,-\omega_2-\omega_1)=S_3(-\omega_1-\omega_2,\omega_2), \tag{4.8}$$

and the identity

$$S_3(-\omega_1,-\omega_2)=\bar{S}_3(\omega_1,\omega_2), \tag{4.9}$$

imply that the values of the bispectrum over the entire plane are completely specified by its values inside the triangle $\triangle$ with vertices $(0,0)$, $(1/2,0)$, $(1/3,1/3)$, (the periodicity is automatically guaranteed in this way). Actually there are 12 equivalent triangle in this sense, see Figure 1.5, page 30, Brillinger and Rosenblatt [27], van Ness [90] and Subba Rao and Gabr [122].

Assumption 72 *Suppose that $\Lambda_1 \subset (0,1/2)$ is a finite union of closed intervals. Let Λ_2 be some finite union of compact domains lying inside the open triangle $\triangle$.*

Let us suppose that the spectrum and the bispectrum of the process Y_t depend on a single parameter ϑ which is not a multiplicative one. (The parameter ϑ is multiplicative if $S_2(\omega,c\vartheta)=cS_2(\omega,\vartheta)$ and $S_3(\omega_1,\omega_2,c\vartheta)=cS_3(\omega_1,\omega_2,\vartheta)$ for any *positive real number c*.) A multiplicative parameter can not be estimated by minimization of (4.10). We consider the estimate ϑ_T for the unknown parameter ϑ obtained from minimization of the functional

$$\begin{aligned} Q_T(\vartheta) &= p_1\frac{B_{1T}}{\beta_1}\sum_{b_{1k}\in\Lambda_1}\left(\frac{S_2(b_{1k},\vartheta)-S_{2T}(b_{1k})}{S_2(b_{1k},\vartheta)}\right)^2 \\ &+ q_1\frac{B_{2T}^2}{\beta_2}\sum_{(b_{2m},b_{2l})\in\Lambda_2}\frac{|S_3(b_{2m},b_{2l},\vartheta)-S_{3T}(b_{2m},b_{2l})|^2}{S_2(b_{2m},\vartheta)\,S_2(b_{2l},\vartheta)S_2(b_{2m}+b_{2l},\vartheta)}, \end{aligned} \tag{4.10}$$

where frequencies b_{1k} are equally spaced in $[0,1]$ by bandwidth B_{1T} as well as frequencies b_{2k} by bandwidth B_{2T}, $p_1\in(0,1)$, $q_1=1-p_1$, and constants β_1, β_2 are the Lebesque measure of Λ_1 and Λ_2, respectively. The role of β_1 and β_2 is that both sums in (4.10) are averaged since the numbers of terms are about to β_1/B_{1T} and β_2/B_{2T}^2. For technical reasons we shall consider an equivalent form of (4.10).

The spectrum is symmetric to 1/2 and the symmetry relations (4.8), (4.9) of the bispectrum define the transformations

$$\begin{aligned} \mathcal{T}_1(\omega_1,\omega_2)&=(\omega_2,\omega_1), & \mathcal{T}_2(\omega_1,\omega_2)&=(\omega_1,1-\omega_2-\omega_1), \\ \mathcal{T}_3(\omega_1,\omega_2)&=(1-\omega_1-\omega_2,\omega_2), & \mathcal{T}_4(\omega_1,\omega_2)&=(1-\omega_1,1-\omega_2), \end{aligned} \tag{4.11a}$$

from the triangle $\triangle$, by mod(1), into the whole square $[0,1]^2$. Denote $\Lambda_1^{\square}$ and $\Lambda_2^{\square}$ the sets with origins Λ_1 and Λ_2, respectively, and with property that they are invariant according to transformations $\mathcal{T}_j$, $j = 1, 2, 3, 4$. Actually $\Lambda_1^{\square} = \Lambda_1 \cup \{\omega = 1 - \lambda \mid \lambda \in \Lambda_1\}$. There will be some advantage of changing the domains of summations in (4.10) into the symmetric ones because the result of summation over set $\Lambda_2^{\square}$ of the expression of the complex valued bispectrum will then be real. Now set

$$Q_T(\vartheta) = \frac{pB_{1T}}{2} \sum_{b_{1k} \in \Lambda_1^{\square}} \left(\frac{S_2(b_{1k}, \vartheta) - S_{2T}(b_{1k})}{S_2(b_{1k}, \vartheta)} \right)^2 + \frac{qB_{2T}^2}{12} \sum_{(b_{2m}, b_{2l}) \in \Lambda_2^{\square}} \frac{|S_3(b_{2m}, b_{2l}, \vartheta) - S_{3T}(b_{2m}, b_{2l})|^2}{S_2(b_{2m}, \vartheta) S_2(b_{2l}, \vartheta) S_2(b_{2m} + b_{2l}, \vartheta)}, \tag{4.12}$$

where $p = p_1/\beta_1$ and $q = q_1/\beta_2$. For a Gaussian time series this method based on the first term of $Q_T(\vartheta)$ for estimating parameters. When a process is non-Gaussian Brillinger [21] suggested applying both the second and the third order periodograms in (4.10).

Let ϑ_0 be the true value of the parameter $\vartheta \in \Theta$ and put

$$Q(\vartheta) = p \int_{\Lambda_1} \left(\frac{S_2(\omega, \vartheta) - S_2(\omega, \vartheta_0)}{S_2(\omega, \vartheta)} \right)^2 d\omega + q \iint_{\Lambda_2} \frac{|S_3(\omega_1, \omega_2, \vartheta) - S_3(\omega_1, \omega_2, \vartheta_0)|^2}{S_2(\omega_1, \vartheta) S_2(\omega_2, \vartheta) S_2(\omega_1 + \omega_2, \vartheta)} d\omega_1 d\omega_2. \tag{4.13}$$

In the next section we shall give conditions under which $Q_T(\vartheta) \to Q(\vartheta)$ and $\vartheta_T \to \vartheta_0$ in probability as $T \to \infty$.

4.2 Consistency and asymptotic variance of the estimate

To prove the consistency of the estimator based on functional $Q_T(\vartheta)$ we use Lemma 1 from the paper by Brillinger [19], which contains general conditions for the consistency of an estimator based on some functional. It will be quoted in the Appendix E. Theorem 4.2 due to Brillinger [18] is also applied. Our theorem concerning the consistency is the following.

Theorem 73 *Suppose that Y_t satisfies Assumptions* 69 -71 *with* $p = 2, 3, \ldots, n, \ldots$, $|S_3(\omega_1, \omega_2, \vartheta)| < \infty$ *and that both*

$$\left(\frac{S_2(\omega, \vartheta) - S_2(\omega, \vartheta_0)}{S_2(\omega, \vartheta)} \right)^2 \quad \text{and} \quad \frac{|S_3(\omega_1, \omega_2, \vartheta) - S_3(\omega_1, \omega_2, \vartheta_0)|^2}{S_2(\omega_1, \vartheta) S_2(\omega_2, \vartheta) S_2(\omega_3, \vartheta)}$$

have finite total variations on Λ_1 and Λ_2 with Assumption 72, *respectively. Suppose moreover that $Q(\vartheta)$ has a unique minimum at ϑ_0 and is continuous*

in ϑ.
Then $Q_T(\vartheta) \xrightarrow{P} Q(\vartheta)$ *as* $T \to \infty$, *where* $Q_T(\vartheta)$ *and* $Q(\vartheta)$ *are defined by* (4.10) *and* (4.13) *respectively and* $\vartheta_T \xrightarrow{P} \vartheta_0$ *as* $T \to \infty$, *where* ϑ_T *is obtained by minimizing* $Q_T(\vartheta)$. *Moreover,* ϑ_T *is an asymptotically unbiased estimate of* ϑ_0.

See the Appendix E for the proof. Under the regularity conditions above ϑ_T tends in probability to the true value ϑ_0. Therefore one can act as if ϑ_T is near ϑ_0 (for a large T) and set down the following Taylor expansion:

$$\frac{\partial}{\partial\vartheta} Q_T(\vartheta_T) = \frac{\partial}{\partial\vartheta} Q_T(\vartheta_0) + (\vartheta_T - \vartheta_0)\frac{\partial^2}{\partial\vartheta^2} Q_T(\vartheta_T^\star),$$

where $|\vartheta_T^\star - \vartheta_0| < |\vartheta_T - \vartheta_0|$. Since ϑ_T minimizes $Q_T(\vartheta)$, it follows that $\frac{\partial}{\partial\vartheta} Q_T(\vartheta_T) = 0$. Thus for a large T

$$\vartheta_T - \vartheta_0 = -\frac{\frac{\partial}{\partial\vartheta} Q_T(\vartheta_0)}{\frac{\partial^2}{\partial\vartheta^2} Q_T(\vartheta_T^\star)}. \tag{4.14}$$

By a straightforward computation we obtain from (4.12):

$$\begin{aligned}
\frac{\partial}{\partial\vartheta} Q_T(\vartheta) &= pB_{1T} \sum_{b_{1k}\in\Lambda_1^{\square}} \left(S_2(b_{1k}, \vartheta) - S_{2T}(b_{1k})\right) A(b_{1k}, \vartheta) \\
&- pB_{1T} \sum_{b_{1k}\in\Lambda_1^{\square}} \left(S_2(b_{1k}, \vartheta) - S_{2T}(b_{1k})\right)^2 \frac{A(b_{1k}, \vartheta)}{S_2(b_{1k}, \vartheta)} \\
&+ \frac{qB_{2T}^2}{6} \sum_{(b_{2m}, b_{2l})\in\Lambda_2^{\square}} \left(S_3(b_{2m}, b_{2l}, \vartheta) - S_{3T}(b_{2m}, b_{2l})\right) B(b_{2m}, b_{2l}, \vartheta) \\
&- \frac{qB_{2T}^2}{12} \sum_{(b_{2m}, b_{2l})\in\Lambda_2^{\square}} \left|S_3(b_{2m}, b_{2l}, \vartheta) - S_{3T}(b_{2m}, b_{2l})\right|^2 \\
&\times \frac{\frac{\partial}{\partial\vartheta}\log\left(S_2(b_{2m}, \vartheta) S_2(b_{2l}, \vartheta) S_2(b_{2m} + b_{2l}, \vartheta)\right)}{S_2(b_{2m}, \vartheta) S_2(b_{2l}, \vartheta) S_2(b_{2m} + b_{2l}, \vartheta)},
\end{aligned}$$

where

$$A(b, \vartheta) = \frac{\frac{\partial}{\partial\vartheta}\log S_2(b, \vartheta)}{S_2(b, \vartheta)}, \tag{4.15}$$

$$B(b_1, b_2, \vartheta) = \frac{\frac{\partial}{\partial\vartheta}\bar{S}_3(b_1, b_2, \vartheta)}{S_2(b_1, \vartheta) S_2(b_2, \vartheta) S_2(b_1 + b_2, \vartheta)}. \tag{4.16}$$

These notations will be used throughout the chapter. Further algebra leads to the expression for the second derivative:

$$\frac{\partial^2}{\partial\vartheta^2}Q_T(\vartheta) = p\int_{\Lambda_1^{\square}} \left(\frac{\partial}{\partial\vartheta}\log S_2(\omega,\vartheta)\right)^2 d\omega$$

$$+p\int_{\Lambda_1^{\square}} (S_2(\omega,\vartheta) - S_{2T}(\omega)) \frac{\left(\frac{\partial}{\partial\vartheta}\right)^2 \log S_2(\omega,\vartheta) - 3\left(\frac{\partial}{\partial\vartheta}\log S_2(\omega,\vartheta)\right)^2}{S_2(\omega,\vartheta)} d\omega$$

$$-p\int_{\Lambda_1^{\square}} (S_2(\omega,\vartheta) - S_{2T}(\omega))^2 \frac{\left(\frac{\partial}{\partial\vartheta}\right)^2 \log S_2(\omega,\vartheta) - 2\left(\frac{\partial}{\partial\vartheta}\log S_2(\omega,\vartheta)\right)^2}{S_2^2(\omega,\vartheta)} d\omega$$

$$+\frac{q}{12}\iint_{\Lambda_2^{\square}} |S_3(\omega_1,\omega_2,\vartheta) - S_{3T}(\omega_1,\omega_2)|^2$$

$$\times \frac{\frac{\partial^2}{\partial\vartheta^2}\log\prod S_2(\omega_k,\vartheta) - \left(\frac{\partial}{\partial\vartheta}\log\prod S_2(\omega_k,\vartheta)\right)^2}{\prod S_2(\omega_k,\vartheta)}$$

$$+2\left(S_3(\omega_1,\omega_2,\vartheta) - S_{3T}(\omega_1,\omega_2)\right)$$

$$\times \frac{\frac{\partial^2}{\partial\vartheta^2}\bar{S}_3(\omega_1,\omega_2,\vartheta) - 2\frac{\partial}{\partial\vartheta}\bar{S}_3(\omega_1,\omega_2,\vartheta)\frac{\partial}{\partial\vartheta}\log\prod S_2(\omega_k,\vartheta)}{\prod S_2(\omega_k,\vartheta)} d\omega_1 d\omega_2$$

$$+\frac{q}{6}\iint_{\Lambda_2^{\square}} \frac{\left|\frac{\partial}{\partial\vartheta}S_3(\omega_1,\omega_2,\vartheta)\right|^2}{\prod S_2(\omega_k,\vartheta)} d\omega_1 d\omega_2 + O(B_T),$$

where, as usual, $\omega_1+\omega_2+\omega_3=0$ and

$$\prod S_2(\omega_k,\vartheta) = \prod_{k=1}^{3} S_2(\omega_k,\vartheta).$$

Denote the denominator of (4.14) by $\Sigma_0^{1/2}(\vartheta_0)$. Now we obtain the asymptotic variance for the estimator ϑ_T:

Lemma 74 *Under the assumptions of Theorem 73*

$$\mathsf{P}\lim_{T\to\infty}\frac{\partial^2}{\partial\vartheta^2}Q_T(\vartheta_T) = \Sigma_0^{1/2}(\vartheta_0)$$

$$= p\int_{\Lambda_1^{\square}} \left(\frac{\partial}{\partial\vartheta}\log S_2(\omega,\vartheta_0)\right)^2 d\omega$$

$$+\frac{q}{6}\iint_{\Lambda_2^{\square}} \frac{\left|\frac{\partial}{\partial\vartheta}S_3(\omega_1,\omega_2,\vartheta_0)\right|^2}{S_2(\omega_1,\vartheta_0)S_2(\omega_2,\vartheta_0)S_2(\omega_1+\omega_2,\vartheta_0)} d\omega_1 d\omega_2. \tag{4.17}$$

Let us turn to the numerator of the expression defined by $\vartheta_T - \vartheta_0$. The following expression is real. It influences the limiting behavior of the estimator ϑ_T

$$Q_{1T}(\vartheta_0) = pB_{1T} \sum_{b_{1k}\in\Lambda_1^{\square}} (S_2(b_{1k}, \vartheta_0) - S_{2T}(b_{1k}))\, A\,(b_{1k}, \vartheta_0)$$

$$+ \frac{qB_{2T}^2}{6} \sum_{(b_{2m}, b_{2l})\in\Lambda_2^{\square}} (S_3(b_{2m}, b_{2l}, \vartheta_0) - S_{3T}\,(b_{2m}, b_{2l}))\, B(b_{2m}, b_{2l}, \vartheta_0).$$

Denote the asymptotic variance of $Q_{1T}(\vartheta_0)$ by

$$\Sigma_1(\vartheta_0) = \lim_{T\to\infty} T\mathrm{var}\,(Q_{1T}(\vartheta_0))\,. \tag{4.18}$$

The statistics $Q_{1T}(\vartheta_0)$ consists of two terms, therefore the variance of it will be given as the sum·

$$\Sigma_1(\vartheta_0) = \Sigma_2(\vartheta_0) + \Sigma_3(\vartheta_0) + 2\Sigma_{23}(\vartheta_0).$$

The asymptotic variance of the second order term $\Sigma_2(\vartheta_0)$ has been given by Rice [106] as well, and may be expressed by

$$\Sigma_2(\vartheta_0) = p^2 \iint_{\Lambda_1^{\square}\times\Lambda_1^{\square}} S_4(\omega, \lambda, -\lambda, \vartheta_0) A\,(\omega, \vartheta_0)\, A(\lambda, \vartheta_0) d\omega d\lambda$$
$$+ 2p^2 \left\|W_1\right\|^2 \int_{\Lambda_1^{\square}} \left(\frac{\partial}{\partial\vartheta} \log S_2(\omega, \vartheta_0)\right)^2 d\omega.$$

Now the asymptotic covariance (4.6) is applied to get the variance of the third order term

$$\Sigma_3(\vartheta_0) = \frac{q^2}{6^2} \iiiint_{\Lambda_2^{\square}\times\Lambda_2^{\square}} S_6\,(\omega_1, \omega_2, -\lambda_1, -\lambda_2, \omega_3, \vartheta_0)\, B(\omega_1, \omega_2, \vartheta_0)$$
$$\times B(-\lambda_1, -\,\lambda_2, \vartheta_0) d\omega_1 d\omega_2 d\lambda_1 d\lambda_2$$
$$+ \frac{q^2 W_{23}}{4} \iiint_{\Lambda_{21}^{\square}\times\Lambda_2^{\square}} S_3(\lambda_1, \lambda_2, \vartheta) B(\lambda_1, \lambda_2, \vartheta_0)$$
$$\times \bar{S}_3(-\lambda_3, \omega_2, \vartheta) \bar{B}(-\lambda_3, \omega_2, \vartheta_0)$$
$$+ S_4(\lambda_1, \lambda_2\,, -\omega_2, \vartheta_0) \bar{B}(-\omega_2 - \lambda_3, \omega_2, \vartheta_0)$$
$$\times B(\lambda_1, \lambda_2, \vartheta_0) S_2(\lambda_3, \vartheta_0) d\lambda_1 d\lambda_2 d\omega_2$$

$$+\frac{q^2W_{20}}{6}\iiint_{\Lambda_{21}^{\square}\times\Lambda_2^{\square}} S_4(\lambda_1,\lambda_2,0,\vartheta_0)B(\lambda_1,\lambda_2,\vartheta_0)\bar{B}(0,\omega_2,\vartheta_0)$$

$$\times S_2(\omega_2,\vartheta_0)d\lambda_1 d\lambda_2 d\omega_2+\frac{q^2W_{20}^2}{4S_2(0,\vartheta_0)}\left|\int_{\Lambda_1^{\square}}\frac{\frac{\partial}{\partial\vartheta}S_3(\omega_1,0,\vartheta_0)}{S_2(\omega_1,\vartheta_0)}d\omega_1\right|^2$$

$$+\frac{q^2\left\|W_2\right\|^2}{6}\iint_{\Lambda_2^{\square}}\frac{\left|\frac{\partial}{\partial\vartheta}S_3(\omega_1,\omega_2,\vartheta_0\right|^2}{S_2(\omega_1,\vartheta_0)S_2(\omega_2,\vartheta_0)S_2(\omega_3,\vartheta_0)}d\omega_1 d\omega_2,$$

where set $\Lambda_{21}^{\square}$ is the orthogonal projection of $\Lambda_2^{\square}$ onto $[0,1]$. The covariance between the second and third order terms is

$$\Sigma_{23}(\vartheta_0)=$$

$$\frac{pq}{6}\iiint_{\Lambda_2^{\square}\times\Lambda_1^{\square}} S_5(\omega,-\omega,-\omega_1,-\omega_2,\vartheta_0)\bar{B}(\omega_1,\omega_2,\vartheta_0)$$

$$\times A(\omega,\vartheta_0)d\omega_1 d\omega_2 d\omega$$

$$+pqW_{12}(\rho)\iint_{\Lambda_{21}^{\square}\times\Lambda_1^{\square}} S_3(-\omega_1,-\omega_2,\vartheta_0)\bar{B}(\omega_1,\omega_2,\vartheta_0)$$

$$\times\frac{\partial}{\partial\vartheta}\log S_2(\omega_1,\vartheta_0)d\omega_1 d\omega_2$$

$$+\frac{pqW_{20}}{2}\iint_{\Lambda_{21}^{\square}\times\Lambda_1^{\square}} S_3(\omega_1,0,\vartheta_0)\bar{B}(\omega_1,0,\vartheta_0)$$

$$\times\frac{\partial}{\partial\vartheta}\log S_2(\omega_2,\vartheta_0)d\omega_1 d\omega_2,$$

where again $\chi_{(A)}$ is the indicator variable of set A. The last integral is zero if the domains are $[0,1]$ and $[0,1]^3$, respectively.

Consider the following functional (depending on the periodograms of the second and the third orders instead of the smoothed periodograms) taken at Fourier frequencies f_k, f_l, $k,l=1,2,\ldots,T$

$$R_T(\vartheta_0)=\frac{p}{2T}\sum_{f_k\in\Lambda_1^{\square}}\left(S_2(f_k,\vartheta_0)-I_{2T}(f_k)\right)A(f_k,\vartheta_0)$$

$$+\frac{q}{6T^2}\sum_{(f_k,f_l)\in\Lambda_2^{\square}}\left(S_3(f_k,f_l,\vartheta_0)-I_{3T}(f_k,f_l)\right)B(f_k,f_l,\vartheta_0).$$

The asymptotic variance $\Sigma_{R1}(\vartheta_0)$ of $R_T(\vartheta_0)$ is the same as $\Sigma_1(\vartheta_0)$, where all constants depending on the weight functions W's are changed to 1.

Lemma 75 *Under the assumptions of Theorem 73 the asymptotic variance*

$$\Sigma_{R1}(\vartheta_0)=\lim_{T\to\infty}T\operatorname{var}\left(Q_{1T}(\vartheta_0)\right), \tag{4.19}$$

of $R_T(\vartheta_0)$ *is given as the sum*

$$\Sigma_{R1}(\vartheta_0) = \Sigma_{R2}(\vartheta_0) + \Sigma_{R3}(\vartheta_0) + 2\Sigma_{R23}(\vartheta_0),$$

with

$$\Sigma_{R2}(\vartheta_0) = p^2 \iint_{\Lambda_1^{\square}\times\Lambda_1^{\square}} S_4(\omega, \lambda, -\lambda, \vartheta_0) A(\omega, \vartheta_0)\, A(\lambda, \vartheta_0) d\omega d\lambda + 2p^2 \int_{\Lambda_1^{\square}} \left(\frac{\partial}{\partial\vartheta} \log S_2(\omega, \vartheta_0)\right)^2 d\omega,$$

$$\begin{aligned}\Sigma_{R3}(\vartheta_0) = \frac{q^2}{6^2} \iiiint_{\Lambda_2^{\square}\times\Lambda_2^{\square}} & S_6(\omega_1, \omega_2, -\lambda_1, -\lambda_2, \omega_3, \vartheta_0) \\ \times \bar{B}(\omega_1, \omega_2, \vartheta_0) & B(-\lambda_1, -\lambda_2, \vartheta_0) d\omega_1 d\omega_2 d\lambda_1 d\lambda_2 \\ + \frac{q^2}{4} \iiint_{\Lambda_{21}^{\square}\times\Lambda_2^{\square}} & S_3(\lambda_1, \lambda_2, \vartheta) B(\lambda_1, \lambda_2, \vartheta_0) \bar{S}_3(-\lambda_3, \omega_2, \vartheta) \\ & \times \bar{B}(-\lambda_3, \omega_2, \vartheta_0) \\ + S_4(\lambda_1, \lambda_2, -\omega_2, \vartheta_0) & \bar{B}(\lambda_3, \omega_2, \vartheta_0) B(\lambda_1, \lambda_2, \vartheta_0) S_2(\omega_2, \vartheta_0) d\lambda_1 d\lambda_2 d\omega_2 \\ + \frac{q^2}{6} \iiint_{\Lambda_{21}^{\square}\times\Lambda_2^{\square}} & S_4(\lambda_1, \lambda_2, 0, \vartheta_0) B(\lambda_1, \lambda_2, \vartheta_0) \\ \times \bar{B}(0, \omega_2, \vartheta_0) & S_2(\omega_2, \vartheta_0) d\lambda_1 d\lambda_2 d\omega_2 \\ + \frac{q^2}{4S_2(0, \vartheta_0)} & \left| \iint_{\Lambda_1^{\square}} \frac{\frac{\partial}{\partial\vartheta} S_3(\omega_1, 0, \vartheta_0)}{S_2(\omega_1, \vartheta_0)} d\omega_1 \right|^2 \\ + \frac{q^2}{6} \iint_{\Lambda_2^{\square}} & \frac{\left|\frac{\partial}{\partial\vartheta} S_3(\omega_1, \omega_2, \vartheta_0\right|^2}{S_2(\omega_1, \vartheta_0) S_2(\omega_2, \vartheta_0) S_2(\omega_3, \vartheta_0)} d\omega_1 d\omega_2,\end{aligned}$$

and

$$\begin{aligned}\Sigma_{R23}(\vartheta_0) = & \\ \frac{pq}{6} \iiint_{\Lambda_2^{\square}\times\Lambda_1^{\square}} & S_5(\omega, -\omega, -\omega_1, -\omega_2, \vartheta_0)\, \bar{B}(\omega_1, \omega_2, \vartheta_0) A(\omega_2, \vartheta_0) d\omega_1 d\omega_2 d\omega \\ + pq \iint_{\Lambda_{21}^{\square}\times\Lambda_1^{\square}} & S_3(-\omega_1, -\omega_2, \vartheta_0)\, \bar{B}(\omega_1, \omega_2, \vartheta_0) \frac{\partial}{\partial\vartheta} \log S_2(\omega_1, \vartheta_0) d\omega_1 d\omega_2 \\ + \frac{pq}{2} \iint_{\Lambda_{21}^{\square}\times\Lambda_1^{\square}} & S_3(\omega_1, 0, \vartheta_0)\, \bar{B}(\omega_1, 0, \vartheta_0) \frac{\partial}{\partial\vartheta} \log S_2(\omega_2, \vartheta_0) d\omega_1 d\omega_2.\end{aligned}$$

There is an interesting case concerning the statistic R_T if one slows down the convergence by bandwidth $B_T = B_{1T} = B_{2T}$ and considers the quantity

$$R_T^\star(\vartheta_0) = \frac{p\sqrt{B_T}}{2} \sum_{b_k \in \Lambda_1^\square} \left(S_2(b_k, \vartheta_0) - I_{2T}(b_k)\right) A(b_k, \vartheta_0)$$
$$+ \frac{q\sqrt{B_T}}{6T} \sum_{(b_k, b_l) \in \Lambda_2^\square} \left[\left(S_3(b_k, b_l, \vartheta_0) - I_{3T}(b_k, b_l)\right) B(b_k, b_l, \vartheta_0)\right].$$

Then the asymptotic variance $\Sigma_{R^\star 1}(\vartheta_0)$ of $R_T^\star(\vartheta_0)$ is the same as $\Sigma_0(\vartheta_0)$, see (4.17).

4.3 Asymptotic normality of the estimate

Theorem 76 *Suppose that conditions with $p = 1, 2, \ldots$ of Theorem 73 are fulfilled. Then the estimator ϑ_T defined by (4.10) is asymptotically Gaussian:*

$$\sqrt{T}(\vartheta_T - \vartheta_0) \xrightarrow{D} \mathcal{N}(0, \Sigma_1(\vartheta_0)/\Sigma_0(\vartheta_0)) \quad \text{as } T \to \infty,$$

where $\Sigma_0(\vartheta_0)$ is defined by (4.17) and $\Sigma_1(\vartheta_0)$, by (4.18).

Theorem 76 follows from Theorem 73, Lemmas 74, 75 and Slutsky's argument, see [106]. For further reference of the central limit theorem we mention Brillinger [20], Rosenblatt [107], Hannan [49], Bradley [15] and Miller [85].

4.4 Asymptotic variance in the case of linear processes

A consequence of (4.19) is that for a linear invertible process

$$Y_t = \left(\sum_{k=0}^{\infty} a_k z^{-k}\right) e_t = a\left(z^{-1}\right) e_t, \qquad a_0 = 1,$$

where z^{-1} is the back shift operator and the noise process e_t is i.i.d. with zero expectation and a finite sixth order moment. Note that the distribution of e_t is assumed not to depend of the unknown parameter ϑ. The asymptotic variance of $R_T(\vartheta_0)$ can be calculated by the skewness and the kurtosis of e_t. Put $\Lambda_1 = (0, 1/2)$, Λ_2 is the entire triangle $\triangle$, moreover

$$\eta = \int_0^{1/2} \left(\frac{\partial}{\partial \vartheta} \log S_2(\omega, \vartheta_0)\right)^2 d\omega,$$

the coefficients of skewness ς and the kurtosis κ of e_t are defined by

$$\varsigma = \frac{\operatorname{Cum}(e_t, e_t, e_t)}{[\operatorname{var}(e_t)]^{3/2}},$$

and

$$\kappa = \frac{\operatorname{Cum}(e_t, e_t, e_t, e_t)}{[\operatorname{var}(e_t)]^2},$$

respectively. Note that κ is different from another used definition of coefficient of kurtosis,namely from $\mathsf{E}(e_t - \mathsf{E}e_t)^4 / [\operatorname{var}(e_t)]^2$ because $\operatorname{Cum}(e_t, e_t, e_t, e_t) = \mathsf{E}(e_t - \mathsf{E}e_t)^4 - 3\sigma_e^4 \neq \mathsf{E}(e_t - \mathsf{E}e_t)^4$, see formula (1.8), p. 13. Note that for an invertible process

$$\begin{aligned} \int_0^{1/2} \left(\frac{\partial}{\partial \vartheta} \log S_2(\omega, \vartheta_0) \right)^2 d\omega &= \frac{1}{2} \int_0^1 \left(\frac{\partial}{\partial \vartheta} \log S_2(\omega, \vartheta_0) \right)^2 d\omega \\ &= \int_0^1 \left| \frac{\frac{\partial}{\partial \vartheta} a\left(z^{-1}\right)}{a\left(z^{-1}\right)} \right|^2_{\vartheta = \vartheta_0} d\omega, \end{aligned} \tag{4.20}$$

because according to the series expansion of $\frac{\partial}{\partial \vartheta} a\left(z^{-1}\right) / a\left(z^{-1}\right)$

$$\int_0^1 \frac{\frac{\partial}{\partial \vartheta} a\left(z^{-1}\right)}{a\left(z^{-1}\right)} d\omega,$$

is real, therefore

$$\begin{aligned} 0 &= \int_0^{1/2} \frac{\partial}{\partial \vartheta} \log S_2(\omega, \vartheta_0) d\omega \\ &= \int_0^1 \frac{\frac{\partial}{\partial \vartheta} a\left(z^{-1}\right)}{a\left(z^{-1}\right)} d\omega \\ &= \int_0^1 \left(\frac{\frac{\partial}{\partial \vartheta} a\left(z^{-1}\right)}{a\left(z^{-1}\right)} \right)^2 d\omega. \end{aligned}$$

It is known that for a linear process the spectral density of nth order is given by

$$S_n(\omega_1, \omega_2, \ldots, \omega_n) = \operatorname{Cum}\Big(\underbrace{e_t, e_t, \ldots, e_t, e_t}_{n} \Big) \prod_1^n a\left(z_k^{-1}\right),$$

where $z_k^{-1} = \exp(-2\pi i \omega_k)$ and again $\sum_1^n \omega_k = 0$. The asymptotic variance of the estimator ϑ_T is now given by $\Sigma_{R1}(\vartheta_0)/\Sigma_0^2(\vartheta_0)$ where

$$\Sigma_{R1}(\vartheta_0) = \left(4p^2 + \left[\frac{q^2}{4}\kappa + \frac{q^2}{2} + pq \right] \varsigma^2 \right) \eta,$$

and the denominator of the variance of ϑ_T is

$$\Sigma_0^{1/2}(\vartheta_0) = \left(2p + \frac{q}{2}\varsigma^2\right)\eta.$$

See the Appendix E, page 241, for the proof. We recall that $p = p_1/\beta_1$ and $q = q_1/\beta_2$, moreover in this particular case $\beta_1 = 1/2$ and $\beta_2 = 1/12$, respectively. The asymptotic variance of the estimator ϑ_T for a linear process is

$$\begin{aligned} T\operatorname{var}(\vartheta_T - \vartheta_0) &\approx \frac{\Sigma_{R1}(\vartheta_0)}{\Sigma_0(\vartheta_0)} \\ &\approx \frac{4p_1^2 + \left[9q_1^2\kappa + 18q_1^2 + 6p_1q_1\right]\varsigma^2}{\left(2p_1 + 3q_1\varsigma^2\right)^2\eta}, \end{aligned} \tag{4.21}$$

where $p_1 \in (0,1)$, $q_1 = 1 - p_1$.

4.4.1 A worked example and simulations

$AR(1)$ process with Gamma distributed residuals

Suppose that the observations are divided into N disjoint stretches with length T, i.e., the overall length of the observation is NT. The variance of the estimate ϑ_T at ϑ_0 will be close to

$$\begin{aligned} &\operatorname{var}(\vartheta_T - \vartheta_0) \\ &\approx \frac{1}{NT}\frac{4p_1^2\|W_1\|^2 + \left[9W_{23}q_1^2\kappa + 18\|W_2\|^2 q_1^2 + 6p_1q_1W_{12}(\rho)\right]\varsigma^2}{\left(2p_1 + 3q_1\varsigma^2\right)^2\eta}. \end{aligned}$$

Consider the AR(1) process

$$Y_t + \vartheta Y_{t-1} = e_t,$$

where e_t i.i.d. with gamma distribution $\Gamma(\alpha,\lambda)$. The characteristic function of $\Gamma(\alpha,\lambda)$ is

$$\left(1 - \frac{it}{\alpha}\right)^{-\lambda},$$

therefore the nth order cumulant of e_t is

$$\operatorname{Cum}\Big(\underbrace{e_t, e_t, \ldots, e_t, e_t}_{n}\Big) = \frac{(n-1)!\lambda}{\alpha^n},$$

$(0! = 1)$ see Example 6, page 10. The skewness ς and the kurtosis κ of e_t follows from this:

$$\varsigma = 2\lambda^{-1/2},\ \kappa = 6\lambda^{-1}.$$

In particular if e_t has χ^2 distribution with k degree of freedom then

$$\varsigma = 2\sqrt{2k}/k, \ \kappa = 12/k.$$

η calculated by (4.20) is

$$\eta = \frac{1}{1-\vartheta^2}.$$

The default window of MATLAB (command SPECTRUM) for the estimation of the spectrum is the Hanning window and the estimation is done by Welch's method. We consider here the Tukey-Hanning window as W_1. It is defined by

$$W_1\left(\omega\right) = 1 - \cos\left(2\pi\left(\omega + 1/2\right)\right),$$

with $\left\|W_1\right\|^2 = 1.5$.

For the estimation of the bispectrum we use the optimal window of *Subba Rao-Gabr*, see Subba Rào and Gabr[121], i.e.,

$$W_2(\omega_1, \omega_2) = \begin{cases} \frac{4\sqrt{3}}{\pi}[1 - 4(\omega_1^2 + \omega_2^2 + \omega_1\omega_2)], & \omega_1^2 + \omega_2^2 + \omega_1\omega_2 < 1/4, \\ 0, & otherwise. \end{cases}$$

For this window

$$\begin{aligned} \left\|W_2\right\|^2 &= \iint_{-\infty}^{\infty} W_2^2(\omega_1, \omega_2) d\omega_1 d\omega_2 = 1.4628, \\ W_{23} &= \iiint_{-\infty}^{+\infty} W_2(\lambda_1, \lambda_2) W_2(\lambda_1 + \lambda_2, \omega) d\lambda_1 d\lambda_2 d\omega = 1.1157, \\ W_{12}\left(1\right) &= \iint_{-\infty}^{+\infty} W_1\left(\omega_1\right) W_2(\omega_1, \omega_2) d\omega_1 d\omega_2 = 1.2565, \end{aligned}$$

where W_1 is the Tukey-Hanning window and W_{12} is considered at $\varrho = 1$ because the bandwidths B_{1T} and B_{2T} are chosen to be equal during the simulation. The variance of the estimate ϑ_T at ϑ_0 under the above conditions is

$$\begin{aligned} &NT \operatorname{var}\left(\vartheta_T - \vartheta_0\right) \\ &\approx \left(1 - \vartheta_0^2\right) \frac{\left\|W_1\right\|^2 p_1^2 \lambda^2 + \left[54W_{23} + 18\left\|W_2\right\|^2 \lambda\right] q_1^2 + 6W_{12}\left(1\right) \lambda p_1 q_1}{\left(p_1\lambda + 6q_1\right)^2}. \end{aligned}$$

It is interesting to consider the surface defined by the expression

$$\begin{aligned} &f\left(p_1, \lambda\right) = \left(1 - \vartheta_0^2\right) \\ &\times \frac{\left\|W_1\right\|^2 p_1^2 \lambda^2 + \left[54W_{23} + 18\left\|W_2\right\|^2 \lambda\right] q_1^2 + 6W_{12}\left(1\right) \lambda p_1 q_1}{\left(p_1\lambda + 6q_1\right)^2}. \end{aligned}$$

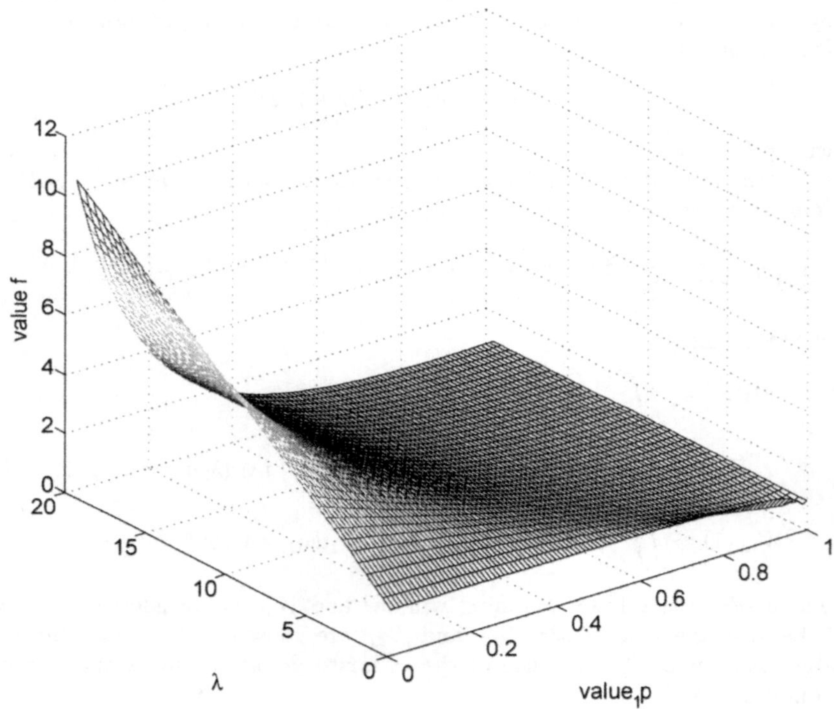

FIGURE 4.1. Surface for the variance according to expression $f(p_1, \lambda)$

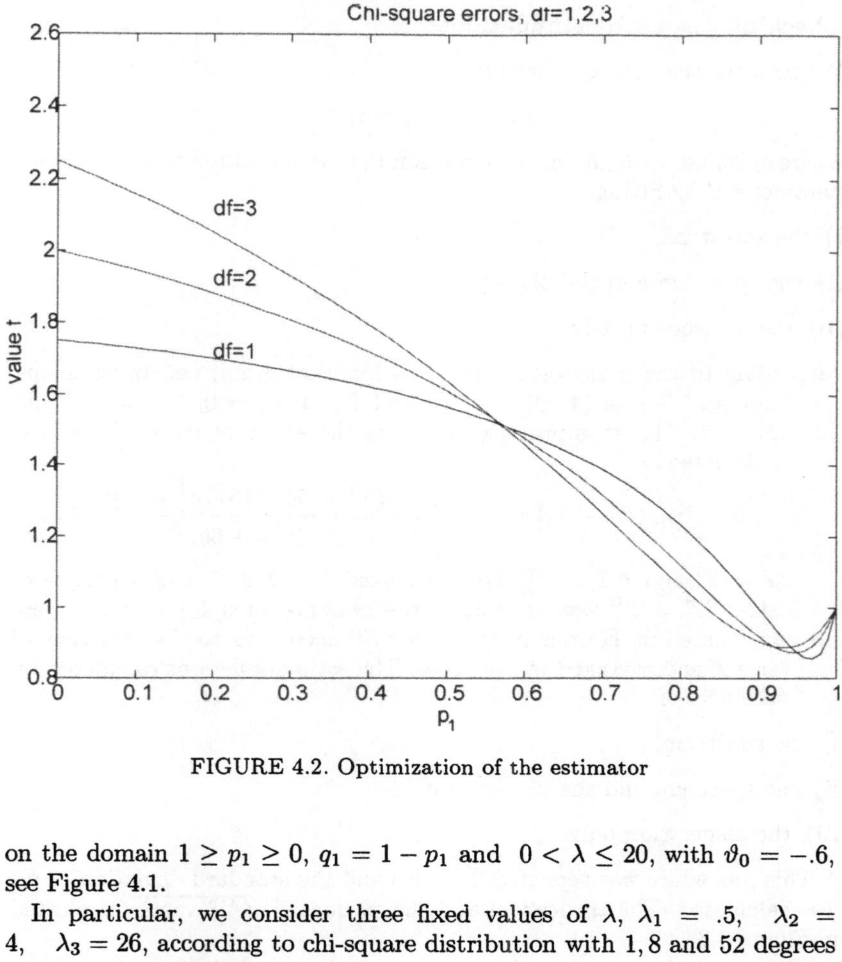

FIGURE 4.2. Optimization of the estimator

on the domain $1 \geq p_1 \geq 0$, $q_1 = 1 - p_1$ and $0 < \lambda \leq 20$, with $\vartheta_0 = -.6$, see Figure 4.1.

In particular, we consider three fixed values of λ, $\lambda_1 = .5$, $\lambda_2 = 4$, $\lambda_3 = 26$, according to chi-square distribution with 1, 8 and 52 degrees of freedom, see Figure 4.2.

It may be seen that there is not a great difference at $p_1 = q_1 = 1/2$, i.e., when the estimation is based on the spectrum and the bispectrum with equal weights. The following table shows the asymptotic variances at $p_1 = 0, p_1 = 1/2$ and $p_1 = 1$.

	$p_1 = 1$	$p_1 = 1/2$	$p_1 = 0$
$df = 1$	0.9600	1.1748	1.3051
$df = 8$	0.9600	1.4062	2.9435
$df = 52$	0.9600	1.2218	13.2416

Checking results by simulations

We consider the autoregressive process

$$Y_t + \vartheta Y_{t-1} = e_t,$$

where e_t is i.i.d. with gamma distribution $\Gamma(1, \lambda)$ and estimate the unknown parameter ϑ by fitting

i) the spectrum,

ii) the spectrum and the bispectrum,

iii) the bispectrum only

according to the periodogram and the biperiodogram, i.e., by changing S_{2T} and S_{3T} in (4.10) into I_{2T} and I_{3T} at $p_1 = 0$, $p_1 = 1/2$ and $p_1 = 1$. The asymptotic variance of the estimate ϑ_T at ϑ_0 is now calculated as

$$NT \operatorname{var}(\vartheta_T - \vartheta_0) \approx \left(1 - \vartheta_0^2\right) \frac{p_1^2\lambda^2 + [54 + 18\lambda]\, q_1^2 + 6\lambda p_1 q_1}{(p_1\lambda + 6q_1)^2}.$$

$T = 2^6$ and $NT = 2^{10}$ For each fixed $\lambda = 1, 2, \dots, 42$ a sample of size $NT = 2^{10}$ was generated, the periodograms I_{2T} and I_{3T} were calculated at Fourier frequencies k/T according to the stretches of size T and averaged by stretches. The estimate ϑ_T was calculated by the help of

i) the spectrum,

ii) the spectrum and the bispectrum,

iii) the bispectrum only

This procedure was repeated 2^7 times and the standard deviation of ϑ_T was calculated. The theoretical and the estimated $\sqrt{NT \operatorname{var} \vartheta_T}$ is plotted in Figure 4.3.

The calculation of the estimator ϑ_T based on the bispectrum and the biperiodogram only is more and more difficult because, as λ is growing, the gamma distribution is getting close to the Gaussian. The bias of the estimator ϑ_T takes the smallest value when p_1 is zero and the largest one when $p_1 = 1$, see Figure 4.4. The average over λ during the simulation was the following:.

	$p_1 = 1$	$p_1 = 1/2$	$p_1 = 0$
averaged bias	0.0115	0.0085	0.0041

Note that the variance of e_t, i.e., λ, can be estimated easily by

$$\lambda_T = \widehat{\sigma}_e^2 = \sum_t (Y_t + \vartheta_T Y_{t-1})^2 .$$

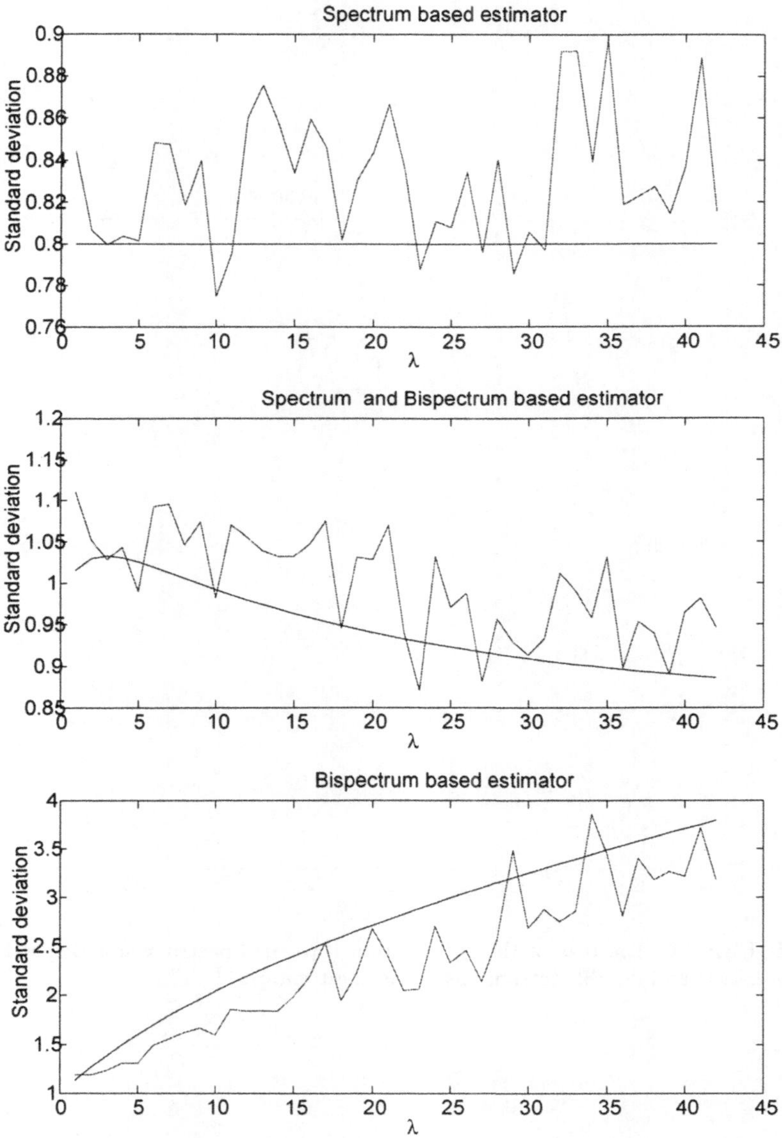

FIGURE 4.3. Theoretical and estimated standard deviation of the estimator ϑ_T

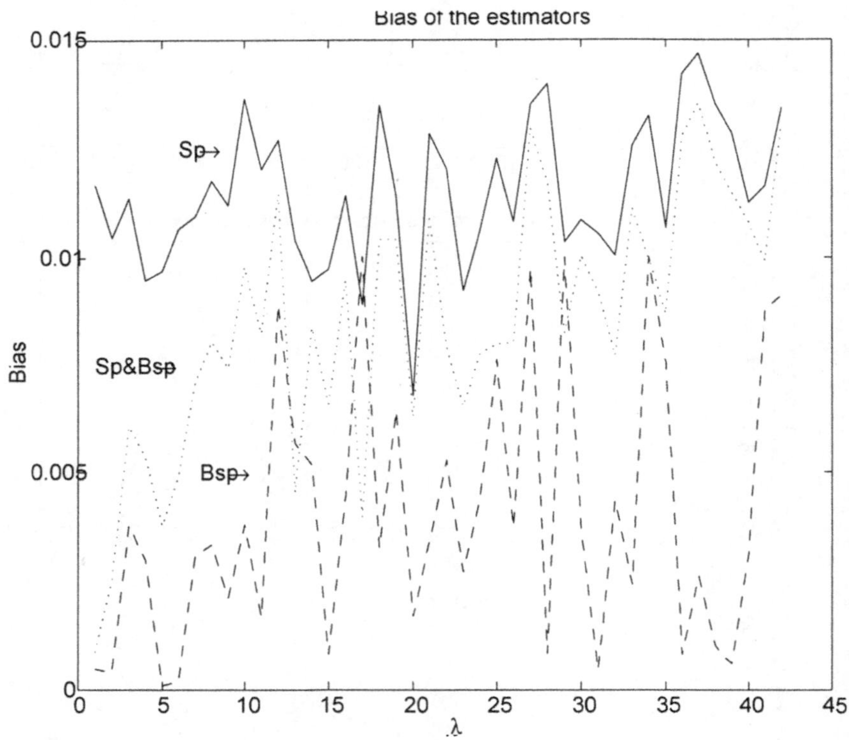

FIGURE 4.4. The bias of the i) Spectrum (Sp), ii) Spectrum and Bispectrum (Sp&Bsp) and iii) Bispectrum (Bsp) based estimators

5

Linearity Test

A stationary Gaussian time series has the following properties: (i) the residual series of the moving average representation is a sequence of independent (and Gaussian) series and (ii) the best predictor, i.e., the conditional expectation of the observation according to the past is linear. Both properties lead to the notion of the linearity of a time series. We follow Hannan's [52] definition that the model is linear if the linear predictor is optimal. This assumption seems to be the minimum requirement. This means that the residual sequence e_t fulfils the following conditions:

$$\mathsf{E}\left(e_t \mid \mathcal{F}_{t-1}\right) = 0, \quad \mathsf{E}\left(e_t^2 \mid \mathcal{F}_{t-1}\right) = \sigma^2, \tag{5.1}$$

where $\mathcal{F}_t$ is the Borel field generated by e_s, $s \leq t$. Subba Rao and Gabr[121] and Hinich[58] assumed that the residual sequence $\{e_t\}$ is independent, which is a stronger assumption than the assumption we have made here. Keenan[68] suggested a time domain test for second order nonlinearity which is analogous to Tukey's one degree of freedom test for nonadditivity. Hong-zhi and Bing[60] used a Kolmogorov-Smirnov type test for linearity. It is based on the best linear prediction error. Linearity test is discussed in [28], [9], [10], [39] and [123].

In view of Hannan's definition of linearity, testing the linearity is equivalent to checking the optimality of the linear predictor. An alternative is that the optimal predictor is of a quadratic type. The test is based on an additive property of the bispectrum of the residual process which characterizes our hypothesis. The distribution of the test statistics is determined from the asymptotic distribution of the estimated bispectrum. The smoothed biperiodogram is used for the bispectrum estimation. The additivity property of the bispectrum requires a particular set of frequencies and these will be used in the construction of the test statistics. We include two examples, one of them is a nonlinear model for which the linear predictor is optimal. The other one is an uncorrelated series for which the linear predictor and the quadratic predictor are explicitly calculated. The procedure is illustrated with simulated and real data. Some papers of this subject include: [137], [136] and [135].

5.1 Quadratic predictor

Finding the best predictor is a long standing problem. The prediction of a stationary time series usually means the following: Take the Hilbert space generated by the elements of the series as vectors in L_2, then find the orthogonal projection of an element in the future to the subspace generated by the past of the series. The precise mathematical formulation and the solution of this problem is given by Kolmogorov [70] in his pioneering work on stationary processes. This predictor is called linear because it is a linear combination of the values from the past. Wiener [145] gave an independent solution for finding the best linear predictor.

It is well known fact that the best least squares predictor is the conditional expectation with respect to the past of the process. Wiener himself, collaborating with Kallianpur and Masani, see [82] and Kallianpur[67], gave a method for finding the best predictor and showed that under certain circumstances the Hilbert space spanned by all the polynomials of the past is the same as the Hilbert space generated by the random variables with second moments, measurable with respect to the σ- algebra generated by the past. Interesting particular cases were considered by Nelsson and Van Ness[89] from computational aspects of a quadratic predictor, by Granger and Andersen[42] for transformed series and by Hida and Kallianpur[56] for squared Gaussian Markov processes in continuous time.

We give a brief survey of the construction of the quadratic predictor as a projection of the value of a time series at time $t+1$ to a space generated by all the linear and quadratic functions of its past. Suppose we are given a zero mean third order stationary time series Y_t. We assume that the fourth order moments exist. The spectrum and bispectrum are given by

$$S_Y(\lambda) = \sum_{k=-\infty}^{\infty} c_{YY}(k) z^{-k},$$

$$B_Y(\lambda_1, \lambda_2) = \sum_{k,j=-\infty}^{\infty} c_{YYY}(k,j) z_1^{-k} z_2^{-j},$$

respectively where $z = e^{i2\pi\lambda}$, $z_1 = e^{i2\pi\lambda_1}$, $z_2 = e^{i2\pi\lambda_2}$, $c_{YY}(k) = \mathsf{E}Y_0Y_k$ and $c_{YYY}(k,j) = \mathsf{E}Y_0Y_kY_j$. The lag one quadratic predictor considered is of the form

$$\hat{Y}_Q(t+1) = \sum_{k=0}^{\infty} a_k Y_{t-k} + \sum_{j,k=0}^{\infty} a_{jk} Y_{t-j} Y_{t-k}, \tag{5.2}$$

and the coefficients $a_k, a_{j,k}$ are chosen such that the mean square error

$$\mathsf{E}\left|Y_{t+1} - \hat{Y}_Q(t+1)\right|^2,$$

is minimum.

The construction of the linear predictor $\hat{Y}_L(t)$ is well known, see Priestley [100], and is based on the spectrum S_Y of the process. One needs only the Szegő assumption, i.e.,

$$\int_0^1 log S_Y(\lambda) d\lambda > -\infty, \tag{5.3}$$

to be satisfied. We further assume that $\log(S_Y(z))$ is analytic in the annulus $\rho < |z| < 1/\rho, (\rho < 1)$. Then the spectral density function has the form

$$S_Y(\omega) = Const. |G(e^{-i2\pi\omega})|^2,$$
$$G(z) = \sum_{u=0}^{\infty} g_u z^u, \ g_0 = 1,$$

and the one step linear predictor of Y_t given $(Y_{t-1}, Y_{t-2}, \dots)$ is

$$\hat{Y}_L(t) = \left(1 - \frac{1}{G(B)}\right) Y_t,$$

where B is the backwardshift operator. The error of the predictor is

$$e_t = Y_t - \hat{Y}_L(t) = G^{-1}(B) Y_t,$$

e_t will also be called an residual process. Note that in this case Y_t has moving average representation

$$Y_t = \sum_{k=0}^{\infty} g_k e_{t-k}.$$

Here $\{e_t\}$ is an uncorrelated sequence, therefore it will not necessarily satisfy assumption (5.1). Let us suppose now that the best linear predictor has already been constructed, we are looking for the best quadratic predictor and would like to check whether it is really better than the linear one. The bispectrum of the residual process e_t will answer the question.

We use Masani and Wiener's[82] definition of the *spectrum of a distribution*, which is often used in measure theory.

By the spectrum of a distribution function F on Euclidean q-space we mean the set of points $\mathbf{x} = (x_1, x_2, \ldots, x_q)$ such that $\mu(I) > 0$ for every open interval I containing $\mathbf{x}$, where μ is the Lebesgue-Stieltjes measure generated by F.

Theorem 77 *Let the time series Y_t be third order stationary with spectral density S_Y. Moreover suppose that the fourth order moments of Y_t exist, and let S_Y satisfy the Szegő condition (5.3) and let each finite dimensional distribution of Y_t have a positive spectrum. Then the necessary and sufficient condition for the equivalence of the linear predictor $\hat{Y}_L(t)$ and the quadratic predictor $\hat{Y}_Q(t)$ is that the bispectrum $B_e(z_1, z_2)$ of the residual process e_t have the form*

$$B_e(z_1, z_2) = H(z_1) + H(z_2) + H(z_1^{-1} z_2^{-1}), \tag{5.4}$$

where

$$H(z) = \sum_{k=0}^{\infty} c_k z^k, \quad c_k = \mathsf{E} e_0 e_k e_k, \tag{5.5}$$

and

$$z = e^{i2\pi\lambda}, \; z_1 = e^{i2\pi\lambda_1}, \; z_2 = e^{i2\pi\lambda_2}.$$

Proof. To prove that the members of the vector system

$$S = \{Y_{t-j}, \; Y_{t-k}Y_{t-l}, \; j = 0, 1, 2..., k, l = 0, 1, 2, ...\},$$

are linearly independent in the Hilbert space H_S generated by S we refer to the paper by Masani and Wiener[82]. Consider any polynomial P from S, i.e.,

$$P(Y_{t-k_1}, ..., Y_{t-k_Q}) = \sum g_r Y_{t-k_r} + \sum g_{r,q} Y_{t-k_r} Y_{t-k_q},$$

then the norm square

$$\|P\|^2 = \int_{R^Q} |P(y_1, ... y_Q)|^2 \, dF_{k_1, ... k_Q}(y_1, ..., y_Q),$$

cannot be zero unless $g_r = g_{r,q} = 0$ because of the positive spectrum of the distribution $F_{k_1, ... k_Q}$. So the subspaces H_{S_1} and H_{S_2} in H_S are disjoint where H_{S_1} and H_{S_2} are the Hilbert spaces generated by $\{Y_{t-k}, k = 0, 1, 2, ...\}$ and $\{Y_{t-j}Y_{t-k}, j, k = 0, 1, 2, ...\}$, respectively.

Now the necessary and sufficient assumption for $\hat{Y}_L(t) = \hat{Y}_Q(t)$ is that the error of the orthogonal projection of Y_t to the Hilbert space H_{S_1} should be orthogonal to H_{S_2} as well, i.e.,

$$\mathsf{E}\left(y_t - \hat{Y}_L(t)\right) Y_{t-j}Y_{t-k} = \mathsf{E}e_tY_{t-j}Y_{t-k} = 0, \ k, j = 1, 2, \ldots \tag{5.6}$$

Under the assumptions of the theorem there exists a linear relationship between the process Y_t and its residual e_t, therefore (5.6) is equivalent to

$$c_{ee}(-j, -k) = \mathsf{E}e_te_{t-j}e_{t-k} = 0, \ j, k = 1, 2, \ldots .$$

In view of the symmetry of the third order moments

$$\begin{aligned} c_{ee}(-j, -k) &= c_{ee}(-k, -j) = c_{ee}(j, j-k) = c_{ee}(j-k, j) \\ &= c_{ee}(k, k-j) = c_{ee}(k-j, k), \end{aligned}$$

the only possibility for $c_{ee}(j, k)$ not to be zero is the case when either $k = j$, or $k = 0$, $j > 0$, or $j = 0$, $k > 0$. Therefore

$$B_e(z_1, z_2) = H(z_1) + H(z_2) + H(z_1^{-1}z_2^{-1}),$$

where

$$H(z) = \sum_{k=0}^{\infty} c_k z^k; \quad c_k = c_{ee}(k, k).$$

■

We note here that the assumption (5.4) is automatically satisfied when the bispectrum of the process Y_t is zero for all frequencies because the bispectrum of the linearly filtered process e_t is given as a product of the bispectrum of the process Y_t and the filter. This implies that the bispectrum of the residual process e_t is also zero. Therefore it may happen that although the residual process is not independent, the best predictor is linear, see an example given below.

To decide whether the bispectrum of the residual process e_t is of the form (5.4) the following theorem is useful. The theorem gives necessary and sufficient conditions for an arbitrary bispectrum to have the additive representation (5.4). The form (5.5) of function $H(z)$ will be guaranteed by the Szegő assumption (5.3). A similar statement has been given by Sakaguchi [110].

Theorem 78 *Let $B(\lambda_1, \lambda_2)$ be a bispectrum which is differentiable with respect to λ_1. Then the following statements are equivalent.*

1. *For any (α, β, γ) the bispectrum $B(\lambda_1, \lambda_2)$ satisfies the relation*

$$\begin{aligned} &B(\alpha, \beta) + B(\gamma, 0) + B(-\alpha + \gamma, -\beta - \gamma) \\ &\quad = B(\beta, \gamma) + B(0, -\alpha - \beta) + B(-\alpha + \gamma, -\gamma), \end{aligned} \tag{5.7}$$

2. *there exists a function* $H(\lambda)$ *such that*

$$B(\lambda_1, \lambda_2) = H(\lambda_1) + H(\lambda_2) + H(-\lambda_1 - \lambda_2), \tag{5.8}$$

where $H(\lambda)$ *is a complex valued function on* $\mathbb{R}$ *and satisfies*

$$H(-\lambda) = H^*(\lambda), \; H(\lambda + k) = H(\lambda), \; k = \pm 1, \pm 2, \dots , \tag{5.9}$$

for any $\lambda \in R$, *where* H^* *denotes the complex conjugate of* H.

See the Appendix F, page 245, for the proof. Note here that it might happen that equation (5.7) is satisfied yet the linear predictor is not optimal. The optimal predictor $\tilde{Y}_t = E(Y_t \mid Y_{t-1}, Y_{t-2}, \dots)$ in general is a functional of the past $(Y_{t-1}, Y_{t-2}, \dots)$, and Theorem 77 is useful only to check whether the quadratic terms are missing from the polynomial expansion of $\tilde{Y}_t$ or not.

The best linear predictor for an uncorrelated series is its mean. If the series is not Gaussian then, even for an uncorrelated series, it is reasonable to check the quadratic predictor. This question is considered in the following example.

5.1.1 *Quadratic predictor for a simple bilinear model*

One of the simplest bilinear models is given by

$$X_t = dX_{t-2}\mathrm{w}_{t-1} + \mathrm{w}_t, \tag{5.10}$$

where w_t is a Gaussian white noise series with $\mathsf{E}\mathrm{w}_t = 0$, and $\mathsf{E}\mathrm{w}_t^2 = \sigma^2$. It is easy to see that X_t is uncorrelated. The best quadratic predictor for X_t is given by

$$X_Q(t+1) = a_{1,2} X_t X_{t-1},$$

where the coefficient $a_{1,2}$ is defined below.

It can be shown that X_t has the representation

$$X_t = \sum_{k=0}^{\infty} d^k \mathrm{w}_{t-2k} \prod_{j=1}^{k} \mathrm{w}_{t-2j+1}, \tag{5.11}$$

in L_2 sense. It is direct to obtain moments of X_t from the representation (5.11), and we can show that the necessary and sufficient condition of stationarity is that $d^2\sigma^2 < 1$. The moments and cumulants are as follows:

$$\begin{aligned}
m_1 &= \mathsf{E}X_t = 0, \\
m_2(k) &= \mathsf{E}X_t X_{t+k} = \delta_{k=0}\sigma_X^2, \\
m_3(k,l) &= \mathsf{E}X_t X_{t+k} X_{t+l} = \delta_{k=1,l=2} m_3(1,2), \\
m_4(0,0,0) &= \mathsf{E}X_t^4 , \\
m_4(0,1,1) &= \mathsf{E}X_t^2 X_{t-1}^2 ,
\end{aligned}$$

where

$$\sigma_X^2 = \frac{\sigma^2}{1-d^2\sigma^2},$$
$$m_3(1,2) = \frac{d\sigma^4}{1-d^2\sigma^2},$$
$$m_4(0,0,0) = \frac{1+d^2\sigma^2}{(1-d^2\sigma^2)(1-3d^4\sigma^4)},$$
$$m_4(0,1,1) = \frac{\sigma^4(1+2d^2\sigma^2)}{(1-d^2\sigma^2)^2}.$$

The existence of the fourth order moment requires $3d^4\sigma^4 < 1$. The spectrum and the bispectrum of the process X_t are given by

$$S(z) = \frac{\sigma^2}{1-d^2\sigma^2},$$
$$B(z_1,z_2) = \frac{d\sigma^4\left(z_1^{-1}z_2 + z_1z_2^{-1} + z_1^{-2}z_2^{-1} + z_1^{-1}z_2^{-2} + z_1^2z_2 + z_1z_2^2\right)}{1-d^2\sigma^2}.$$

Construction of the quadratic predictor

Consider the Hilbert space $\mathcal{H}_Q$ generated by the variables $X_{t-k}, k \geq 1$, and $X_{t-k}X_{t-l}, \quad k,l \geq 1$ and the predictor given by the orthogonal projection of X_t onto $\mathcal{H}_Q$. The only element of $\mathcal{H}_Q$ which is not orthogonal to X_t is $X_{t-1}X_{t-2}$. Therefore the best quadratic predictor is

$$X_Q(t) = a_{1,2}X_{t-1}X_{t-2},$$

where

$$a_{1,2} = d\frac{1-d^2\sigma^2}{1+2d^2\sigma^2}.$$

The variance of the best linear predictor, which is σ_X^2, is reduced to

$$\mathsf{E}\left|X_t - X_Q(t)\right|^2 = \sigma_X^2 - \frac{d^2\sigma^4}{1+2d^2\sigma^2}.$$

The maximum of the reduction relatively to σ_X^2 is reached at $d^2\sigma^2 = \left(\sqrt{3}-1\right)/2$, i.e., when the process is normalized the maximal reduction is 0.134. When we have considered different models and simulations, the quadratic predictor compared to the linear we frequently reduced the variance by a small percentage.

5.2 The test statistics

Our object here is to construct a statistic for testing whether the process is linear or not. Theorem 77 gives a necessary condition for the equality of the best linear and quadratic predictors. We must check whether the bispectrum of the residual process has the additive form (5.7). The following hypotheses will be considered:

H_0: $\mathsf{E}\left(\hat{Y}_L(t) - \hat{Y}_Q(t)\right)^2 = 0$;

H_1: $\mathsf{E}\left(\hat{Y}_L(t) - \hat{Y}_Q(t)\right)^2 > 0$.

Note that if the process is linear, i.e., the residual sequence e_t satisfies the assumption (5.1), then H_0 follows and if H_1 is true, the process cannot be linear. First we fit a linear model for Y_t and consider the residual series $\hat{e}_t$ obtained from the fitted linear model. Let us denote the estimate of the bispectrum of $\hat{e}_t$ by $B_e^T(\omega_1, \omega_2)$, where the length of our time series is T. The smoothed biperiodogram will be used for the estimation of the bispectrum of $\hat{e}_t$. Recall that the biperiodogram of $\hat{e}_t$ is

$$I_{eee}^T(\omega_1, \omega_2) = \frac{1}{T} d_T(\omega_1) d_T(\omega_2) d_T(-\omega_1 - \omega_2),$$

where

$$d_T(\omega) = \sum_{t=0}^{T-1} \hat{e}_t e^{-i2\pi\omega t},$$

is the Fourier transform of the residual series $\hat{e}_t$.

Let $W(\omega_1, \omega_2)$ be a nonnegative bounded continuous weight function such that

$$\iint_{-\infty}^{+\infty} W(\omega_1, \omega_2) d\omega_1 d\omega_2 = 1.$$

We choose $W(\omega_1, \omega_2)$ to be 0 for $|\ \omega_j\ | > 1/2, \quad j = 1, 2$ and also $W(\omega_1, \omega_2)$ to satisfy the symmetry conditions

$$\begin{aligned} W(\omega_1, \omega_2) &= W(\omega_2, \omega_1), \\ &= W(\omega_1, \omega_3), \end{aligned}$$

where $\omega_3 = -\omega_1 - \omega_2 \bmod (1)$ and therefore $\omega_3 \in [-1/2, 1/2]$. Now the smoothed biperiodogram is defined by

$$B_e^T(\lambda_1, \lambda_2) = \frac{1}{T^2 b_T^2} \sum_{s_1, s_2 = 1}^{T-1} W\left(b_T^{-1}\left(\lambda_1 - \frac{s_1}{T}\right), b_T^{-1}\left(\lambda_2 - \frac{s_2}{T}\right)\right) I_{eee}^T\left(\frac{s_1}{T}, \frac{s_2}{T}\right),$$

where b_T denotes a sequence of scale parameters such that $b_T > 0$, $b_T \to 0$, $b_T^2 T \to \infty$, as $T \to \infty$. The estimate of B_e^T actually involves a weighting of $(2Tb_T + 1)^2$ biperiodogram ordinates in the neighborhood of (λ_1, λ_2). If we assume that e_t is strictly stationary and

$$\sum_{s=-\infty}^{\infty} (1 + |s|) |\mathrm{Cum}(e_s, e_0)| < \infty,$$

$$\sum_{s_1, s_2=-\infty}^{\infty} (1 + |s_j|) |\mathrm{Cum}(e_{s_1}, e_{s_2}, e_0)| < \infty, \quad j = 1, 2,$$

then, see Brillinger[19], Rosenblatt [107], $\{B_e^T\}$ is consistent and the entries are asymptotically jointly normal with covariances

$$\begin{aligned}\lim_{T\to\infty} Tb_T^2 \mathrm{Cov}\left(B_e^T(\lambda_1, \lambda_2), B_e^T(\mu_1, \mu_2)\right) \\ &= \sigma_e^6 W_{01}^2 (\delta_{\mu_1=0} + \delta_{\mu_2=0} + \delta_{\mu_3=0})(\delta_{\lambda_1=0} + \delta_{\lambda_2=0} + \delta_{\lambda_3=0}) \\ &+ \sigma_e^6 W_2^2 \left[\delta_{\lambda_1=\mu_1}(\delta_{\lambda_2=\mu_2} + \delta_{\lambda_2=\mu_3})\right. \\ + \delta_{\lambda_1=\mu_2}(\delta_{\lambda_2=\mu_1} + \delta_{\lambda_2=\mu_3}) &+ \left.\delta_{\lambda_1=\mu_3}(\delta_{\lambda_2=\mu_1} + \delta_{\lambda_2=\mu_2})\right], \qquad (5.12)\end{aligned}$$

where as usual $\mu_3 = -\mu_1 - \mu_2$, $\lambda_3 = -\lambda_1 - \lambda_2$, δ is the Kronecker symbol and "=" of the frequencies is mod (1). Also

$$W_2^2 = \iint_{-\infty}^{\infty} W^2(\omega_1, \omega_2) d\omega_1 d\omega_2, \quad W_{01} = \int_{-\infty}^{\infty} W(0, \omega) d\omega.$$

The expression (5.12) is valid in general for stationary processes and depends on the spectral density. Now because e_t is uncorrelated with a constant spectrum, the asymptotic covariance does not depend on the frequencies. It follows from (5.12) that in different points the bispectrum estimates are asymptotically independent inside the triangle with vertices $(0, 0), (0, 1/2), (1/3, 1/3)$, which will be referred to in the sequel as Δ. On the boundary of Δ they are correlated, see (5.12). If $\lambda_1 \neq \lambda_2$, $\lambda_1\lambda_2 \neq 0$ and $\lambda_1 \neq -2\lambda_2$, the variance of the bispectrum estimate is

$$\lim_{T\to\infty} Tb_T^2 \mathrm{Var} B_e^T(\lambda_1, \lambda_2) = \sigma_e^6 W_2^2, \qquad (5.13)$$

which implies

$$\lim_{T\to\infty} Tb_T^2 \mathrm{Var}\, \mathrm{Re}\, B_e^T(\lambda_1, \lambda_2) = \sigma_e^6 W_2^2/2,$$

and

$$\lim_{T\to\infty} Tb_T^2 \mathrm{Var}\, \mathrm{Im}\, B_e^T(\lambda_1, \lambda_2) = \sigma_e^6 W_2^2/2.$$

If $0 < \omega_1 < 1/2$, then

$$\lim_{T\to\infty} Tb_T^2 \mathrm{Var} B_e^T(\omega_1, 0) = \sigma_e^6 \left(W_2^2 + W_{01}^2\right). \qquad (5.14)$$

It is well known that the bispectrum satisfies the symmetry conditions, see Section 1.5.3, page 26 and it is periodic

$$B(\omega_1, \omega_2) = B(\omega_1 + k, \omega_2 + j), \quad k, j = 1, 2, \ldots. \tag{5.15}$$

Therefore every frequency in the plane can be transformed into the triangle $\triangle$ in view of the symmetry and (5.15) without changing the value of the bispectrum except for complex conjugation. The transformations $\mathcal{T}_k$, $k = 1, 2, 3, 4$, (see (4.11a), page 161) map by mod(1) the whole square $[0,1]^2$ into the triangle $\triangle$. Now, let us choose and fix a triplet (α, β, γ) such that the map by $\mathcal{T}_k$ of the following six points

$$\begin{aligned}&(\alpha, \beta), \quad (\gamma, 0), \quad (-\alpha+\gamma, -\beta-\gamma), \\ &(\beta, \gamma), \quad (0, -\alpha-\beta), \quad (-\alpha+\gamma, -\gamma),\end{aligned} \tag{5.16}$$

be different inside the triangle $\triangle$. We define the statistic

$$\begin{aligned}Q^T(\alpha, \beta, \gamma) &\doteq B_e^T(\alpha, \beta) + B_e^T(\gamma, 0) + B_e^T(-\alpha+\gamma, -\beta-\gamma) \\ &- B_e^T(\beta, \gamma) - B_e^T(0, -\alpha-\beta) - B_e^T(-\alpha+\gamma, -\gamma),\end{aligned} \tag{5.17}$$

with its asymptotic expectation

$$\begin{aligned}Q(\alpha, \beta, \gamma) = B_e(\alpha, \beta) + B_e(\gamma, 0) + B_e(-\alpha+\gamma, -\beta-\gamma) \\ - B_e(\beta, \gamma) - B_e(0, -\alpha-\beta) - B_e(-\alpha+\gamma, -\gamma).\end{aligned}$$

Under hypothesis H_0 the value of $Q(\alpha, \beta, \gamma)$ is zero. This is a consequence of Theorem 77. The variance of ReQ^T and ImQ^T can be shown from (5.13) and (5.14) to be

$$\operatorname{Var} Q^T(\alpha, \beta, \gamma) \sim \frac{6\sigma_e^6 W_2^2}{T\, b_T^2}. \tag{5.18}$$

We standardize the statistics Q and under H_0 the statistics are asymptotically independent and Gaussian. Let

$$R_1^T(\alpha, \beta, \gamma) = \operatorname{Re} Q^T(\alpha, \beta, \gamma) \sqrt{\frac{T\, b_T^2}{3\sigma_e^6 W_2^2}},$$

$$R_2^T(\alpha, \beta, \gamma) = \operatorname{Im} Q^T(\alpha, \beta, \gamma) \sqrt{\frac{T\, b_T^2}{3\sigma_e^6 W_2^2}}.$$

We divide the entire set of observation into K separate stretches of length T. For each fixed triplet (α, β, γ) we get $2K$ statistics

$$R_1^{T,1}(\alpha, \beta, \gamma), \ldots, R_1^{T,K}(\alpha, \beta, \gamma),$$

$$R_2^{T,1}(\alpha,\beta,\gamma),\ldots,R_2^{T,K}(\alpha,\beta,\gamma),$$

which are asymptotically independent with the same distribution as $R_1^T(\alpha,\beta,\gamma)$ and $R_2^T(\alpha,\beta,\gamma)$ hence we have

$$M_1^{T,K}(\alpha,\beta,\gamma)=\frac{R_1^{T,1}(\alpha,\beta,\gamma)+\cdots+R_1^{T,K}(\alpha,\beta,\gamma)}{\sqrt{K}},$$

$$M_2^{T,K}(\alpha,\beta,\gamma)=\frac{R_2^{T,1}(\alpha,\beta,\gamma)+\cdots+R_2^{T,K}(\alpha,\beta,\gamma)}{\sqrt{K}},$$

with the following expectation and variance

$$\mathsf{E}\, M_1^{T,K}(\alpha,\beta,\gamma)\sim \operatorname{Re}Q(\alpha,\beta,\gamma)\sqrt{\frac{K\;Tb_T^2}{3\sigma_e^6W_2^2}}, \tag{5.19}$$

$$\mathsf{E}M_2^{T,K}(\alpha,\beta,\gamma)\sim \operatorname{Im}Q(\alpha,\beta,\gamma)\sqrt{\frac{K\;T\;b_T^2}{3\sigma_e^6W_2^2}}, \tag{5.20}$$

and

$$\operatorname{Var}M_1^{T,K}(\alpha,\beta,\gamma)\sim 1,\quad \operatorname{Var}M_2^{T,K}(\alpha,\beta,\gamma)\sim 1.$$

The expectations in (5.19) and (5.20) are equal to zero under H_0. Let us now consider a set of triplets

$$(\alpha_1,\beta_1,\gamma_1),\ldots,(\alpha_n,\beta_n,\gamma_n).$$

For each triplet $(\alpha_i,\beta_i,\gamma_i)$ we have ordered six pairs of frequencies by (5.16). Now consider the $6n$ pairs of frequencies

$$(\alpha_i,\beta_i),\ \ (\gamma_i,0),\ \ (-\alpha_i+\gamma_i,-\beta_i-\gamma_i), \tag{5.21}$$

$$(\beta_i,\gamma_i),\ \ (0,-\alpha_i-\beta_i),\ \ (-\alpha_i+\gamma_i,-\gamma_i),\ i=1...n, \tag{5.22}$$

which are distinct. A possible selection of these triplets is given in the next section. Note that if $j\neq k$, then the statistics $Q^T(\alpha_j,\beta_j,\gamma_j)$ and $Q^T(\alpha_k,\beta_k,\gamma_k)$ are uncorrelated, and it follows from the asymptotic normality that $Q^T(\alpha_j,\beta_j,\gamma_j)$ and $Q^T(\alpha_k,\beta_k,\gamma_k)$ are asymptotically independent.

Now we are in a position to construct the test statistics

$$\mathcal{G}_n^{T,K}=\left[M_1^{T,K}(\alpha_1,\beta_1,\gamma_1)\right]^2+\cdots+\left[M_1^{T,K}(\alpha_n,\beta_n,\gamma_n)\right]^2$$
$$+\left[M_2^{T,K}(\alpha_1,\beta_1,\gamma_1)\right]^2+\cdots+\left[M_2^{T,K}(\alpha_n,\beta_n,\gamma_n)\right]^2. \tag{5.23}$$

If the linear predictor is optimal (versus the quadratic one) the statistics $\mathcal{G}_n^{T,K}$ (5.23) according to the appropriate frequencies $(\alpha_k, \beta_k, \gamma_k)$, $k = 1, 2, \ldots n$ has an appropriate central χ^2 distribution with $2n$ degrees of freedom. Otherwise, if (5.7) does not hold, the distribution is shifted to the right, i.e., if at least one $Q(\alpha, \beta, \gamma)$ is different from zero, the noncentrality parameter of the test statistics $\mathcal{G}_n^{T,K}$ tends to infinity. The test is consistent under H_1.

5.3 Comments on computing the test statistics

We compute the values of the bispectrum for the frequencies defined by $(k/T, l/T), k, l = -T/2 \ldots T/2$. The problem is to find as many triplets

$$(\alpha_1, \beta_1, \gamma_1) = (k_1/T, l_1/T, m_1/T),$$

$$\vdots$$

$$(\alpha_n, \beta_n, \gamma_n) = (k_n/T, l_n/T, m_n/T),$$

as possible under the assumption that the following $6n$ points and their transformation inside the triangle $\triangle$ are different

$$(k_i/T, l_i/T), \quad (m_i/T, 0),$$

$$((m_i - k_i)/T, -(l_i + m_i)/T), (l_i/T, m_i/T),$$

$$(0, -(k_i + l_i)/T), \quad ((m_i - k_i)/T, -m_i/T) \quad, i = 1 \ldots n. \tag{5.24}$$

Test statistics (5.23) are computed at points (5.24) without transformation. It is clear that the maximal number of triplets is less than $T/4$. For example when $T = 64$, we get 7 triplets which satisfy all the properties necessary for the computation. They are

$$(17,\ 27,\ 30)\ (17,\ 21,\ 10)\ (17,\ 24,\ 27)\ (18,\ 27,\ 14)$$
$$(18,\ 21,\ 24)\ (19,\ 30,\ 1)\ (21,\ 27,\ 9)\ .$$

The bispectrum is estimated via an FFT and smoothing. The residual series has a zero mean, therefore the biperiodogram is zero at $(0, 0)$ and along the axes. It has no information about the value of the bispectrum there. Therefore we applied the smoothing such that the biperiodogram was considered in the entire plane first. Though the data has a zero mean, the FFT of each subset is not necessary zero at frequency zero. It provides an estimate for the Fourier transform at zero. There is another point in using stretches. If one works with scale parameter B_T close to $\sqrt{T}$ then the noncentrality parameter, and so the power of our test, depends on the number of stretches explicitly, see (5.19, 5.20). If the sample size is small, the overlapping stretches with tapering are suggested. The computation starts by fitting a linear model and, this can be done by a standard method using the AIC criteria, [100], and then fit an AR model to the data using the LS procedure.

5.4 Simulations and real data

5.4.1 Homogeneous bilinear realizable time series with Hermite degree-2

An example of the non-Gaussian time series for which the best predictor is linear against all the polynomial ones is the homogeneous bilinear (realizable) time series with Hermite degree-2, see Section 3.6.2, page 140. We consider briefly an example (for details see Terdik and Máth [136]).

Example 79 *Let the process Y_t be given by the state space equations*

$$\sum_{k=0}^{P_1} a_k^{(1)} X_{t-k}^{(1)} = \mathrm{w}_t,$$

$$\sum_{k=0}^{P_2} a_k^{(2)} X_{t-k}^{(2)} = \sum_{m=1,n=0}^{R,S} c_{m,m+n} X_{t-m-n}^{(1)} \mathrm{w}_{t-m} + const.,$$

$$Y_t = X_{t+1}^{(2)}, \tag{5.25}$$

where w_t is Gaussian white noise series.

The polynomials $\alpha_{21}(z), \alpha_{22}(z)$ and $\gamma(z,v)$ corresponding to model (5.25) are given by

$$\alpha_{21}(z) = \sum_{k=0}^{P_1} a_k^{(1)} z^{-k}; \quad a_0^{(1)} = 1,$$

$$\alpha_{22}(z) = \sum_{k=0}^{P_2} a_k^{(2)} z^{-k}; \quad a_0^{(2)} = 1,$$

$$\gamma(z,v) = \sum_{m=1,n=0}^{R,S} c_{m,m+n} z^{-n} v^{-m}.$$

The process Y_t is *homogeneous, bilinear realizable and with Hermite degree-2*. It is *separable* if the polynomial γ is the product of two polynomials of a single variable, i.e.,

$$\gamma(z,v) = \gamma_1(z)\gamma_0(v),$$

where

$$\gamma_0(v) = \sum_{m=1}^{R} c_{m,m} v^{-m}.$$

As the spectrum and the bispectrum for bilinear realizable processes with Hermite degree-2 are explicitly given, see Theorem 65, p. 142 and Theorem 66 p. 66, it follows that

Theorem 80 *If the homogeneous bilinear realizable and Hermite degree-2 process (5.25) is separable and the roots of γ_0 are inside the unit circle, then the best linear predictor is the best quadratic one as well.*

Proof. In this case the spectrum has the following form, (see Terdik and Meaux[138]),

$$\sigma^4 \left|\frac{\gamma_0(z_1)}{\alpha_{22}(z_1)}\right|^2 \left[1 + \int_0^1 \left|\frac{\gamma_1(z)}{\alpha_{21}(z)}\right|^2 d\lambda\right] = \sigma_e^2 \left|\frac{\gamma_0(z_1)}{\alpha_{22}(z_1)}\right|^2,$$

where σ_e^2 is the variance of the residual series of the best linear predictor. If we assume that the roots of γ_0 are inside the unit circle, the residual series has the following form

$$e_t = \frac{\alpha_{22}(B)}{\gamma_0(B)} Y_t,$$

where B is the backward shift operator, i.e., $BY_t = Y_{t-1}$ and the bispectrum of the residual series is also simple, that is

$$\Psi(z_1, z_2) = 6\sigma^6 (h(z_1) + h(z_2) + h(z_1^{-1} z_2^{-1})).$$

Here

$$h(z_1) = 1/3 + \int_0^1 \frac{|\gamma_1(z)|^2 \gamma_1(z z_1^{-1})}{|\alpha_{21}(z)|^2 \alpha_{21}(z z_1^{-1})} d\lambda,$$

As $\Psi(z_1, z_2)$ satisfies the necessary and sufficient condition of Theorem 78, the proof is completed. ■

We show in addititon to the Theorem 80 that if the homogeneous bilinear realizable and Hermite degree-2 process is separable and the roots of γ_0 are inside the unit circle, then the best linear predictor is optimal as compared to all polynomial predictors. We can show that the time series Y_t has the Wiener-Itô spectral representation of the form

$$Y_t = \iint_0^1 (z_1 z_2)^t \frac{\gamma_0(z_1 z_2)\gamma_1(z_1)}{\alpha_{22}(z_1 z_2)\alpha_{21}(z_1)} W(d\lambda_1, d\lambda_2), \tag{5.26}$$

where $z = e^{i2\pi\lambda}$, $z_1 = e^{i2\pi\lambda_1}$, $z_2 = e^{i2\pi\lambda_2}$ (see Terdik and Subba Rao [139] for more details). Therefore the spectral density of Y_t is given by

$$S_Y(z) = \sigma_e^2 \left| \frac{\gamma_0(z)}{\alpha_{22}(z)} \right|^2, \quad z = e^{i2\pi\lambda}.$$

Process Y_t with the above spectral density can be written in the form

$$Y_t = \frac{\gamma_0(B)}{\alpha_{22}(B)} e_t,$$

where $\alpha_{22}(B)$ and $\gamma_0(B)$ are polynomials in the shift operator B. The residual process e_t has the spectral representation

$$e_t = \iint_0^1 (z_1 z_2)^t \frac{\gamma_1(z_1)}{\alpha_{21}(z_1)} W(d\lambda_1, d\lambda_2).$$

This representation follows from (5.26) if one applies the operator $\alpha_{22}(B)/\gamma_0(B)$ to the series Y_t. It can also be shown that e_t satisfies the equations

$$\begin{aligned} \alpha_{21}(B) X_t^{(3)} &= \gamma_1(B) \mathrm{w}_t, \\ e_t &= X_t^{(3)} \mathrm{w}_t - \sigma^2, \end{aligned} \tag{5.27}$$

as well. Now $X_t^{(3)}$ is a Gaussian time series with moving average representation

$$X_t^{(3)} = \frac{\gamma_1(B)}{\alpha_{21}(B)} \mathrm{w}_t = \mathrm{w}_t + \phi(\mathrm{w}_{t-1}, \mathrm{w}_{t-2}, \ldots),$$

therefore

$$e_t = \mathrm{w}_t^2 - \sigma^2 + \mathrm{w}_t \phi(\mathrm{w}_{t-1}, \mathrm{w}_{t-2}, \ldots).$$

It is easy to see that for any functions ϕ_1 and ϕ_2

$$\mathsf{E}\left[(\mathrm{w}_t^2 - \sigma^2 + \mathrm{w}_t \phi_1(\mathrm{w}_{t-1}, \mathrm{w}_{t-2} \ldots)) \phi_2(\mathrm{w}_{t-1}, \mathrm{w}_{t-2}, \ldots) \right] = 0.$$

Particularly, for all $n = 1, 2, \ldots$, $k_1, k_2, \ldots, k_n = 1, 2, \ldots$, $i_1, i_2, \ldots, i_n = 1, 2, \ldots$ we have

$$\mathsf{E} e_t (Y_{t-i_1}^{k_1} Y_{t-i_2}^{k_2} \ldots Y_{t-i_n}^{k_n} - \mathsf{E} Y_{t-i_1}^{k_1} Y_{t-i_2}^{k_2} \ldots Y_{t-i_n}^{k_n}) = 0,$$

i.e., e_t is orthogonal not only to the linear Hilbert space generated by the past of process Y_t but to the Hilbert space generated by all of the polynomials of the past of process Y_t. Moreover, denote $\mathcal{F}_t$ the Borel field generated by e_s, $s \leq t$ and, $\mathcal{F}_t^{\mathrm{w}}$ the Borel field generated by w_s, $s \leq t$. So $\mathcal{F}_t \subseteq \mathcal{F}_t^{\mathrm{w}}$, and therefore

$$\mathsf{E}\left(e_t \mid \mathcal{F}_{t-1}\right) = \mathsf{E}\left(\mathsf{E}\left(e_t \mid \mathcal{F}_{t-1}^{\mathrm{w}}\right) \mid \mathcal{F}_{t-1}\right) = 0,$$

similarly

$$\mathsf{E}\left(e_t^2 \mid \mathcal{F}_{t-1}\right) = \sigma_e^2.$$

i.e., Y_t is linear.

5.4.2 Results of simulations

We have compared all our results with Hinich's linearity test . The same set of triplets given earlier were used. The series w_t is a zero mean Gaussian white noise with variance $\sigma^2 = 2$. The number of replications was 100 for each model. We estimated the bispectrum using the smooth function of *Subba Rao-Gabr*, see Subba Rao and Gabr [121], i.e.,

$$\begin{aligned} W(\omega_1, \omega_2) &= \frac{4\sqrt{3}}{\pi}[1 - 4(\omega_1^2 + \omega_2^2 + \omega_1\omega_2)], \quad && \omega_1^2 + \omega_2^2 + \omega_1\omega_2 < 1/4, \\ &= 0 && \textit{otherwise}. \end{aligned}$$

For this window

$$\iint_{-\infty}^{\infty} W^2(\omega_1, \omega_2) d\omega_1 d\omega_2 = 1.4628.$$

In addition of the models of this section we also consider the models proposed by Keenan [68] and by Hong-zhi and Bing [60].The following models were employed:

1. An autoregressive model of order 2 given by the equation

$$X_t = 0.5 \cdot X_{t-1} - 0.4 \cdot X_{t-2} + \mathrm{w}_t.$$

   This is a Gaussian stationary linear model. The linearity is accepted at level .05, 95 percent by Hinich test and 97 percent by the test (5.23), see the figures below.

2. A series of Hermite polynomials of order 2, i.e.,

$$X_t = \mathrm{w}_t^2 - \sigma^2.$$

   This model will be referred to as $H2$. X_t is an independent (non-Gaussian) series so it is linear not only by our definition of linearity but by the stronger one as well. The Hinich test does not work properly because linearity is rejected 94 times out of 100 at level .05. Our test gives fully satisfactory results accepting linearity at 91 percent.

3. A bilinear model

$$X_t = 0.4 \cdot X_{t-1} + 0.65 \cdot X_{t-2} + 0.5 \cdot X_{t-1}\mathrm{w}_{t-1} + \mathrm{w}_t.$$

It will be called model B. It is stationary and nonlinear. Both tests rejected linearity at $93 - 95$ percent.

4. A model denoted by SB is a simple bilinear model

$$X_t = 0.65 \cdot X_{t-2} \cdot \mathrm{w}_{t-1} + \mathrm{w}_t.$$

It is a stationary and nonlinear model. The Hinich test rejected linearity of this model at 96 and our test at 97 percent.

5. A model denoted by $BH2S$ is a homogeneous bilinear model with Hermite degree-2. It is given by the equation (5.25) with the polynomials

$$\begin{aligned}\alpha_{21}(z) &= 1 - 0.5 \cdot z^{-1} + 0.4 \cdot z^{-2},\\ \alpha_{22}(z) &= 1 + 0.3 \cdot z^{-1} - 0.1 \cdot z^{-2},\\ \gamma(z,v) &= 1 - 0.4 \cdot z^{-1} - 0.7 \cdot v^{-1} + 0.28 \cdot z^{-1}v^{-1}.\end{aligned}$$

Polynomial $\gamma(z,v)$ is separable therefore the process is linear in our sense but, as we have shown, its residual series is not independent, see (5.27). The linearity of this model is rejected by the Hinich test at 99% which is correct because the model is not linear by the stronger definition of linearity. It is accepted by our test in 87% of the times.

6. A model denoted by $BH2NS$ is a homogeneous bilinear model, see (5.25), with the polynomials

$$\begin{aligned}\alpha_{21}(z) &= 1 - 0.5 \cdot z^{-1} + 0.4 \cdot z^{-2},\\ \alpha_{22}(z) &= 1 + 0.3 \cdot z^{-1} - 0.1 \cdot z^{-2},\\ \gamma(z,v) &= 1 + 0.4 \cdot z^{-1} + 0.7 \cdot v^{-1} - 0.3 \cdot z^{-1}v^{-1}.\end{aligned}$$

The difference from the previous model is that polynomial $\gamma(z,v)$ is nonseparable, therefore the process is nonlinear. Our test is able to make a difference between models $BH2S$ and $BH2NS$, i.e., between the separable and nonseparable bilinear homogeneous models. The Hinich test gives a correct decision in this case, rejecting linearity at 100 percent. Our test accepts linearity in 22% of the times.

For these particular models we plotted the sorted p-values of both tests based on 100 simulations. The length of each simulation was $n = 2^9$. We chose $B_T = T^{\varepsilon - 1/2}$ with $\varepsilon = .01$, and $T = 128$ so $TB_T = 12$. The Figures 5.1 and 5.2 contain the graphs of the sorted p-values $p \in [0,1]$ $(y - axis)$ versus the number of simulation $k = 1, 2, \ldots 100$ $(x - axis)$, according to the models listed above, respectively. The crossing of the line *Level* .05 shows the number of simulations rejected at level .05.

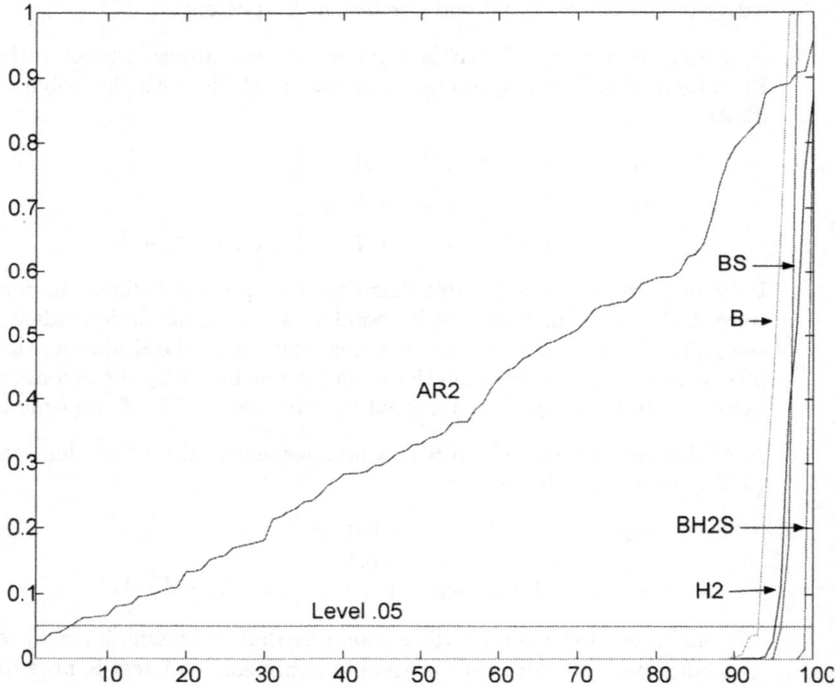

FIGURE 5.1. Results of Hinich Test

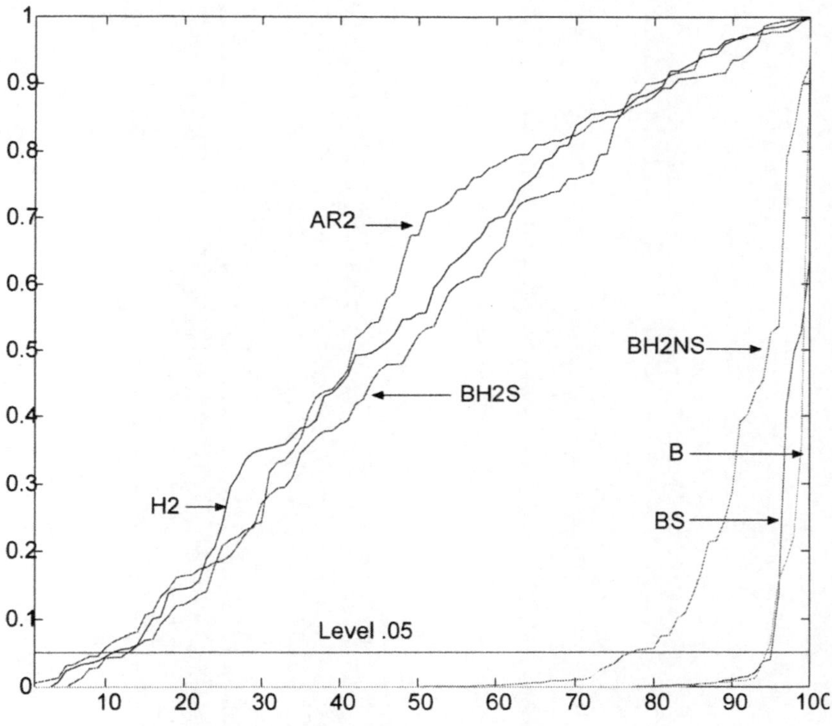

FIGURE 5.2. Results of the New Linearity Test

6
Some Applications

6.1 Testing linearity

We have used the linearity test for real data. The linear residual process was calculated by fitting an AR model using the LS method. The order P of the autoregression was determined by AIC criteria.

6.1.1 Geomagnetic Indices

aa indices of geomagnetic activity (1868-1987).

The data are monthly and the total number of observations is $n = 1428$, see [6], [7], [8], and [132]. This time series has an eleven-year cycle approximately. The 11-year periodicity was partially removed first by calculating and subtracting the seasonal means. The estimated AIC order of the series is $P = 12$. The size of the observations allows us to put the stretch length T either 2^7 or 2^8; the window bandwidths TB_T are 12 and 15, respectively. The p-values according to the different stretch lengths are 0.2039 and 0.3096. In both cases the p-values are greater than 0.05. The Hinich linearity test gives *zero* and $4.3640 * 10^{-7}$, respectively for the stretch lengths 2^7 and 2^8. Therefore our conclusion is that the series is linear in our general sense, but it can not be put into a moving average representation with any i.i.d. noise process.

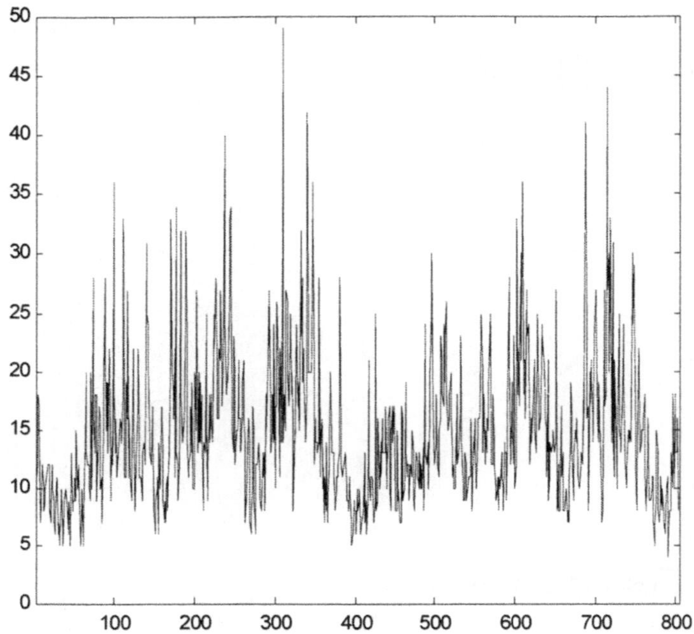

FIGURE 6.1. Geomagnetic planetary indices (ap), monthly averages between 1932 and 1998

ap indices of geomagnetic activity (1932-1998).

The following table shows the results of linearity tests of the Hinich test (H-test) and our test of Chapter 5.

	Gaussianity	Linearity, $T = 2^7$	Linearity, $T = 2^8$
H-test	0	0	0
TM-test	–	0.0637	0.0461

There is a basic difference between the results of two tests. Therefore our conclusion is that the series of *ap* indices is weakly linear. The bilinear fitting of order $P = 1, 2, 3, 4$, (see p. 206) for this time series strengthens the weakly linearity because all the estimated bilinear parameters can be considered zero,.see [37] for some modeling of *ap*. The spectrum for the ap-indices clearly shows the well known periodicity for 11 years and 6 months. The bispectrum is estimated at Fourier frequencies of 2^{-6} with Subba Rao–Gabr optimal window of size 8.

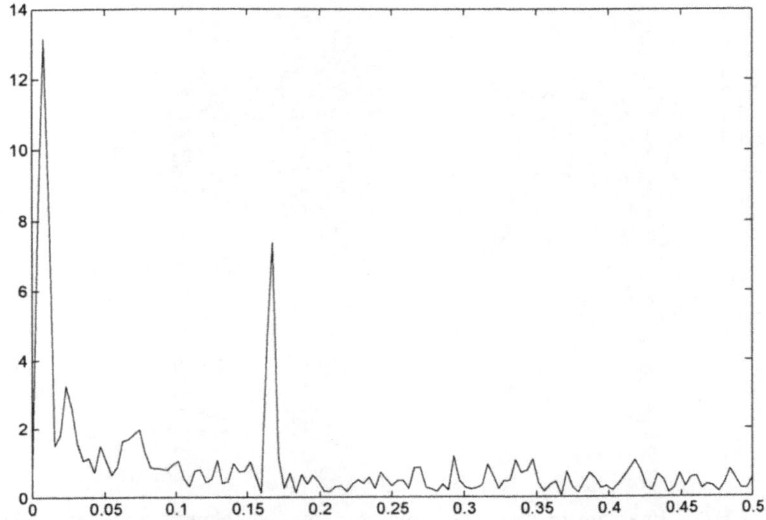

FIGURE 6.2. Spectrum for geomagnetic planetary indices (ap)

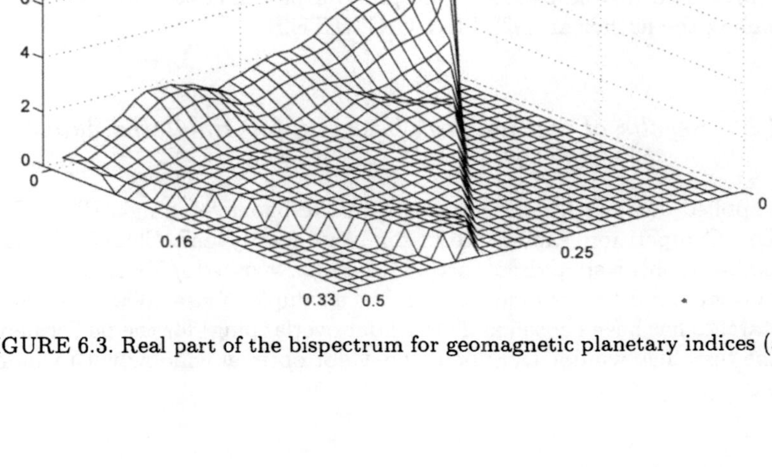

FIGURE 6.3. Real part of the bispectrum for geomagnetic planetary indices (ap)

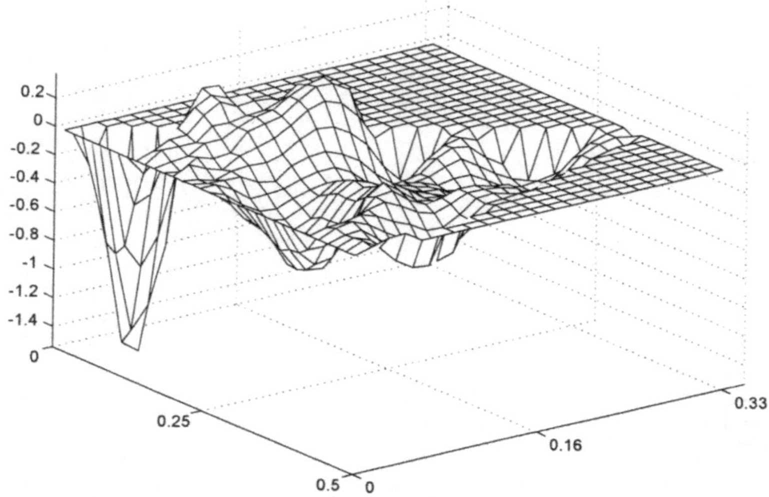

FIGURE 6.4. Imaginary part of the bispectrum for geomagnetic planetary indices (ap)

Wolfer sunspot numbers (1700-1987)

This is a well known time series in the literature, $n = 288$. The number of observations does not allow larger stretch lengths than $T = 2^6$ with window bandwidth $TB_T = 8$. The estimated AIC order of autoregression is 9. The p-values according to both our test and the Hinich test are 0.0442 and 0.0102, respectively. So both tests reject the linearity of the series at level 0.05.

A documented MATLAB program has been developed to use the linearity test presented in this chapter. The program is available upon request by e-mailing the author at TERDIK@CIC.KLTE.HU.

6.1.2 Results of testing weak linearity for simulated data at WUECON

We applied our test for the data used in the paper "A Single-Blind Controlled Competition Among Tests for Nonlinearity and Chaos" [9]. It is available at this web address: http://wueconb.wustl.edu/~barnett/. These data correspond to five models; each has a sample of size 2000. The length of starches has been chosen as 2^6 (without overlapping) for the performance of the test. The window is Subba Rao–Gabr optimal window with window size 8.

1. Model I is the deterministic, chaotic Feigenbaum recursion of the form:

$$Y_t = 3.57Y_{t-1}\left(1 - Y_{t-1}\right),$$

where the initial condition was set at $Y_0 = .7$.
Result of the test: p value is 0; linearity is **strongly rejected.**

2. Model II is a *GARCH* process of the following form:

$$h_t = 1 + .8h_{t-1} + .1h_{t-1}\mathrm{w}_{t-1}^2,$$
$$Y_t = \sqrt{h_t}\mathrm{w}_t,$$

with $h_0 = 1$ and $Y_0 = 0$.
Result of the test: p value is 0.0591; linearity is **accepted.**

3. Model III is a nonlinear moving average process of the following form:

$$Y_t = \mathrm{w}_t + .8\mathrm{w}_{t-1}\mathrm{w}_{t-2}.$$

Result of the test: p value is 0; linearity is **strongly rejected.**

4. Model IV is an *ARCH* process of the following form:

$$Y_t = \mathrm{w}_t\sqrt{(1 - .5Y_t^2)},$$

with the value of the initial observation set at $Y_0 = 0$.
Result of the test: p value is 0.0019; linearity is **strongly rejected.**

5. Model V is an *ARMA* model of the form:

$$Y_t = .8Y_{t-1} + .15Y_{t-2} + \mathrm{w}_t + .3\mathrm{w}_{t-1},$$

with $Y_0 = 1$ and $Y_1 = .7$.
Result of the test: p value is 0.6989; linearity is **strongly accepted.**

6.1.3 GARCH model fitting

We now go a bit further in studying the data of Model II. Both the real and imaginary part of the biperiodogram of this data set is zero (see Figure 6.5), but the square of the series Y_t^2 has some bispectrum, Figure 6.6, which is not constant. Therefore the process Y_t cannot be a white noise.

A quick Withle estimate of the parameters are $\widehat{a_0} = 0.8020$, $\widehat{d_0} = 0.099$, and $\widehat{c_0} = 0.9846$ respectively [35]. They are reasonably close to the theoretical parameters $a_0 = 0.8$, $d_0 = 0.1$, and $c_0 = 1$. Note that here $\sigma^2 = 1$ by the definition of the model. This is based on a comparison of the theoretical spectrum of the *GARCH* process and estimated spectrum of the data (see Figure 6.7).

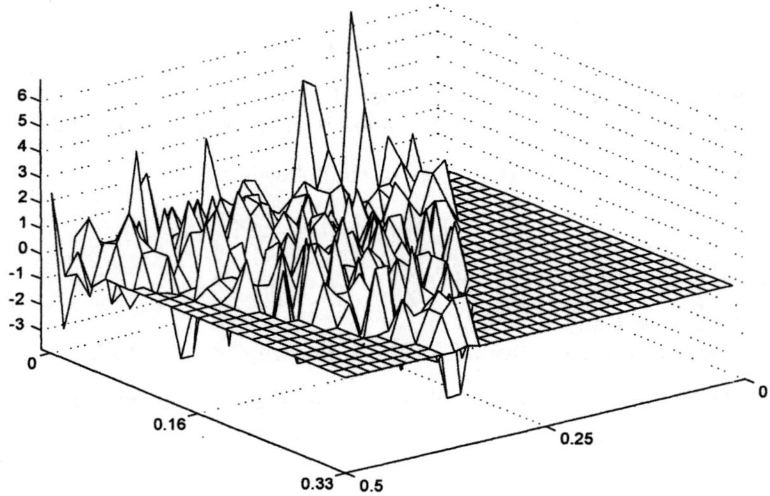

FIGURE 6.5. Real part of the biperiodogram of the data set for Model II

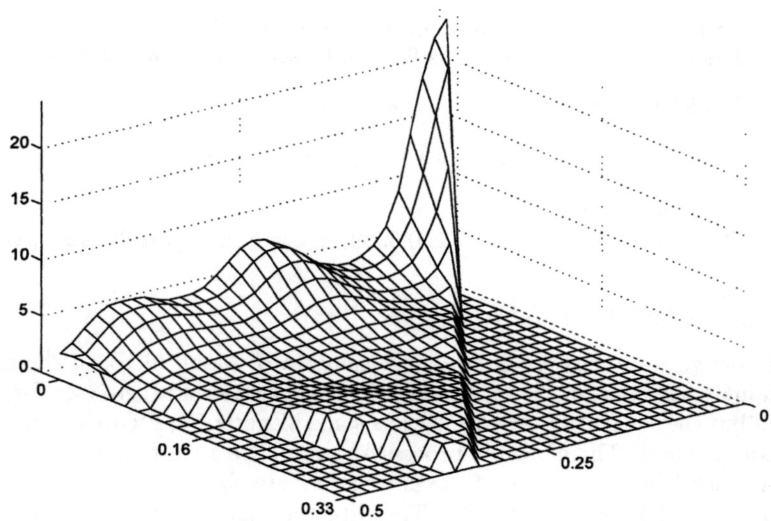

FIGURE 6.6. Real part of the smoothed biperiodogram of the square of the data for Model II

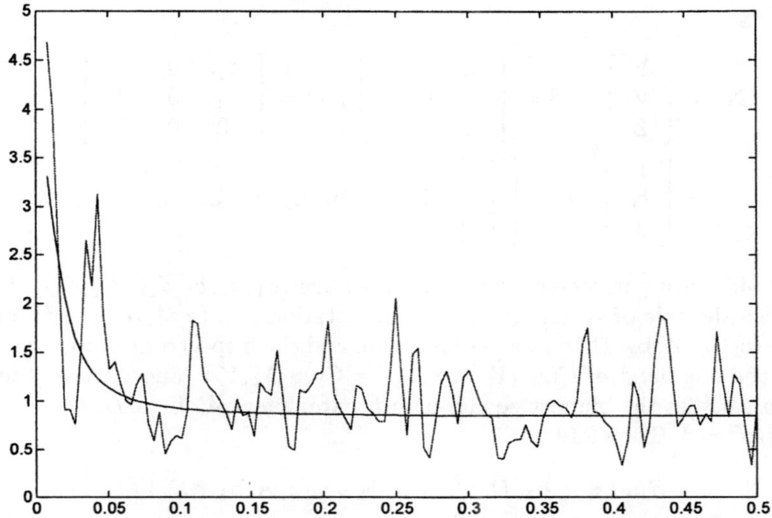

FIGURE 6.7. The estimated spectrum of the data and theoretical spectrum of square of the GARCH process with the estimated parameters

6.2 Bilinear fitting

6.2.1 Parameter estimation for bilinear processes

We have already shown (see [131]) that for the process

$$Y_t = aY_{t-1} + dY_{t-1}\mathrm{w}_{t-1} + \mathrm{w}_t,$$

the theoretical and the estimated spectrum, cross-spectrum, and bispectrum are useful for the parameter estimation. The estimator based on the spectrum and the cross-spectrum between Y_t and Y_t^2 needs the fourth order cumulant spectrum, and, in particular, the spectrum of Y_t^2 as one of the norming factor.

Now we are able to estimate the parameters of a general univariate bilinear model, say

$$\begin{aligned} Y_t &= a_1Y_{t-1} + a_2Y_{t-2} + d_{11}Y_{t-1}\mathrm{w}_{t-1} \\ &\quad + d_{12}Y_{t-2}\mathrm{w}_{t-1} + d_{22}Y_{t-2}\mathrm{w}_{t-2} + \mathrm{w}_t + b_1\mathrm{w}_{t-1} + f_0. \end{aligned}$$

where $\mathsf{E}\mathrm{w}_t = 0$ and $\mathsf{E}\mathrm{w}_t^2 = \sigma^2$. The model can be rewritten into the state space form as

$$\begin{aligned} \mathbf{X}_t &= A\mathbf{X}_{t-1} + D\mathbf{X}_{t-1}\mathrm{w}_{t-1} + \mathbf{b}_1\mathrm{w}_{t-1} + \mathbf{b}_2H_2(\mathrm{w}_{t-1}), \\ Y_t &= \mathbf{c}'\mathbf{X}_t + \mathrm{w}_t, \end{aligned}$$

Taking

$$\mathbf{X}_t = \begin{bmatrix} Y \\ V \\ Z \end{bmatrix}_t, \quad A = \begin{bmatrix} a_1 & 1 & 0 \\ a_2 & 0 & 0 \\ 1 & 0 & 0 \end{bmatrix}, \quad D = \begin{bmatrix} d_{11} & 0 & d_{12} \\ d_{22} & 0 & 0 \\ 0 & 0 & 0 \end{bmatrix},$$

$$\mathbf{b} = \begin{bmatrix} 1 \\ b_1 \\ 0 \end{bmatrix}, \quad \mathbf{c} = \begin{bmatrix} 1 \\ 0 \\ 0 \end{bmatrix}, \quad \mathbf{b}_1 = A\mathbf{b}, \;\; \mathbf{b}_2 = D\mathbf{b}.$$

The unknown parameters to be estimated are $(a_1, a_2, b_1, d_{11}, d_{12}, d_{22}, \sigma^2)$. Notice the role of variance σ^2. Use the notation $S_{2,Y}(z, A, \mathbf{b}_1, D, \sigma^2)$ and $S_{3,Y}(z_1, z_2, A, \mathbf{b}_1, D, \sigma^2)$ for the spectrum and the bispectrum, respectively. On the one hand $\sigma^2 \operatorname{Cum}(Y_t/\sigma, Y_s/\sigma) = \operatorname{Cum}(Y_t, Y_s)$, and on the other, Y_t/σ satisfies the bilinear equation with parameters A, $\mathbf{b}_1$, σD, and $\mathsf{E}w_t^2/\sigma^2 = 1$. Therefore

$$S_{2,Y}(z; A, \mathbf{b}_1, D, \sigma^2) = \sigma^2 S_{2,Y/\sigma}(z, A, \mathbf{b}_1, \sigma D, 1),$$
$$S_{3,Y}(z_1, z_2, A, \mathbf{b}_1, D, \sigma^2) = \sigma^3 S_{3,Y/\sigma}(z_1, z_2, A, \mathbf{b}_1, \sigma D, 1).$$

Similarly

$$\operatorname{Var}(Y_t \,|A, \mathbf{b}_1, D, \sigma^2) = \sigma^2 \operatorname{Var}(Y_t/\sigma \,|A, \mathbf{b}_1, \sigma D, 1).$$

The first step of the estimation is the standardization of series Y_t, i.e., consider the series

$$\hat{Y}_t = \frac{Y_t - \mathsf{E}Y_t}{\operatorname{std}(Y_t)},$$

the spectrum and the bispectrum according to the series $\hat{Y}_t$ are

$$\begin{aligned} S_{2,\hat{Y}}(z) &= \frac{1}{\operatorname{Var}(Y_t \,|A, \mathbf{b}_1, D, \sigma^2)} S_{2,Y}(z, A, \mathbf{b}_1, D, \sigma^2) \\ &= \frac{1}{\operatorname{Var}(Y_t/\sigma \,|A, \mathbf{b}_1, \sigma D, 1)} S_{2,Y/\sigma}(z, A, \mathbf{b}_1, \sigma D, 1), \end{aligned}$$

similarly

$$\begin{aligned} S_{3,\hat{Y}}(z_1, z_2) &= \frac{1}{[\operatorname{Var}(Y_t \,|A, \mathbf{b}_1, D, \sigma^2)]^{3/2}} S_{3,Y}(z_1, z_2, A, \mathbf{b}_1, D, \sigma^2) \\ &= \frac{1}{[\operatorname{Var}(Y_t/\sigma \,|A, \mathbf{b}_1, \sigma D, 1)]^{3/2}} S_{3,Y/\sigma}(z_1, z_2, A, \mathbf{b}_1, \sigma D, 1). \end{aligned}$$

Next estimate $S_{2,\hat{Y}}(z)$ and $S_{3,\hat{Y}}(z_1, z_2)$ by some consistent estimator, e.g., by the smoothed periodogram and biperiodogram. Then estimate the unknown parameters contained in the matrices A and σD by the minimization of the functional $Q_T(\vartheta)$ (see equation (4.10), p. 161; the explicit expressions for the theoretical spectrum and bispectrum are given in Section 3.3.3, p. 96). Suppose that the estimates $\widehat{A}$ and $\widehat{\sigma D}$ are given; then the estimate $\widehat{\sigma^2}$ of σ^2 is calculated by

$$\widehat{\sigma^2} = \frac{\widehat{\operatorname{Var}(Y_t)}}{\operatorname{Var}(Y_t/\sigma \left| \widehat{A}, \widehat{\mathbf{b}_1}, \widehat{\sigma D}, 1 \right.)}.$$

The nominator is the estimated variance of Y_t, and the denominator is the theoretical variance given with the help of the estimated parameters $\widehat{A}$ and $\widehat{\sigma D}$. Finally the estimate of all the unknown parameters is

$$\left[\widehat{A}, \widehat{\mathbf{b}_1}, \frac{\widehat{\sigma D}}{\sqrt{\widehat{\sigma^2}}}, \widehat{\sigma^2} \right].$$

The particular parameters for the simulation were chosen as the linear part $a_1 = -.3$, $a_2 = -.1$, $b_1 = .4$; the coefficients of the bilinear part $d_{11} = .3$, $d_{12} = -.2$, $d_{22} = .15$; and the variance of the noise $\sigma^2 = 2$. The sample size of the simulated series was $n = 2^{12}$, and the number of frequencies for the spectra was 2^6. The demeaned data were broken into L overlapping stretches of length $K = 2^6$. The estimates of the parameters were computed by putting the estimated spectrum and the biperiodogram into the functional $Q_T(\vartheta)$ (see equation (4.10), p. 161). There were five outliers out of 186 simulations. After removing the outliers we summarized the mean and the standard deviation of the estimates of 181 simulations in Table 6.1. When the Subba Rao–Gabr optimal window was used for smoothing the bispectrum, there were no outliers. The mean and the standard deviation of the estimates of 188 simulations are given in Table 6.2. Both the tables show that the method works well for bilinear time series although not all the assumptions of Theorem 73 are fulfiled.

Table 6.1; Results of the estimation of the seven parameters of simulated bilinear time series by spectrum, and biperiodogram

	N	Range	Min.	Max.	Mean	Std.	Skew.	Kurt.
A1	181	.42	-.58	-.15	-.3142	.0795	-.291	.006
A2	181	.36	-.28	.08	-.0827	.0522	-.025	1.111
B1	181	.46	.18	.64	.3937	.0901	.039	-.269
D11	181	.22	.22	.44	.3095	.0386	.483	.258
D12	181	.19	-.33	-.14	-.2112	.0337	-.520	.608
D22	181	.24	.06	.30	.1433	.0330	.672	2.129
SIG	181	1.03	1.16	2.19	1.8233	.1789	-.249	.196

Table 6.2; Results of the estimation of the seven parameters of simulated bilinear time series by the spectrum, and bispectrum

	N	Range	Min.	Max.	Mean	Std.	Skew.	Kurt.
A1	188	.35	-.51	-.17	-.3311	.0715	-.124	-.349
A2	188	.25	-.20	.05	-.0690	.0447	-.011	.314
B1	188	.41	.20	.61	.3899	.0815	.149	-.299
D11	188	.20	.23	.43	.3074	.0362	.590	.149
D12	188	.19	-.32	-.13	-.2027	.0320	-.628	1.029
D22	188	.19	.06	.25	.1275	.0272	.847	1.813
SIG	188	.84	1.35	2.18	1.8701	.1600	-.338	.012

6.2.2 *Bilinear fitting for real data*

We fitted the following particular bilinear model;

$$\sum_{m=0}^{P} a_m Y_{t-m} = \sum_{m=0}^{P-1} b_m \mathrm{w}_{t-m} + \sum_{m=1}^{P-1} \sum_{n=0}^{P-m} d_{m,m+n} Y_{t-m-n} \mathrm{w}_{t-m}. \tag{6.1}$$

In engineering, it is usual to fit an $ARMA$ model with orders P and $P-1$. If a model already contains P and $P-1$ shifts of the observation and the errors respectively, then the lower triangular bilinear model with these shifts is exactly above one. The capacity of computer (PC) in hand allowed only the orders $P = 1, 2, 3, 4$ because the number of parameters is

$$\frac{P^2 + 5P - 4}{2} + 1,$$

and for $P = 4$, the time of estimation is counted by days. In some cases the estimated parameters imply the use of subset bilinear fitting, which needs even more processor time. Therefore these cases are skipped. The selection criteria between models was the minimal variance of the noise process.

The S&P 500 Index

We consider the data set of the S&P 500 Index (see p. xiii), between 04/21/82 and 10/02/98. The daily closingdata were differenced once, and these values were put in the equation (6.1) as Y_t. It has sample size $N = 4158$ with $\mathsf{Var}(Y_t) = 27.1743$. It is strongly nonlinear with significance *zero*. The estimated variances $\widehat{\sigma^2}$ for different orders P are the following:

P	1	2	3	4
$\widehat{\sigma^2}$	25.2099	1.4912	10.0169	23.4001

Therefore the estimated order $P = 2$. The estimated parameters of chosen bilinear model are $\widehat{a_1} = -0.4627$, $\widehat{a_2} = 0.1863$, $\widehat{b_1} = -2.6156$, $\widehat{d_{11}} = 0.4378$, $\widehat{d_{12}} = 0.1105$, $\widehat{\sigma^2} = 1.4912$.

IBM stock prices

The daily closing prices, between January 2, 1968 and May 29, 1998, were differenced once. The result is the series Y_t. The number of the observation is $N = 7653$ and the $\mathsf{Var}(Y_t) = 0.6048$. It is strongly nonlinear with significance *zero*. One can hardly choose a bilinear model by the following estimates of variances:

P	1	2	3	4
$\widehat{\sigma^2}$	0.5271	0.5414	0.4608	0.4294

Therefore, similar the Section 6.1.3 , p. 201, we fit a GARCH(1,1) model to the data. A Withle estimate of the GARCH parameters [35].according to the model

$$\begin{aligned} h_{0,t} &= c_0 + a_0 h_{0,t-1} + d_0 h_{0,t-1} \mathrm{w}_{t-1}^2, \\ Y_t &= \sqrt{h_{0,t}} \mathrm{w}_t. \end{aligned} \tag{6.2}$$

are $\widehat{a_0} = 0.6784$, $\widehat{d_0} = 0.1539$, $\widehat{c_0} = 0.1014$. Please note that here $\sigma^2 = 1$.

Magnetic field data (MFD)

This data set contains the daily average Y_t of BT (T component) in RTN units (nanoteslas) between February 17, 1992 and June 30, 1997 with sample size $N = 1961$ (see p. xiv). It is strongly nonlinear with significance *zero* and $\mathsf{Var}(Y_t) = 0.3279$. The estimated variances

P	1	2	3	4
$\widehat{\sigma^2}$	0.2765	0.4122	0.4839	0.4364

give back, what is known from the theory of Block equation, that $P = 1$. The estimated parameters of chosen bilinear model are $\widehat{a_1} = -0.5421$, $\widehat{d_{11}} = 0.0541$ and $\widehat{\sigma^2} = 0.2765$.

We list here the estimated value $\widehat{d_{11}}$ at different order P for different data sets:

P	1	2	3	4
S&P	0.0392	0.4378	0.1462	0.0331
IBM	−0.2270	−0.1976	−0.1977	−0.2558
MFD	0.0541	0.0581	0.0129	0.0022

It is seen that the estimates basically do not change for the IBM data. This might suggest a subset bilinear model. To spare computing time, we did not check this.

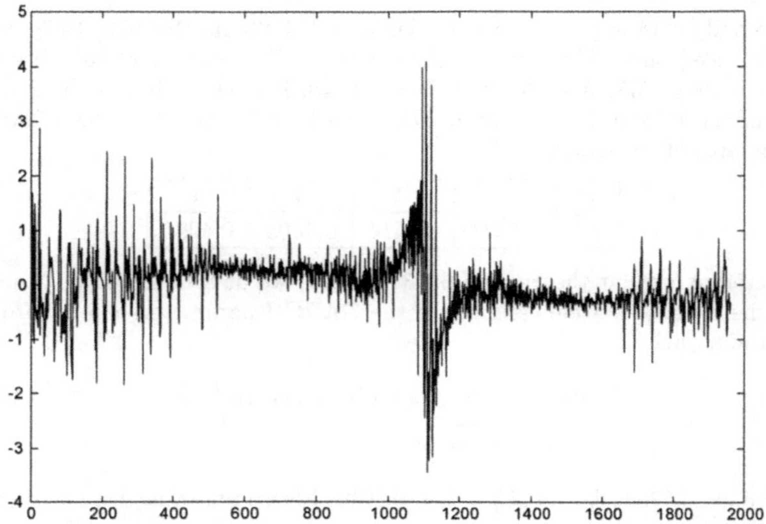

FIGURE 6.8. Magnetic field data, BT component in RTN (nT), between February 17, 1992 and June 30, 1997

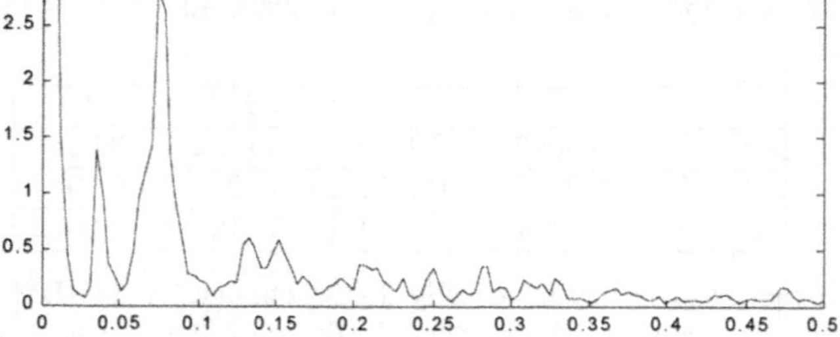

FIGURE 6.9. Spectrum of magnetic field data

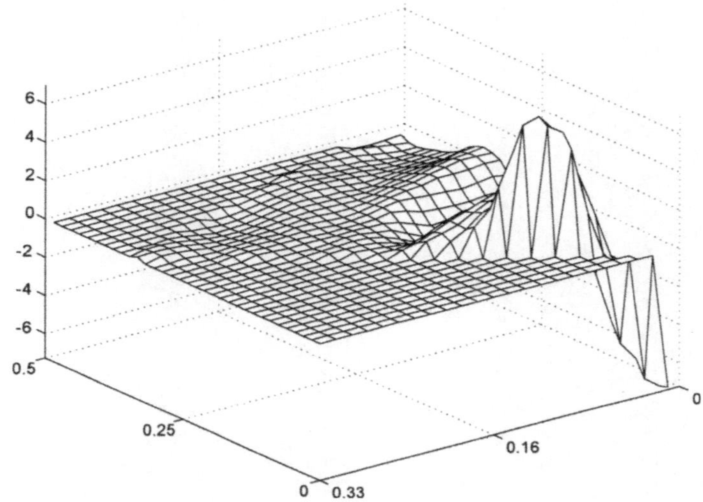

FIGURE 6.10. Estimated bispectrum (real part) of magnetic field data

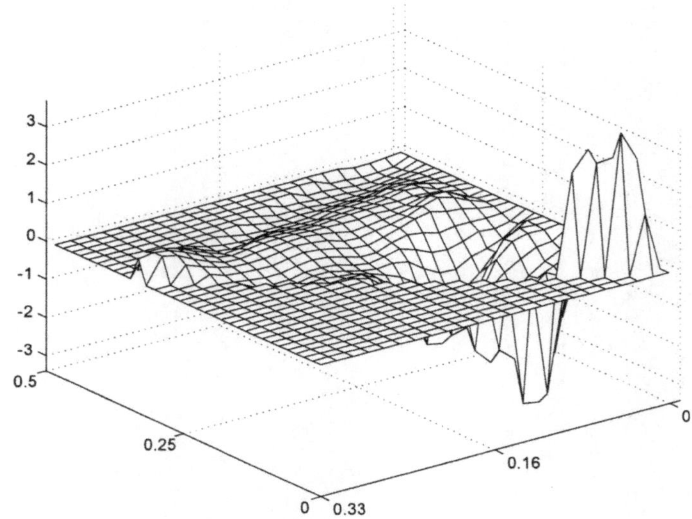

FIGURE 6.11. Estimated bispectrum (imaginary part) of magnetic field data

Appendix A
Moments

We introduce here the moment in a similar fashion as it was used for the definition of cumulants, the corresponding theorems of probability see [36]. Consider some random variable $X_{(1:n)}$. The characteristic function of X is defined by

$$\phi(\lambda_{(1:n)}) = \mathsf{E}\exp\left(i\left(\lambda_{(1:n)}, X_{(1:n)}\right)\right),$$

where $\lambda_{(1:n)}$ are real entries. The characteristic function always exists. Moreover $\mathsf{E}\,|X_j|^{2M} < \infty$, $j \in \{1, 2, \dots n\}$ for some M if and only if all the partial derivatives

$$\frac{\partial^{\Sigma k_{(1:n)}}}{\partial \lambda_{(1:n)}^{k_{(1:n)}}} \phi(\lambda_{(1:n)}),$$

up to order $\Sigma k_{(1:n)} \leq 2M$ exist, are continuous and

$$\phi(\lambda_{(1:n)}) = \sum_{\Sigma k_{(1:n)} \leq 2M} \frac{i^{\Sigma k_{(1:n)}}}{k_{(1:n)}!} \mathsf{E} X_{(1:n)}^{k_{(1:n)}} \lambda_{(1:n)}^{k_{(1:n)}} + o(|\lambda_{(1:n)}|^{2M}), \qquad \text{(A.1)}$$

around zero, where $k_{(1:n)}$ denotes the vector $(k_1, k_2, \dots, k_n)$ with nonnegative entries. One might consider the equation (A.1) as the definition of the moments. Notice that there is no such assumption that all the components of $X_{(1:n)}$ should be different. Therefore an equivalent definition of the mth, $m \leq 2M$, order moment of variable X can be given as follows. Put $X_{(1:m)} = (X, X, \dots, X)$ with the characteristic function

$$\phi(\lambda_{(1:n)}) = \mathsf{E}\exp\left(iX \sum \lambda_{(1:n)}\right).$$

Now the mth order derivative of

$$\phi_1(\lambda) = \mathsf{E}\exp(i\lambda X),$$

exists and is continuous at zero if and only if the derivative

$$\frac{\partial^m}{\partial\lambda_{(1:m)}}\phi(\lambda_{(1:m)}),$$

exists and is continuous at zero. In general there is no need for other than the first order partial derivatives of the characteristic function with respect to each variable. To show this let us start with some random variable $X_{(1:2M)}$ with characteristic function

$$\phi(\lambda_{(1:2M)}) = \mathsf{E}\exp\left(i\left(\lambda_{(1:2M)}, X_{(1:2M)}\right)\right).$$

Suppose that the derivative

$$\frac{\partial^{2M}}{\partial\lambda_{(1:2M)}}\phi(\lambda_{(1:2M)}),$$

exists and is continuous, then the each moment of order m up to $2M$ are defined by the equation

$$\phi(\lambda_{(1:m)}) = \sum_{\substack{\Sigma k_{(1:m)}\leq m \\ 0\leq k_j\leq 1}} i^{\Sigma k_{(1:m)}}\mathsf{E}X_{(1:m)}^{k_{(1:m)}}\lambda_{(1:m)}^{k_{(1:m)}} + o(|\lambda_{(1:m)}|^m).$$

The moment of order m is considered to be the case when $\Sigma k_{(1:m)} = m$, where $k_{(1:m)} = \vec{1}$. The $(m_1, m_2)^{th}$ order moment of the random variables X and Y, i.e., $\mathsf{E}X^{m_1}Y^{m_2}$ is a particular case when the first m_1 components of $X_{(1:m_1+m_2)}$ are equal to X and the rest m_2 components are equal to Y. An interesting property of this definition of the moment $\mathsf{E}X_{(1:m)}^{\vec{1}}$ is that it is multilinear in the entries of $X_{(1:m)}$. It is symmetric in the components of $X_{(1:m)}$ as well. We repeat once again that the definition of the mth order moment of the random variable $X_{(1:m)}$ can be considered to be

$$\mathsf{E}X_{(1:m)}^{\vec{1}} = \left.\frac{(-i\partial)^m}{\partial\lambda_{(1:m)}^{\vec{1}}}\phi_{X_{(1:m)}}(\lambda_{(1:m)})\right|_{\lambda_{(1:m)}=0}.$$

In particular if the entries of $X_{(1:n)}$ are independent

$$\mathsf{E}X_{(1:n)}^{k_{(1:n)}} = \left(\mathsf{E}X_{(1:n)}\right)^{k_{(1:n)}}. \tag{A.2}$$

Appendix B

Proofs for the Chapter Stationary Bilinear Models

B.1 Proofs for Subsection 3.2.3

In this Appendix we present some results concerning matrix theory, tensor–product of matrices and elementary algebra, cited earlier. Denote by $\mathbf{C}$ the set of complex numbers, by $\mathbf{C}_\circ$ the closed unit disc in $\mathbf{C}$ and by x, y, z complex numbers. Moreover, set

$$\nu(P) = \max\{\ |x|\ |x \in \mathbf{C}, P(x) = 0\},$$

where P is a polynomial with real coefficients.

The following corollary and two propositions are used in the proof of Theorem 37.

Proposition 81 *Let $a_i \in \mathbf{R}$, $i = 1, \ldots, n$ and*

$$P(x) = x^n - a_1 x^{n-1} - \ldots - a_n.$$

Then $\nu(P) < 1$ iff $\sum_{i=1}^n |a_i| < 1$.

Proof. See exercises 16, 17 in [97] tom. I. §3.

Corollary 82 *Let $a_i \in \mathbf{R}$, $i = 0, \ldots, n$, $|a_0| < 1$ and*

$$P(x) = x^n - (1 - a_0^2)^{-1} \sum_{i=1}^{n} a_i^2 x^{n-i}.$$

Then $\nu(P) < 1$ iff $\sum_{i=0}^n a_i^2 < 1$.

Proposition 83 *Let $a_i, b_i \in R$, $i = 0, \ldots, n$, $|a_0|, |b_0| < 1$ and*

$$P(x) = x^n - (1 - a_0 b_0)^{-1}(a_1 b_1 x^{n-1} + \ldots + a_n b_n),$$
$$Q(x) = x^n - (1 - a_0^2)^{-1}(a_1^2 x^{n-1} + \ldots + a_n^2),$$
$$R(x) = x^n - (1 - b_0^2)^{-1}(b_1^2 x^{n-1} + \ldots + b_n^2).$$

If $\nu(Q) < 1$ and $\nu(R) < 1$ then $\nu(P) < 1$.

Proof. Use Proposition 81, Corollary 82 and the Cauchy–Schwarz inequality. ■

Afterwards we generalize these results to polynomials with special matrix coefficients. Define the closed convex cone K and M in $\mathbf{C}^{m^2}$ and $\mathbf{R}^{m^2 \times m^2}$, respectively, by

$$K = \left\{ v \in \mathbf{C}^{m^2} \middle| v = \sum_{i=1}^{n} v_i \otimes \overline{v}_i, \quad v_i \in \mathbf{C}^m i = 1, \ldots, n \right\},$$
$$M = \left\{ M \in \mathbf{R}^{m^2 \times m^2} \middle| M = \sum_{i=1}^{\infty} M_i^{\otimes 2}, \quad M_i \in \mathbf{R}^{m \times m} \right\}. \tag{B.1}$$

Moreover, denote by $B_A(x, y)$ the bilinear form $\langle Ax, y \rangle$, where $A \in \mathbf{R}^{m^2 \times m^2}$, $x, y \in \mathbf{C}^{m^2}$ and $\langle , \rangle$ denotes the usual inner product in $\mathbf{C}^{m^2}$.

K and M have properties as follows:

Proposition 84 *(i) Let $v \in \mathbf{C}^{m^2}$. Then there exist $U_i \in \mathbf{C}^{m \times m}$ $(i = 1, 2)$, $U_i U_i^* = I$ and $S = diag(s_1, \ldots, s_m)$, $s_1 \geq \ldots \geq s_m \geq 0$ such that*

$$v = (\overline{U}_2 \otimes U_1)\ Vec(S),$$

where $Vec(S) \in C^{m^2}$, $Vec(S)^T = (s_{(1)}^T, \ldots, s_{(m)}^T)$, $s_{(i)}$ is the ith column of S and * denotes the Hermite–transpose.
(ii) If $u, v \in K$ and $A \in M$, then $\langle u, v \rangle, B_A(u, v) \in \mathbf{R}_+$.
(iii) Let $\sum_{i=1}^{\infty} G_i^{\otimes 2}$ and $\sum_{i=1}^{\infty} F_i^{\otimes 2}$ be convergent, where $G_i, F_i \in R^{m \times m}$ $(i = 1, 2, \ldots)$. Then for all $u, v \in \mathbf{C}^{m^2}$, $\|u\| = \|v\| = 1$ there exist $u_i, v_i \in K$, $\|u_i\| = \|v_i\| = 1$, $i = 1, 2$ and $G, F \in M$ so that

$$|B_M(u, v)| \leq \frac{1}{2}(B_F(u_1, v_1) + B_G(u_2, v_2)), \tag{B.2}$$

where $M = \sum_{i=1}^{\infty} G_i \otimes F_i$
(iv)Let $A \in M$, then there exist $u, v \in K$, $\|u\| = \|v\| = 1$ such that the maximum of $|B_A|$ on $\mathbf{C}_\circ \times \mathbf{C}_\circ$ is $B_A(u, v)$.

Proof.
(ii) can be proved by elementary algebra.
(iii) Let $u = (\overline{U}_2 \otimes U_1)\,\mathrm{Vec}(S^{(1)})$, $v = (\overline{V}_2 \otimes V_1)\,\mathrm{Vec}(S^{(2)})$, U_i, V_i and $S^{(i)}$, $i = 1, 2$, possess properties described in (i), moreover $\mathrm{tr}[(S^{(i)})^2] = 1$,

which is equivalent to $\|u\| = \|v\| = 1$. Denote by e_k the kth unit vector in $\mathbf{R}^m$, then

$$\begin{aligned}
|B_M(u,v)| &\le \sum_{i=1}^{\infty}\sum_{k,l=1}^{m} s_k^{(1)} s_l^{(2)} |\langle G_i\overline{U}_2 e_k, \overline{V}_2 e_l\rangle\langle F_i U_1 e_k, V_1 e_l\rangle| \\
&\le \sum_{i=1}^{\infty}\sum_{k,l=1}^{m} s_k^{(1)} s_l^{(2)} \frac{1}{2}\left(|\langle G_i\overline{U}_2 e_k, \overline{V}_2 e_l\rangle|^2 + |\langle F_i U_1 e_k, V_1 e_l\rangle|^2\right) \\
&= \frac{1}{2}\left(B_G(u_1,v_1) + B_F(u_2,v_2)\right),
\end{aligned}$$

where

$$\begin{aligned}
&G = \sum_{i=1}^{\infty} G_i^{\otimes 2}, \ F = \sum_{i=1}^{\infty} F_i^{\otimes 2}, \\
&u_i = (\overline{U}_i \otimes U_i)\mathrm{Vec}(S^{(1)}), \ v_i = (\overline{V}_i \otimes V_i)\mathrm{Vec}(S^{(2)}).
\end{aligned}$$

(iv) The existence of the maximum and its location is well-known from calculus. The location of the maximum in K follows from (B.2). ∎

Denote by $\rho(A)$ the spectral radius of the matrix A.

Proposition 85 *(i) Let $\sum_{k=0}^{\infty} A_k$ be convergent, where $A_k \in M$, see (B.1), $k = 0,1,\dots$. Then*

$$P(x) = \sum_{k=0}^{\infty} A_k x^k,$$

converges for every $x \in \mathbf{C}_\circ$ and

$$\rho(P(x)) \le \rho(P(1)) \qquad \text{for all } x \in \mathbf{C}_\circ. \tag{B.3}$$

(ii) Let $\sum_{k=0}^{\infty} F_k^{\otimes 2}$ and $\sum_{k=0}^{\infty} G_k^{\otimes 2}$ be convergent, where $F_k, G_k \in \mathbf{R}^{m\times m}$, $k = 0,1,\dots$. Then

$$S(x) = \sum_{k=0}^{\infty} F_k \otimes G_k x^k,$$

converges on $\mathbf{C}_\circ$ and

$$\rho(S(x)) \le \max\left\{\rho\left(\sum_{k=0}^{\infty} F_k^{\otimes 2}\right), \rho\left(\sum_{k=0}^{\infty} G_k^{\otimes 2}\right)\right\}, \quad \text{for all } x \in \mathbf{C}_\circ.$$

Proof.

(i) First we show the convergence of $P(x)$ on $\mathbf{C}_\circ$. From our assumption it follows that $B_{P(1)}$ exists. Let $B_x = \sum_{k=0}^{\infty} B_{A_k} x^k$ if $x \in \mathbf{C}_\circ$. Then by Proposition 84 (iii) we have

$$|B_x(u,v)| \le \frac{1}{2}\left(B_{P(1)}(u_1,v_1) + B_{P(1)}(u_2,v_2)\right) < \infty.$$

Therefore B_x exists and is finite on the whole $\mathbf{C}^{m^2} \times \mathbf{C}^{m^2}$ and it is direct to see that B_x is a bilinear form, so it determines uniquely a matrix $R(x)$ in $\mathbf{C}^{m^2 \times m^2}$ so that $B_{R(x)} = B_x$ for all $x \in \mathbf{C}_\circ$. We prove that $R(x) = P(x)$ on $\mathbf{C}_\circ$. Let $u, v \in \mathbf{C}^{m^2}$. Then

$$\left| B_{R(x)-\sum_{k=0}^{n} A_k x^k}(u,v) \right| \leq \max_{i=1,2} \sum_{k=n+1}^{\infty} B_{A_k}(u_i, v_i) \to 0, \quad \text{if } n \to \infty,$$

shows that $\sum_{k=0}^{n} A_k u x^k \xrightarrow{w} R(x)u$ for all $u \in \mathbf{C}^{m^2}$, $x \in \mathbf{C}_\circ$. But in finite dimensional space weak convergence is equivalent to the usual convergence.

Next we prove (B.3). Let $n \in \mathbf{N}$ arbitrary and $u, v \in \mathbf{C}^{m^2}$, $\|u\| = \|v\| = 1$. Then for all $x \in \mathbf{C}_\circ$

$$\begin{aligned} |B_{P^n(x)}(u,v)| &\leq \sum_{k_1,\ldots,k_n=0}^{\infty} |x|^{k_1+\ldots+k_n} |B_{A_{k_1}\cdots A_{k_n}}(u,v)| \\ &\leq \sum_{k_1,\ldots,k_n=0}^{\infty} \frac{1}{2} \sum_{i=1}^{2} B_{A_{k_1}\cdots A_{k_n}}(u_i, v_i) \leq \max_{i=1,2} B_{P^n(1)}(u_i, v_i), \end{aligned}$$

where $u_i, v_i \in K$ are the same as in Proposition 84 (iii). So that by Proposition 84 (iv), and since $P^n(1) \in M$,

$$\begin{aligned} \max_{\|u\|=\|v\|=1} |B_{P^n(x)}(u,v)| &\leq \max_{\|u\|=\|v\|=1;\ u,v \in K} |B_{P^n(1)}(u,v)| \\ &= \max_{\|u\|=\|v\|=1} |B_{P^n(1)}(u,v)|, \end{aligned}$$

i.e., $\|P^n(x)\| \leq \|P^n(1)\|$ for all $x \in \mathbf{C}_\circ$ and $n \in \mathbf{N}$. The spectral radius theorem, see [108] confirms our assertion.

(ii) It can be proved as (i). ■

Remark 86 *One can prove more than we did, namely, $\rho(P(|x|))$ is a monotone increasing function in $|x|$.*

Now we prove a theorem on maximal eigenvalue of matrices in M.

Theorem 87 *If $A \in M$, then the spectral radius $\rho(A)$ of A is one of the nonnegative eigenvalues of A.*

Proof. If $\rho(A) = 0$, the theorem is trivial. Let $\rho(A) \neq 0$ and $B = \frac{1}{\rho(A)} A$. Then $B \in M$, $\rho(B) = 1$. Assume that 1 is not an eigenvalue of B. We prove that this assumption leads to a contradiction.

It is easy to see that

$$K \supseteq BK \supseteq B^2 K \supseteq \ldots .$$

Let $K^\infty = \bigcap_{n=0}^{\infty} B^n K$. Then K^∞ is a closed convex cone. We use the Jordan–decomposition of B, i.e., if $\lambda_1, \ldots, \lambda_p$ are eigenvalues of B with multiplicities $m_1, \ldots, m_p$ then there exist subspaces $V_1, \ldots, V_p$ such that the dimension of V_j is m_j, B is invariant on V_j, $\mathbf{C}^{m^2} = \bigoplus_{j=1}^{p} V_j$ and $N_j = B - \lambda_j I$ is nilpotent on V_j (see [48] §58, Theorem 2).

We prove that there exists a vector v different from 0 in K such that $-v \in K$, which leads to a contradiction with Proposition 84. (ii). The proof is broken down into three steps.

1. Let $|\lambda_j| = 1$ if $j = 1, \ldots, q$, $|\lambda_j| \neq 1$ if $j = q+1, \ldots, p$ ($q \in \{1, \ldots, p\}$).Then we prove that $K^\infty \subset \bigoplus_{j=1}^{q} V_j$.

2. There exist $v \in K^\infty$ $v \neq 0$.

3. There exists $w \in K^\infty \setminus \{0\}$ such that $-w \in K^\infty$ as well.

Proof, step by step:

1. Let $w \in K$ be arbitrary and $w = \sum_{j=1}^{p} w_j$, $w_j \in V_j$. Moreover, let n_j be the index of nilpotent linear mapping N_j on V_j and $n_0 = \max_{1 \le j \le p}\{n_j\}$. Then, if $n > n_0$,

$$
\begin{aligned}
B^n w &= \sum_{j=1}^{p} B^n w_j = \sum_{j=1}^{q} B^n w_j + \sum_{j=q+1}^{p} (\lambda_j I + N_j)^n w_j \\
&= \sum_{j=1}^{q} B^n w_j + \sum_{j=q+1}^{p} \left(\lambda_j^n w_j + \binom{n}{1} \lambda_j^{n-1} N_j w_j + \ldots \right. \\
&\left. + \binom{n}{n_0 - 1} \lambda_j^{n-n_0+1} N_j^{n_0-1} w_j \right),
\end{aligned}
$$

but

$$
\lim_{n \to \infty} \binom{n}{k} \lambda^{n-k} = 0, \quad \text{if } |\lambda| < 1, \text{ and } k \text{ is fixed.}
$$

2. We prove step 2. indirectly. Assume that $K^\infty = \{0\}$. Let $\{u_i, i = 1, \ldots, s\}$ be a maximal independent system in K. Then there exists $n_0 \in N$ such that $B^{n_0} u_i = 0$ for all $i \in \{1, \ldots, s\}$. Assuming the contrary we have for some $i \in \{1, \ldots, s\}$ $B^n u_i \neq 0$ for all $n \in \mathbf{N}$. Put the sequence $\{w_n \,|\, w_n = \frac{1}{\|B^n u_i\|} B^n u_i\}$. Then $w_n \in B^n K$, $\|w_n\| = 1$ for all $n \in \mathbf{N}$ and there exists an accumulation point w of this sequence. Then $w \in B^n K$ for all $n \in \mathbf{N}$ so that $w \in K^\infty$, $\|w\| = 1$ and this contradicts the assumption $K^\infty = \{0\}$. Hence we have $B^{n_0} K = \{0\}$ and applying Proposition 84 and the spectral radius theorem it is easy to prove that $\rho(B) = 0$, which is a contradiction.

3. By 1. and 2. there exists $v \in K^\infty$ so that $v = \sum_{j=1}^{q} v_j$, $v_j \in V_j$ ($j \in \{1, \ldots, q\}$) and

$$
B^n v = \sum_{j=1}^{q} \left(\lambda_j^n v_j + \binom{n}{1} \lambda_j^{n-1} N_j v_j + \ldots + \binom{n}{n_0 - 1} \lambda_j^{n-n_0+1} N_j^{n_0-1} v_j \right),
$$

where $n > n_0 = \max_{1\le j\le q}\{n_j\}$. Put an increasing sequence $\{n_k, k = 1, 2, \ldots\}$ in $\mathbf{N}$ such that $n_1 > 2n_0$ and

$$\mu_j = \lim_{k\to\infty} \lambda_j^{n_k - n_0 + 1},$$

exists for all $j \in \{1, \ldots, q\}$. The existence of this subsequence is evident. Since

$$\lim_{k\to\infty} \binom{n_k}{l} \Big/ \binom{n_k}{n_0 - 1} = 0,$$

for all $l < n_0 - 1$, and K^∞ is closed, an $u \in K^\infty \setminus \{0\}$ can be determined by

$$u = \lim_{k\to\infty} \frac{1}{\binom{n_k}{n_0-1}} B^{n_k} v = \sum_{j=1}^{q} u_j,$$

where

$$u_j = \mu_j N_j^{n_0-1} v_j, \qquad j = 1, \ldots, q,$$

and $Bu_j = \lambda_j u_j$.

Let $\lambda_j = e^{2\pi i \varphi_j}$ where $\varphi_j \in (0,1)$ $j = 1, \ldots, q$ so that the first r, $0 \le r \le q$ eigenvalues are irrational, i.e., $\varphi_j = \frac{p_j}{q_j}$, $p_j, q_j \in \mathbf{N}$, p_j and q_j are coprime, $j = r+1, \ldots, q$. Let $n_0 = \prod_{j=r+1}^{q} q_j$, then

$$\sum_{l=0}^{n_0-1} B^l u = \sum_{j=1}^{r} \sum_{l=0}^{n_0-1} B^l u_j + \sum_{j=r+1}^{q} \underbrace{\frac{e^{2\pi i n_0 \varphi_j} - 1}{e^{2\pi i \varphi_j} - 1}}_{=0} u_j$$

$$= \sum_{j=1}^{r} \sum_{l=0}^{n_0-1} B^l u_j \in K^\infty.$$

If every φ_j is rational ($r = 0$), put $w = u$. Then

$$-w = \sum_{l=1}^{n_0-1} B^l u \in K^\infty,$$

satisfies our assertion. If $r \ne 0$ then let

$$u^{(1)} = \sum_{j=1}^{r} \sum_{l=0}^{n_0-1} B^l u_j \in K^\infty,$$

and then

$$B^n u^{(1)} = \sum_{j=1}^{r} e^{2\pi i n \varphi_j} u_j^{(1)}.$$

It is known from number theory, that $\{e^{2\pi in\varphi}, n = 1, 2, \dots\}$ is everywhere dense on the unit disc if φ irrational, is an immediate consequence of Kronecker's theorem, see [53] Theorem 442. In view of this fact there exist a sequence $\{n_k, k = 1, 2, \dots\}$ in $\mathbf{N}$ such that

$$\tau_j = \lim_{k\to\infty} e^{2\pi in_k\varphi_j} \quad j = 1, \dots, r,$$

exists and $\tau_r = -1$.
Case 1: if $\tau_j = -1$ for all $j \in \{1, \dots, r\}$, then

$$-u^{(1)} = \sum_{j=1}^{r} \tau_j u_j^{(1)} \in K^\infty,$$

i.e., $w = u^{(1)}$ satisfies our assertion.
Case 2: if $\tau_j \neq -1$ for some $j \in \{1, \dots, r\}$ then let

$$u^{(2)} = u^{(1)} + \sum_{j=1}^{r} \tau_j u_j^{(1)}.$$

Then $u^{(2)} \in K^\infty$, $u^{(2)} \in \bigoplus_{j=1}^{r-1} V_j$ and $u^{(2)} \neq 0$. Let us suppose that $u^{(i)}$, $1 \leq i \leq r$ is constructed, where $u^{(i)} \in K^\infty \setminus \{0\}$ and $u^{(i)} \in \bigoplus_{j=1}^{r-i+1} V_j$. Then in a similar way we have such either $-u^{(i)} \in K^\infty$ or there exists $u^{(i+1)} \in K^\infty \setminus \{0\}$ that $u^{(i+1)} \in \bigoplus_{j=1}^{r-i} V_j$. But this procedure is completed in finite steps in the first case to verify our assertion. ∎

Remark 88 *In a similar way it can be seen that there exists an eigenvector in K^∞ with eigenvalue* 1.

Now we are in a position to prove a generalization of Corollary 82 of Proposition 81 in terms are used in Theorem 45.

Proposition 89 *Let $A_i \in M$, $i = 1, \dots, n$ and*

$$P(x) = det(Ix^n - A_1x^{n-1} - \dots - A_n).$$

Then the following statements are equivalent:

(i) $\nu(P) < 1$
(ii) $\rho\left(\sum_{i=1}^{n} A_i\right) < 1$.

Proof. Let

$$V(z) = A_1 z + \dots + A_n z^n, \qquad z \in \mathbf{C}.$$

Then

$$P(x) = x^{nm^2} \det(I - V(x^{-1})), \qquad \text{for all } x \in \mathbf{C} \setminus \{0\}, \tag{B.4}$$

and $V(1) = \sum_{i=1}^{n} A_i$.
(ii) $\Longrightarrow$ (i)
By Proposition 85 $\rho(V(x^{-1})) \leq \rho(V(1))$, for all $|x| \geq 1$, so that $P(x) \neq 0$ if $|x| \geq 1$, which is equivalent to (i).
(i) $\Longrightarrow$ (ii)
Let us assume that $\rho\left(V(1)\right) \geq 1$ and define

$$f : [1, \infty) \to \mathbf{R}_{+} : x \mapsto \rho\left(V(x^{-1})\right).$$

f is continuous and $\lim_{x\to\infty} f(x) = 0$, $f(1) \geq 1$, so there exists $\lambda \in [1, \infty)$ such that $f(\lambda) = 1$. By Theorem 87 $\rho\left(V(\lambda^{-1})\right) = 1$ is an eigenvalue of matrix $V(\lambda^{-1})$, hence by (B.4) $P(\lambda) = 0$, i.e., $\nu(P) \geq 1$, which is a contradiction. ■

Proposition 90 *Let $A, B \in M$, $\rho(A) < 1$. Then $\rho((I - A)^{-1}B) < 1$ iff $\rho(A + B) < 1$.*

Proof. Consider the equation

$$\det\left(\lambda^{-1}I - (A + B)\right) = \det\left(\lambda^{-1} - A\right)\det\left(I - \left(\lambda^{-1}I - A\right)^{-1}B\right), \tag{B.5}$$

$\lambda \in \mathbf{C} \setminus \{0\}$, and let

$$V(\lambda) = (\lambda^{-1}I - A)^{-1}B = \sum_{k=0}^{\infty} A^k B \lambda^{k+1},$$

then $V(1) = (I - A)^{-1}B$.
$\rho((I - A)^{-1}B) < 1 \Longrightarrow \rho(A + B) < 1$.
By Propositions 85 $\rho(V(\lambda)) \leq \rho(V(1))$ for all $\lambda \in \mathbf{C}_{\circ}$. This fact and (B.5) verifies our statements.
$\rho(A + B) < 1 \Longrightarrow \rho((I - A)^{-1}B) < 1$ can be proved as (i) $\Longrightarrow$ (ii) in Proposition 89. ■

Corollary 91 *Let $A_i \in M$ $(i = 0, 1, \ldots, n)$, $\rho(A_0) < 1$ and*

$$P(x) = det(Ix^n - (I - A_0)^{-1}(A_1 x^{n-1} - \ldots - A_n)).$$

Then $\nu(P) < 1$ iff $\rho(\sum_{i=0}^{n} A_i) < 1$.

Proof. Apply Proposition 89, 90. ■

Proposition 92 *Let $F_i, G_i \in \mathbf{R}^{m\times m}$ $(i = 0, 1, \ldots n)$, $\rho(F_0) < 1$, $\rho(G_0) < 1$ and*

$$P(x) = det(Ix^n - (I - F_0 \otimes G_0)^{-1}(F_1 \otimes G_1 x^{n-1} + \ldots + F_n \otimes G_n)).$$

Then $\nu(P) < 1$ if $\max\{\rho(\sum_{i=0}^{n} F_i^{\otimes 2}), \rho(\sum_{i=0}^{n} G_i^{\otimes 2})\} < 1$.

Proof. Let

$$V(z) = F_0 \otimes G_0 + F_1 \otimes G_1 z + \ldots + F_n \otimes G_n z^n.$$

Then

$$P(x) = x^{nm^2} \det^{-1}(I - F_0 \otimes G_0)\det(I - V(x^{-1})).$$

By Proposition 85 (ii) $\rho(V(x^{-1})) \leq \max\{\rho(\sum_{i=0}^n F_i^{\otimes 2}), \rho(\sum_{i=0}^n G_i^{\otimes 2})\}$, which shows the truth of the statement. ■

Corollary 93 *Let F_i, G_i, P the same as in Proposition 92 and*

$$Q(x) = det(Ix^n - (I - F_0^{\otimes 2})^{-1}(F_1^{\otimes 2}x^{n-1} + \ldots + F_n^{\otimes 2})),$$
$$R(x) = det(Ix^n - (I - G_0^{\otimes 2})^{-1}(G_1^{\otimes 2}x^{n-1} + \ldots + G_n^{\otimes 2})).$$

Then $\nu(P) < 1$ if $\nu(Q) < 1$ and $\nu(R) < 1$.

Proof. It is evident by Proposition 92 and Corollary 91. ■

Finally, the following lemma can be proved easily by the definition of the determinant

Lemma 94 *Let $A_{ij} \in \mathbf{R}^{m \times m}$, $i, j = 1, \ldots, n$*

$$A = \begin{bmatrix} A_{11} & \cdots & A_{1n} \\ \vdots & \ddots & \vdots \\ A_{n1} & \cdots & A_{nn} \end{bmatrix},$$

and $A_{ij} = 0$ if $i < j$. Then

$$det(A) = \prod_{i=1}^{n} det(A_{ii}).$$

Appendix C

Proofs for Section 3.6.1

We consider first the expected value of the product, when t is fixed:

$$\eta_k = e_{t-k-1}^2 \prod_{j=1}^{k-2} e_{t-j} \,.$$

The variable η_k is a product of k terms therefore $E\,\eta_k = 0$ if the product contains an odd number of elements because of the Gaussianity of e_t.

The even order moments of Gaussian variables can be expressed by the covariances so we can prove the following.

Lemma 95 *For every* $h \in (0, 1/2)$

$$E\,\eta_{2K} \geq C^K(K) \cdot (2K-1)!!,$$

where $C(k)$ *denotes the covariance function of* e_t, *i.e.,*

$$C(k) = Ee_t e_{t+k} = \sigma_e^2 \frac{-h(h)_k}{(-h)_{k+1}} \,.$$

Note that $(h)_k = h(h+1)(h+2)\ldots(h+k-1)$.

Proof. $C(k) > 0$ for all k. The variance of e_t is

$$\sigma_e^2 = \sigma_w^2 \frac{\Gamma(1-2h)}{\Gamma^2(1-h)} \,.$$

Now $Ee_t = 0$ and e_t is Gaussian, therefore

$$Ee_{t-2K-1}^2 \prod_{j=1}^{2K-2} e_{t-j} = \sum_{(2K-1)!!}^{*} \prod_{j=1}^{K} Ee_{n_j} e_{m_j},$$

where the summation $\sum^*$ is taken over all the different partitions containing pairs only,

$$\{(n_j, m_j), \quad j = 1, 2, \ldots, K\}$$

of the set $\mathcal{N}_{2K-1} \cup \{2K-1\}$, and their number is

$$(2K-1)!! = (2K-1)(2K-3)\ldots 1, \tag{C.1}$$

where the following notation is applied:

$$\mathcal{N}_K = \{1, 2, \ldots, K\}.$$

$C(k)$ is monotone decreasing and positive, therefore

$$E\eta_{2K} \geq E\prod_{j=1}^{2K} e_{t-j}.$$

Indeed, if one changes at least one index $2K-1$ into index $2K$ in the product

$$\prod_{j=1}^{K} Ee_{n_j}e_{m_j},$$

then one gets a smaller term, i.e., for every $1 \leq j \leq 2K-1$

$$\begin{aligned} Ee_{2K-1}e_j &= \operatorname{Cov}(e_{2K-1}, e_j) \\ &= C(2K-1-j) \geq C(2K-j) = Ee_{2K}e_j. \end{aligned}$$

Now we are going to show by induction that the inequality

$$E\prod_{j=1}^{2K} e_{t-j} \geq (2K-1)!!\, C^K(K), \tag{C.2}$$

is true, where $(2K-1)!!$ is (C.1).

Take $K = 2$, and all the possible products of covariances are

$$C^2(1),\ C^2(2),\ C(1)C(3).$$

It is easy to see that

$$C^2(1) \geq C(1)C(3) \geq C^2(2),$$

for every $h \in (0, 1/2)$.

Suppose now that (C.2) is true for $K = 2, 3, \ldots, N$. Let $K = N+1$ and let M denote the maximum of the distances $|n_j - m_j|$ for a partition

$$\left\{(n_j, m_j), n_j, m_j \in \mathcal{N}_{2(N+1)}, \quad j = 1, 2, \ldots, N+1\right\}.$$

It is clear that $M \leq 2N+1$. Separate $C(M)$ from the product according to the partition, and we get by induction

$$\prod_{j=1}^{N+1} \mathsf{E} e_{n_j} e_{m_j} = C(M) \prod_{j=1}^{N} \mathsf{E} e_{n_j} e_{m_j} \geq C(2N+1) C^N(N).$$

Consider the ratio

$$\frac{C(2N+1)C^N(N)}{C^{N+1}(N+1)} = \prod_{j=0}^{N-1} \frac{(h+N+1+j)(-h+N+1)}{(-h+N+2+j)(h+N)} \geq 1,$$

because

$$\frac{(h+N+1+j)(-h+N+1)}{(-h+N+2+j)(h+N)} \geq 1,$$

for all $j = 0, 1, 2, \ldots, N-1$ and every $h \in (0, 1/2)$. Q.E.D.

Proof. The proof of Theorem 64 is based on the above Lemma. We shall prove that for every $d \neq 0$

$$\lim_{n \to \infty} \mathsf{E} \left(e_t + \sum_{k=1}^{n} d^k e_{t-k}^2 \prod_{j=1}^{k-1} e_{t-j} \right) = \infty.$$

By the Lemma

$$\mathsf{E} \left(e_t + \sum_{k=1}^{2N-2} d^k e_{t-k}^2 \prod_{j=1}^{k-1} e_{t-j} \right) \geq \sum_{K=1}^{N} d^{2K} (2K-1)!!\, C^K(K).$$

The ratio test for the convergence of a series can be applied for large K, i.e.,

$$d^2 \frac{(2K+1)!!\, C^{K+1}(K+1)}{(2K-1)!!\, C^K(K)} \sim d^2(2K+1) \frac{(K+1)^{(k+1)(2h-1)}}{K^{K(2h-1)}}$$
$$\sim d^2(2K+1)^{2h} \exp(2h - 1 + \ln 2) > 1.$$

We have used the property that $C(K) \sim K^{2h-1}$ for large K. Q.E.D. ■

Appendix D

Cumulants and Fourier transforms for GARCH(1,1)

The GARCH model was defined in Section 3.7. We repeat here equations 3.77 and 3.78

$$V_t = aV_{t-1} + dV_{t-1}H_2(\mathrm{w}_{t-1}) + dH_2(\mathrm{w}_{t-1}),$$
$$Y_t = V_t + V_tH_2(\mathrm{w}_t) + H_2(\mathrm{w}_t) + 1,$$

and the notation of restricted Fourier transforms of cumulants of jointly stationary time series X_t, Y_t, and V_t, i.e.,

$$S^+_{X,Y}(z) = \sum_{s=1}^{\infty} \operatorname{Cum}(X_{t+s}, Y_t)\, z^{-s},$$
$$S^+_{X,Y,V}(z_1, z_2) = \sum_{k=1}^{\infty}\sum_{l=1}^{\infty} \operatorname{Cum}(X_{t+k+l}, Y_{t+l}, V_t) z_1^{-k-l} z_2^{-l}.$$

The cumulants necessary for the bispectrum of Y_t are:

1.

$$\operatorname{Cum}(V_{t+s}H_{2,t+s}, H_{2,t}) = 0, \quad \text{for all } s,$$

2.

$$\operatorname{Cum}(V_{t+s}, H_{2,t}) = \delta_{\{s>0\}} a^{s-1} 2d,$$

with Fourier transform

$$S^+_{V,H_2}(z) = \frac{2d}{z-a},$$

3.

$$\mathrm{Cum}\,(Q_{t+k}, H_{2,t}) = \left(a^2 + 2d^2\right) \mathrm{Cum}\,(Q_{t+k-1}, H_{2,t}) \\ + 4d^2\, \mathrm{Cum}\,(V_{t+k-1}, H_{2,t})\,,$$

with Fourier transform

$$S^{+}_{Q,H_2}(z) = \frac{4d\,(a+2d)\,\mathrm{Var}\,(V_t) + 8d^2}{z - (a^2 + 2d^2)} \\ + \frac{4d^2 S^{+}_{V,H_2}(z)}{z - (a^2 + 2d^2)},$$

4.

$$\mathrm{Cum}\,(V_{t+l}, Q_t H_{2,t}) = \delta_{\{l>0\}} 2a^{l-1}\left[2d\, \mathrm{Cum}\,(Q_t, V_t) + 2d\, \mathrm{Var}\,(V_t)\right],$$

5.

$$\mathrm{Cum}\left(V_{t+l}, V_t H^2_{2,t}\right) = \begin{cases} 2a^{l-1}\,(2a + 8d)\,\mathrm{Var}\,(V_t) & \text{if} \quad l > 0, \\ 4a^{-l}\,\mathrm{Var}\,(V_t) & \text{if} \quad l \le 0, \end{cases}$$

6.

$$\mathrm{Cum}\left(V_{t+l}, H^2_{2,t}\right) = \delta_{\{l>0\}} 8a^{l-1} d,$$

7.

$$\mathrm{Cum}\left(V_{t+l}, Q_t H^2_{2,t}\right) = \begin{cases} a^{l-1}\left[(2a + 8d)\,\mathrm{Cum}\,(Q_t, V_t) + 8d\,\mathrm{Var}\,(V_t)\right] & \text{if} \quad l > 0, \\ 2\,\mathrm{Cum}\,(Q_{t-l}, V_t) & \text{if} \quad l \le 0, \end{cases}$$

8.

$$\mathrm{Cum}\,(V_{t+l}, V_t H_{2,t}) = \delta_{\{s>0\}} a^{s-1} 2d \mathrm{Var}\,(V_t)\,,$$

with Fourier transform

$$S^{+}_{V,VH_2}(z) = \frac{2d\,\mathrm{Var}\,(V_t)}{z - a},$$

9.

$$\mathrm{Cum}\,(V_{t+k}, V_{t+k}, V_t H_{2,t}) = \left(a^2 + 2d^2\right) \mathrm{Cum}\,(Q_{t+k-1}, V_t H_{2,t}) \\ + 4d^2\, \mathrm{Cum}\,(V_{t+k}, V_t H_{2,t})\,,$$

with Fourier transform

$$S^{+}_{Q,VH_2}(z) = \frac{a\left[4d\,(a+2d)\,\mathrm{Cum}\,(Q_t, V_t) + 4d\,(a+4d)\,\mathrm{Var}\,(V_t)\right]}{z - (a^2 + 2d^2)} \\ + \frac{4d^2 S^{+}_{V,VH_2}(z)}{z - (a^2 + 2d^2)},$$

10.

$$\mathrm{Cum}\left(V_{t+k}, V_{t+k}, H_{2,t}\right) = \mathrm{Cum}\left(Q_{t+k}, H_{2,t}\right),$$

11.

$$\mathrm{Cum}\left(V_{t+k+l}, V_{t+k}, V_t H_{2,t}\right) = a^l\, \mathrm{Cum}\left(V_{t+k}, V_{t+k}, V_t H_{2,t}\right),$$

with Fourier transform according to $z_1^{-k-l} z_2^{-k} = z_1^{-l} z_3^{k}$, $k.l \geq 1$

$$S^+_{V,V,VH_2}\left(z_1, z_2\right) = \frac{a}{z_1 - a} S^+_{Q,VH_2}\left(z_3^{-1}\right),$$

12.

$$\mathrm{Cum}\left(V_{t+k+l}, V_{t+k}, H_{2,t}\right) = a^l\, \mathrm{Cum}\left(V_{t+k}, V_{t+k}, H_{2,t}\right)$$

$$S^+_{V,V,H_2}\left(z_1, z_2\right) = \frac{a}{z_1 - a} S^+_{Q,H_2}\left(z_3^{-1}\right),$$

13.

$$\mathrm{Cum}\left(V_{t+k+l}, V_{t+k} H_{2,t+k}, V_t\right)$$
$$= a^{l-1} 2d \left[\mathrm{Cum}\left(V_{t+k}, V_{t+k}, V_t\right) + \mathrm{Cum}\left(V_{t+k}, V_t\right)\right],$$

with Fourier transform according to $z_1^{-k-l} z_2^{-k} = z_1^{-l} z_3^{k}$, $k.l \geq 1$

$$S^+_{V,VH_2,V}\left(z_1, z_2\right) = \frac{2d}{z_1 - a}\left[S^+_{Q,V}\left(z_3^{-1}\right) + \frac{\mathrm{Var}\left(V_t\right)}{z_3^{-1} - a}\right],$$

14.

$$\mathrm{Cum}\left(V_{t+k+l}, V_{t+k} H_{2,t+k}, V_t H_{2,t}\right)$$
$$= a^{l-1} 2d \left[\mathrm{Cum}\left(V_{t+k}, V_{t+k}, V_t H_{2,t}\right) + a^{k-1}\, \mathrm{Var}\left(V_t\right)\right],$$

with Fourier transform according to $z_1^{-k-l} z_2^{-k} = z_1^{-l} z_3^{k}$, $k.l \geq 1$

$$S^+_{V,VH_2,VH_2}\left(z_1, z_2\right) = \frac{2d}{z_1 - a}\left[S^+_{Q,VH_2}\left(z_3^{-1}\right) + \frac{\mathrm{Var}\left(V_t\right)}{z_3^{-1} - a}\right],$$

15.

$$\mathrm{Cum}\left(V_{t+k+l}, V_{t+k} H_{2,t+k}, H_{2,t}\right)$$
$$= a^{l-1} 2d \left[\mathrm{Cum}\left(V_{t+k}, V_{t+k}, H_{2,t}\right) + (a + 4d)\, \mathrm{Cum}\left(V_{t+k-1}, H_{2,t}\right)\right],$$

with Fourier transform according to $z_1^{-k-l} z_2^{-k} = z_1^{-l} z_3^{k}$, $k.l \geq 1$

$$S^+_{V,VH_2,H_2}\left(z_1, z_2\right) = \frac{2d}{z_1 - a}\left[S^+_{Q,H_2}\left(z_3^{-1}\right) + \frac{2d\,(a + 4d)}{z_3^{-1} - a}\right],$$

16.

$$\begin{aligned}\operatorname{Cum}\left(V_{t+k+l}, H_{2,t+k}, V_t\right) &= a^{l-1} 2d \operatorname{Cum}\left(V_{t+k}, V_t\right) \\ &= a^{l+k-1} 2d \operatorname{Var}\left(V_t\right),\end{aligned}$$

with Fourier transform according to $z_1^{-k-l} z_2^{-k} = z_1^{-l} z_3^k$, $k.l \geq 1$

$$S^+_{V,H_2,V}\left(z_1, z_2\right) = \frac{a 2d \operatorname{Var}\left(V_t\right)}{\left(z_1 - a\right)\left(z_3^{-1} - a\right)},$$

17.

$$\begin{aligned}\operatorname{Cum}\left(V_{t+k+l}, H_{2,t+k}, V_t H_{2,t}\right) &= a^{l-1} 2d \operatorname{Cum}\left(V_{t+k}, V_t H_{2,t}\right) \\ &= a^{l+k-2} 4d^2 \operatorname{Var}\left(V_t\right),\end{aligned}$$

with Fourier transform according to $z_1^{-k-l} z_2^{-k} = z_1^{-l} z_3^k$, $k.l \geq 1$

$$S^+_{V,H_2,VH_2}\left(z_1, z_2\right) = \frac{4d^2 \operatorname{Var}\left(V_t\right)}{\left(z_1 - a\right)\left(z_3^{-1} - a\right)},$$

18.

$$\begin{aligned}\operatorname{Cum}\left(V_{t+k+l}, H_{2,t+k}, H_{2,t}\right) &= a^{l-1} 2d \operatorname{Cum}\left(V_{t+k}, H_{2,t}\right) \\ &= a^{l+k-2} 4d^2.\end{aligned}$$

with Fourier transform according to $z_1^{-k-l} z_2^{-k} = z_1^{-l} z_3^k$, $k.l \geq 1$

$$S^+_{V,H_2,H_2}\left(z_1, z_2\right) = \frac{4d^2}{\left(z_1 - a\right)\left(z_3^{-1} - a\right)}.$$

19.

$$\begin{aligned}&\operatorname{Cum}\left(V_{t+l} H_{2,t+l}, \left(Y_t - 1\right)^2\right) \\ &= \begin{cases} 0 & \text{if} \quad l > 0, \\ 12 \operatorname{Cum}\left(Q_t, V_t\right) + 20 \operatorname{Var}\left(V_t\right) & \text{if} \quad l = 0, \\ 3 \operatorname{Cum}\left(V_t H_{2,t}, Q_{t-l}\right) + 4 \operatorname{Cum}\left(V_t H_{2,t}, V_{t-l}\right) & \text{if} \quad l < 0, \end{cases}\end{aligned}$$

20.

$$\begin{aligned}&\operatorname{Cum}\left(H_{2,t+l}, \left(Y_t - 1\right)^2\right) \\ &\quad = \begin{cases} 0 & \text{if} \quad l > 0, \\ 12 \operatorname{Var}\left(V_t\right) + 8 & \text{if} \quad l = 0, \\ 3 \operatorname{Cum}\left(H_{2,t}, Q_{t-l}\right) + 4 \operatorname{Cum}\left(H_{2,t}, V_{t-l}\right) & \text{if} \quad l < 0. \end{cases}\end{aligned}$$

Appendix E
Proofs for the Chapter Non-Gaussian Estimation

It is easy to see that the first one of the following approximations is valid, and the second one will be proved just because of the frequent use for the formulae of asymptotic expressions of cumulants. They are also needed in the proof of Theorem 73 as well as Lemma 98.

Lemma 96 *Let f_1 and f_2 be functions from $[0,\, 1]$ and $[0,\, 1]^2$ to R respectively, with finite total variations on $[0, 1]$ and $[0, 1]^2$. Suppose further that $f_2(x, 0)$ and $f_2(0, y)$ are of bounded variation in x and y, respectively. Then*

$$\left| \int_0^1 f_1(x)dx - \frac{1}{n}\sum_{k=1}^{n} f(k/n) \right| \le \frac{1}{n}\mathsf{Var}\,(f_1; [0,\, 1])\,,$$

$$\left| \int_0^1 \int_0^1 f_2(x,y)dxdy - \frac{1}{n^2}\sum_{k.r=1}^{n} f_2(k/n,\, r/n) \right| \le \frac{1}{n^2}\mathsf{Var}\,(f_2; [0,\, 1]^2)\,,$$

$$+ \frac{1}{n}\int_0^1 \mathsf{Var}_x\,(f_2(\cdot, y); [0,\, 1])\, dy$$

$$+ \frac{1}{n}\int_0^1 \mathsf{Var}_y\,(f_2(x, \cdot); [0,\, 1])\, dx.$$

Proof. See [57] for the reference of bounded variation of functions. We recall now that the function $f_2(x, y)$ on the rectangle $[0,\, 1]^2$ is of bounded variation if for any subdivision $\mathcal{A}$ of $[0,\, 1]^2$

$$\sum_{[a_1,a_2]\times[b_1,b_2]\in\mathcal{A}} |f_2(a_1, b_1) - f_2(a_1, b_2) - f_2(a_2, b_1) + f_2(a_2, b_2)| < \infty,$$

and the least upper bound according to the subdivisions is called a total variation of f_2 and denoted by $\operatorname{Var}\left(f_2;[0,\,1]^2\right)$. Now

$$\left|\int_0^1\int_0^1 f_2(x,y)dxdy-\frac{1}{n^2}\sum_{k.r=1}^{n} f_2(k/n,\,r/n)\right|$$
$$=\left|\int_0^{1/n}\int_0^{1/n}\sum_{k.r=1}^{n}\left(f_2(g_{k-1}+x,g_{r-1}+y)-f_2(g_k,g_r)\right)dxdy\right|$$
$$\leq\frac{1}{n^2}\operatorname{Var}\left(f_2;[0,\,1]^2\right)+\frac{1}{n}\int_0^1\operatorname{Var}_x\left(f_2(\cdot,y);[0,\,1]\right)dy$$
$$+\frac{1}{n}\int_0^1\operatorname{Var}_y\left(f_2(x,\cdot);[0,\,1]\right)dx,$$

where $g_k=k/n, \operatorname{Var}_y\left(f_2(x,\cdot);[0,\,1]\right)$ denotes the total variation of f_2 on $[0,\,1]$ when x is fixed. Note that if f_2 is differentiable, the right side can be replaced by

$$\int_0^1\int_0^1\frac{1}{n^2}\left|\frac{\partial^2}{\partial x\partial y}f_2(x,y)\right|+\frac{1}{n}\left|\frac{\partial}{\partial x}f_2(x,y)\right|+\frac{1}{n}\left|\frac{\partial}{\partial y}f_2(x,y)\right|dxdy.$$

■

Moreover, some properties of total variations in general as

$$\operatorname{Var}(fg)\leq c_1\operatorname{Var}(f)+c_2\operatorname{Var}(g),\ \text{ with some contants } c_1,c_2,$$
$$\operatorname{Var}(cf)=|c|\operatorname{Var}(f),$$
$$\operatorname{Var}(f\,(cx))=\operatorname{Var}(f),$$

will be of interest below.

The following result of Brillinger, see Brillinger[20] Theorem 4.3.2, will be frequently applied. If Assumption 69 is satisfied, then

$$\operatorname{Cum}\left(d_T(\omega_1),d_T(\omega_2),\ldots,d_T(\omega_k)\right)=\delta_{\sum\omega_j}TS_k\left(\omega_1,\ldots,\omega_{k-1}\right)+O\left(1\right), \tag{E.1}$$

and the error term is uniform in $\omega_1,\omega_2,\ldots,\omega_k$.

Proof for covariance (4.4), **page 159.** We are going to prove (4.4), i.e., the formula

$$\operatorname{Cov}\left(S_{2T}(\lambda),S_{3T}(\omega_1,\,\omega_2)\right)=\frac{1}{T}S_5\left(\lambda,-\omega_1,\,-\omega_2,-\lambda\right)$$
$$+\frac{W_{12}\left(\rho\right)}{TB_{2T}}\sum_{k=1}^{3}S_2\left(\omega_k\right)\bar{S}_3\left(\omega_{k+1},\,\omega_{k+2}\right)\left(\delta_{\lambda+\omega_k}+\delta_{\lambda-\omega_k}\right)$$
$$+\frac{W_{20}}{TB_{2T}}\sum_{k=1}^{3}S_2\left(\omega_k\right)S_3\left(\lambda,0\right)\delta_{\omega_{k+1}}+O\left(B_{2T}^2T^{-1}\right)+O(T^{-2}B_{1T}^{-2}),$$

Constants W_{12} and W_{20} are defined by

$$W_{12}(\rho) = \iint_{-\infty}^{+\infty} W_1(\omega_1) W_2(\rho\omega_1, \omega_2) d\omega_1 d\omega_2,$$
$$W_{20} = \int_{-\infty}^{+\infty} W_2(\omega_1, \omega_1) d\omega_1.$$

The periodograms are

$$I_{2T}(\omega) = \frac{1}{T} d_T(\omega)\overline{d_T(\omega)},$$
$$I_{3T}(\omega_1, \omega_2) = \frac{1}{T} d_T(\omega_1) d_T(\omega_2)\overline{d_T(\omega_1 + \omega_2)}.$$

Start by

$$\operatorname{Cov}(I_{2T}(\lambda), I_{3T}(\omega_1, \omega_2)) = \frac{1}{T^2}\operatorname{Cum}\left(d_T(\lambda)\overline{d_T(\lambda)}, \overline{d_T(\omega_1)d_T(\omega_2)}d_T(\omega_1+\omega_2)\right).$$

Now if $EX_k = EY_j = 0$, then

$$\begin{aligned}\operatorname{Cum}(X_1X_2, Y_1Y_2Y_3) &= \operatorname{Cum}(X_1, X_2, Y_1, Y_2, Y_3) \\ &+ \sum_{k=1}^{3} \operatorname{Cum}(X_1, Y_k)\operatorname{Cum}(X_2, Y_{k+1}, Y_{k+2}) \\ &+ \sum_{k=1}^{3} \operatorname{Cum}(X_2, Y_k)\operatorname{Cum}(X_1, Y_{k+1}, Y_{k+2}) \\ &+ \sum_{k=1}^{3} \operatorname{Cum}(X_1, X_2, Y_k)\operatorname{Cum}(Y_{k+1}, Y_{k+2}),\end{aligned}$$

see Subsection 1.3.2, page 14. The indices of Y_k are considered mod $(3)+1$. Therefore by (E.1)

$$\begin{aligned}&\operatorname{Cov}(I_{2T}(\lambda), I_{3T}(\omega_1, \omega_2)) = \frac{1}{T}S_5(\lambda, -\lambda, -\omega_1, -\omega_2) \\ &+ \sum_{k=1}^{3} \delta_{\lambda-\omega_k} S_2(\lambda) S_3(-\lambda, -\omega_{k+1}) \\ &+ \sum_{k=1}^{3} \delta_{\lambda+\omega_k} S_2(\lambda) S_3(\lambda, -\omega_{k+1}) + \sum_{k=1}^{3} \delta_{\omega_k} S_2(\omega_{k+1}) S_3(\lambda, 0) \\ &\qquad + O(T^{-1}),\end{aligned} \tag{E.2}$$

as usual $\sum_1^3 \omega_j = 0$ and the indices are meant mod $(3)+1$. Now we recall that

$$S_{2T}(\omega) = \frac{1}{T}\sum_k W_{1T}(\omega - f_k) I_{2T}(f_k),$$

$$S_{3T}(\omega_1, \omega_2) = \frac{1}{T^2}\sum_k\sum_r W_{2T}(\omega_1 - f_k, \omega_2 - f_r) I_{3T}(f_k, f_r).$$

There are three typical terms of $\mathsf{Cov}\,(S_{2T}(\lambda), S_{3T}(\omega_1, \omega_2))$; one is

$$\begin{aligned}
&\frac{1}{T^3}\sum_k\sum_{m,n} W_{1T}(\lambda - f_k,) W_{2T}(\omega_1 - f_m, \omega_2 - f_n) \\
&\qquad \times \frac{1}{T^2}\mathsf{Cum}\left(d_T(f_k), \bar{d}_T(f_k), \bar{d}_T(f_n), \bar{d}_T(f_m), d_T(f_n + f_m)\right) \\
&= \frac{1}{T^4}\sum_k\sum_{m,n} W_{1T}(\lambda - f_k) W_{2T}(\omega_1 - f_m, \omega_2 - f_n) \qquad \text{(E.3)} \\
&\qquad \times S_5\left(f_k, -f_k, -f_m, -f_n\right) + O\left(T^{-2}\right) \\
&= \frac{1}{T}\iiint_0^1 W_{1T}(\lambda - \beta_1) W_{2T}(\omega_1 - \alpha_1, \omega_2 - \alpha_2) \qquad \text{(E.4)} \\
&\qquad \times S_5\left(\beta_1, -\beta_1, -\alpha_1, -\alpha_2\right) d\alpha_1 d\alpha_2 d\beta_1 + O\left(T^{-2} B_{1T}^{-1}\right) \\
&= \frac{1}{T}\iiint_{-\infty}^{+\infty} W_1(\beta_1) W_2(\alpha_1, \alpha_2) \\
&\times S_5\left(\lambda - \beta_1 B_{1T}, \beta_1 B_{1T} - \lambda, \alpha_1 B_{2T} - \omega_1, \alpha_2 B_{2T} - \omega_2\right) d\alpha_1 d\alpha_2 d\beta_1 \\
&\qquad + O\left(T^{-2} B_{1T}^{-1}\right) \\
&= \frac{1}{T} S_5\left(\lambda, -\lambda, -\omega_1, -\omega_2\right) + O\left(T^{-1} B_{2T}\right) + O\left(T^{-2} B_{1T}^{-1}\right), \qquad \text{(E.5)}
\end{aligned}$$

The equation (E.3) follows from the general asymptotical property (E.1) of cumulants of discrete Fourier transform. The equation (E.4) based on the Lemma 96, finaly the equation (E.5) based on the classical Theorem of Taylor.

(If we put, in addition to Assumption 71 on page 157,

$$\iint_{-\infty}^{+\infty} \omega_1 W_2(\omega_1, \omega_2) d\omega_1 d\omega_2 = 0.$$

the we have a stronger result namely, an $O\left(T^{-1}B_{2T}^{2}\right)$ term instead of $O\left(T^{-1}B_{2T}\right)$. Since W_1 is even, the inequality

$$\begin{aligned}
&\iiint_{-\infty}^{+\infty} W_1(\beta_1)W_2(\alpha_1,\alpha_2) \\
&\quad\times\left|S_5\left(\lambda-\beta_1B_{1T},\beta_1B_{1T}-\lambda,\alpha_1B_{2T}-\omega_1,\,\alpha_2B_{2T}-\omega_2\right)\right. \\
&\qquad\left.-S_5\left(\lambda,-\lambda,-\omega_1,-\omega_2\right)\right| d\alpha_1 d\alpha_2 d\beta_1 \\
&\quad\le\frac{B_{2T}^2}{2}\left[\int_{-\infty}^{+\infty} W_1(\beta_1)\left|\beta_1\right|^2 d\beta_1\right. \\
&\qquad\left.+\iint_{-\infty}^{+\infty} W_2(\alpha_1,\alpha_2)\left(\left|\alpha_1\right|+\left|\alpha_2\right|\right)^2 d\alpha_1 d\alpha_2\right]+O\left(B_{2T}^3\right),
\end{aligned}$$

implies an $O\left(T^{-1}B_{2T}^{2}\right)$ term.)

Put

$$\rho=\lim_{T\to\infty}\frac{B_{1T}}{B_{2T}},$$

and remember that the quantities $W_{12}\left(\rho\right)$ and W_{20} are defined above.

Another term according to the second and third terms of $(E.2)$ is

$$\begin{aligned}
&\frac{1}{T^3}\sum_k\sum_{m,n} W_{1T}(\lambda-f_k,\,)W_{2T}(\omega_1-f_m,\,\omega_2-f_n) \\
&\quad\times\frac{1}{T^2}\mathsf{Cum}\left(d_T(f_k),\bar{d}_T(f_m)\right)\mathsf{Cum}\left(\bar{d}_T(f_k),\bar{d}_T(f_n),d_T(f_n+f_m)\right) \\
&=\frac{1}{T^3}\sum_k\sum_{m,n} W_{1T}(\lambda-f_k)W_{2T}(\omega_1-f_m,\,\omega_2-f_n) \\
&\quad\times S_2\left(f_k\right)S_3\left(-f_m,-f_n\right)\delta_{f_k-f_m}+O\left(T^{-2}\right) \\
&=\frac{1}{T}\iint_0^1 W_{1T}(\lambda-\alpha_1)W_{2T}(\omega_1-\alpha_1,\,\omega_2-\alpha_2)S_2\left(\alpha_1\right) \\
&\quad\times S_3\left(-\alpha_1,-\alpha_2\right)d\alpha_1 d\alpha_2+O\left(T^{-2}B_{1T}^{-2}\right), \\
&=\frac{1}{TB_{2T}}\iint_{-\infty}^{+\infty} W_1(\alpha_1)W_2\left(\frac{\omega_1-\lambda}{B_{2T}}-\frac{B_{1T}}{B_{2T}}\alpha_1,\alpha_2\right) \\
&\quad\times S_2\left(\lambda-\alpha_1B_{1T}\right)S_3\left(\alpha_1B_{1T}-\omega_1,\,\alpha_2B_{2T}-\omega_2\right)d\alpha_1 d\alpha_2 \\
&\quad+O\left(T^{-2}B_{2T}^{-2}\right) \\
&=\frac{\delta_{\{0<\rho<\infty\}}W_{12}\left(\rho\right)}{TB_{2T}}\left[S_2\left(\lambda\right)S_3\left(-\omega_1,-\omega_2\right)+O\left(B_{2T}\right)\right]\delta_{\omega_1-\lambda} \\
&\quad+O\left(T^{-2}B_{2T}^{-2}\right),
\end{aligned}$$

where $B_{1T} \leq B_{2T}$ and $0 < \rho < \infty$ are assumed, ρ is defined above. A row of $(E.2)$ implies

$$\frac{1}{T^3}\sum_k\sum_{m,n} W_{1T}(\lambda - f_k)W_{2T}(\omega_1 - f_m, \omega_2 - f_n)$$
$$\times \frac{1}{T^2}\mathsf{Cum}\left(\bar{d}_T(f_m), \bar{d}_T(f_n)\right)\mathsf{Cum}\left(d_T(f_k), \bar{d}_T(f_k), d_T(f_m+f_n)\right)$$
$$= \frac{1}{T^3}\sum_k\sum_{m,n} W_{1T}(\lambda - f_k)W_{2T}(\omega_1 - f_m, \omega_2 - f_n)$$
$$\times S_2(-f_m)\, S_3(f_k, 0)\, \delta_{f_m+f_n} + O\left(T^{-2}\right)$$
$$= \frac{1}{T}\iint_0^1 W_{1T}(\lambda - \alpha_2)W_{2T}(\omega_1 - \alpha_1, \omega_2 + \alpha_1)S_2(\alpha_1)\, S_3(\alpha_2, 0)\, d\alpha_1 d\alpha_2$$
$$+ O(T^{-2}B_{1T}^{-2})$$
$$= \frac{1}{TB_{2T}}\iint_{-\infty}^{+\infty} W_1(\alpha_2)W_2(\alpha_1, \frac{\omega_1+\omega_2}{B_{2T}} - \alpha_1)$$
$$\times S_2(\alpha_1 B_{1T} - \omega_1)\, S_3(\lambda - \alpha_2 B_{1T}, 0)\, d\alpha_1 d\alpha_2$$
$$+ O(T^{-2}B_{1T}^{-2})$$
$$= \frac{W_{20}}{TB_{2T}}\left[S_2(\omega_1)\, S_3(\lambda, 0) + O(B_{2T})\right]\delta_{\omega_1+\omega_2}$$
$$+ O(T^{-2}B_{1T}^{-2}).$$

■

Proof of (4.6) If $EX_k = EY_j = 0$ then

$$\mathsf{Cum}(X_1X_2X_3, Y_1Y_2Y_3) = \mathsf{Cum}(X_1, X_2, X_3, Y_1, Y_2, Y_3)$$
$$+ \sum_{l=1,k=1}^{3} \mathsf{Cum}(X_l, Y_k)\,\mathsf{Cum}(X_{l+1}, X_{l+2}, Y_{k+1}, Y_{k+2})$$
$$+ \sum_{l=1}^{3}\mathsf{Cum}(X_l, X_{l+1})\,\mathsf{Cum}(X_{l+2}, Y_1, Y_2, Y_3)$$
$$+ \sum_{k=1}^{3}\mathsf{Cum}(X_1, X_2, X_3, Y_k)\,\mathsf{Cum}(Y_{k+1}, Y_{k+2})$$
$$+ \sum_{l=1,k=1}^{3} \mathsf{Cum}(X_l, X_{l+1}, Y_k)\,\mathsf{Cum}(X_{l+2}, Y_{k+1}, Y_{k+2})$$
$$+ \sum_{l=1,k=1}^{3} \mathsf{Cum}(X_l, X_{l+1})\,\mathsf{Cum}(X_{l+2}, Y_k)\,\mathsf{Cum}(Y_{k+1}, Y_{k+2}),$$

see Subsection 1.3.2, page 14. We consider only one of the terms of (4.6) to give an example of its proof, say

$$\begin{aligned}
&\frac{1}{T^4}\sum_{k,r}\sum_{m,n} W_{2T}(\omega_1 - f_k,\, \omega_2 - f_r)W_{2T}(\lambda_1 - f_m,\, \lambda_2 - f_n)\\
&\times \frac{1}{T^2}\mathsf{Cum}\left(d_T(f_k), d_T(f_r), \bar{d}_T(f_n)\right)\\
&\times \mathsf{Cum}\left(\bar{d}_T(f_m), \bar{d}_T(f_k + f_r), d_T(f_n + f_m)\right)\\
&= \frac{1}{T^4}\sum_{k,r}\sum_{m,n} W_{2T}(\omega_1 - f_k,\, \omega_2 - f_r)W_{2T}(\lambda_1 - f_m,\, \lambda_2 - f_n)\\
&\times S_3\left(f_k, f_r\right) S_3\left(-f_m, -f_k - f_r\right)\delta_{f_k+f_r-f_n} + O\left(T^{-2}\right)\\
&= \frac{1}{T}\iiint_{[0,1]^3} W_{2T}(\omega_1 - \alpha_1,\, \omega_2 - \alpha_2)W_{2T}(\lambda_1 - \beta_1,\, \lambda_2 + \alpha_3)\\
&\times S_3\left(\alpha_1, \alpha_2\right) S_3\left(-\beta_1, \alpha_3\right) d\alpha_1 d\alpha_2 d\beta_1 + O\left(T^{-2}B_{2T}^{-2}\right)\\
&= \frac{1}{TB_{2T}}\iiint_{-\infty}^{+\infty} W_2(\alpha_1, \alpha_2)W_2(\beta_1, \tfrac{\lambda_2+\omega_3}{B_{2T}} - \alpha_3)\\
&\times S_3\left(\omega_1 - \alpha_1 B_{2T},\, \omega_2 - \alpha_2 B_{2T}\right)\\
&\times S_3\left(-\lambda_1 + \beta_1 B_{2T}, \omega_3 + \alpha_3 B_{2T}\right) d\alpha_1 d\alpha_2 d\beta_1 + O\left(T^{-2}B_{2T}^{-2}\right)\\
&= \frac{W_{23}}{TB_{2T}}\left[S_3\left(\omega_1, \omega_2\right) S_3\left(-\lambda_1, \omega_3\right) + O\left(B_{2T}\right)\right]\delta_{\lambda_2+\omega_3}\\
&+ O\left(T^{-2}B_{2T}^{-2}\right),
\end{aligned}$$

where the abbreviation $\alpha_3 = -\alpha_1 - \alpha_2$ is applied as usual and

$$W_{23} = \iiint_{-\infty}^{+\infty} W_2(\lambda_1, \lambda_2)W_2(\lambda_1 + \lambda_2, \omega)d\lambda_1 d\lambda_2 d\omega.$$

Proof of Lemma 75

We shall indicate the proof of (4.18). More precisely, consider the terms in

$$\Sigma_1(\vartheta_0) = \Sigma_2(\vartheta_0) + \Sigma_3(\vartheta_0) + 2\Sigma_{23}(\vartheta_0).$$

Note that the symmetry of the bispectrum implies that the quantity

$$\sum_{(b_{2m},\, b_{2l})\in\Lambda_2^{\square}} \left(S_3(b_{2m},\, b_{2l},\, \vartheta_0) - S_{3T}\left(b_{2m},\, b_{2l}\right)\right)\frac{\frac{\partial}{\partial\vartheta}\bar{S}_3(b_{2m},\, b_{2l},\, \vartheta_0)}{S_2\left(b_{2m}\right)S_2(b_{2l})S_2(b_{2m} + b_{2l})},$$

is real. In the approximation of covariances (4.3), (4.4) and (4.6) we have seen for example that

$$\frac{1}{T^4}\sum_{k,r}\sum_{m,n} W_{2T}(\omega_1 - f_k, \omega_2 - f_r) W_{2T}(\lambda_1 - f_m, \lambda_2 - f_n) \tag{E.6}$$
$$\times \frac{1}{T^2}\operatorname{Cum}\left(d_T(f_k), d_T(f_r), \bar{d}_T(f_n)\right)$$
$$\times \operatorname{Cum}\left(\bar{d}_T(f_m), \bar{d}_T(f_k + f_r), d_T(f_n + f_m)\right)$$
$$= \frac{W_{23}}{TB_{2T}}\left[S_3(\omega_1, \omega_2) S_3(-\lambda_1, \omega_3) + O(B_{2T})\right]\delta_{\lambda_2+\omega_3} + O\left(T^{-2}B_{2T}^{-2}\right).$$

Therefore the variance

$$\Sigma_3(\vartheta_0) = T\operatorname{Var}\left(\frac{qB_{2T}^2}{6}\sum_{(b_{2m}, b_{2l})\in\Lambda_2^{\square}}\left(S_3(b_{2m}, b_{2l}, \vartheta_0) - S_{3T}(b_{2m}, b_{2l})\right)\right.$$
$$\left.\times B(b_{2m}, b_{2l}, \vartheta_0)\right),$$

consists of different types of terms, one of which according to (E.6) is

$$W_{23}B_{2T}^3 \sum_{(b_{2m}, b_{2l})\in\Lambda_2^{\square}}\ \sum_{(a_{2j}, a_{2k})\in\Lambda_2^{\square}} \left[S_3(b_{2m}, b_{2l})\overline{S}_3(a_{2j}, b_{2m} + b_{2l})\right.$$
$$\left.\times B(b_{2m}, b_{2l}, \vartheta_0)\overline{B}(a_{2j}, a_{2k}, \vartheta_0) + O(B_{2T})\right]\delta_{a_{2k}-b_{2m}-b_{2l}} + O\left(T^{-2}B_{2T}\right),$$

which tends to

$$W_{23}\iiint_{\Lambda_{21}^{\square}\times\Lambda_2^{\square}} S_3(\lambda_1, \lambda_2, \vartheta)B(\lambda_1, \lambda_2, \vartheta_0)\bar{S}_3(\omega_1, -\lambda_3, \vartheta)$$
$$\times \bar{B}(\omega_1, -\lambda_3, \vartheta_0)d\lambda_1 d\lambda_2 d\omega_1,$$

as $T \to \infty$.

Proof of Theorem 73 is essentially checking the assumptions of the following lemma.

Lemma 97 *([19]) If*

1. *Θ is locally compact, complete, separable, metric;*
2. *$Q_T(\omega, \vartheta)$ is real valued, Borel measurable for $(\omega, \vartheta) \in \Omega \times \Theta$ and all T;*
3. *$Q(\vartheta)$ is real valued lower semicontinuous $Q(\vartheta) > Q(\vartheta')$ for all $\vartheta \neq \vartheta'$;*
4. *$Q_T(\omega, \vartheta') = Q(\vartheta') + o_P(1)$, $Q_T(\omega, \vartheta) \geq Q(\vartheta) + o_P(1)$, $\vartheta \neq \vartheta'$, $T \to \infty$;*

5. *given* γ, $h > 0$, $\vartheta_1 \neq \vartheta'$, *there exists a neighborhood* U_1 *of* ϑ_1 *and there exists* T_0 *such that*

$$\mathbf{P}\left(Q_T(\omega,\vartheta_1) - \inf_{\vartheta\in U_1} Q_T(\omega,\vartheta) > \gamma\right) < h \text{ for } T \geq T_0;$$

6. *for each* ω *and* T *there exists* $\hat{\vartheta}$ *such that*

$$Q_T\left(\omega,\hat{\vartheta}\right) = \inf_{\vartheta\in U_1} Q_T(\omega,\vartheta);$$

7. *given* $h > 0$, *there exists a compact set* $C \subset \Theta$ *and* T_0 *such that* $\mathbf{P}\left(\hat{\vartheta} \notin C\right) < h$, $T \geq T_0$;

Then

$$\hat{\vartheta} = \vartheta' + o_P(1), \text{ as } T \to \infty.$$

It would be possible to prove that the variance of $Q_T(\vartheta) - Q(\vartheta)$ tends to zero under our assumptions but it seems to be more elegant to use the following lemma with assumptions of higher order moments. The first statement of the lemma can be found in Brillinger [18] and the second one is shown to be analogous to it.

Lemma 98 *Suppose that Assumptions 69-71 are true for* $p = 2, 3, \ldots, n, \ldots$ *and*
$|S_3(\omega_1,\omega_2,\vartheta)| < \infty$. *Let* C_T, V_T, $T = 1, 2, \ldots$ *be sequences of positive integers,* $C_T \to 0$ *and* V_T *is increasing.*
Then for any $\epsilon > 0$

$$\sup_{p=0,\ldots,V_T} |S_{2T}(C_T p) - S_2(C_T p)| = o_p(V_T^{\epsilon} B_{2T}^{-1/2} T^{-1/2})$$

$$\sup_{q,p=0,\ldots,V_T} |S_{3T}(C_T q,\, C_T p) - S_3(C_T q,\, C_T p)| = o_p(V_T^{\epsilon} B_{2T}^{-1} T^{-1/2}).$$

Proof. In general it can be seen that

$$\operatorname{Cum}\Bigg(\underbrace{S_{3T}(\lambda_1,\,\lambda_2),\ldots,S_{3T}(\lambda_1,\,\lambda_2)}_{K}\Bigg) = O\left(B_{2T}^{-K} T^{-K/2}\right),$$

because the largest term is obtained when the product of cumulants of possible pairs of $d_T(\lambda_j)$ are taken, compare with (4.6), page 160. Therefore

$$\mathsf{E}\,|S_{3T}(\lambda_1,\,\lambda_2) - S_3(\lambda_1,\,\lambda_2)|^{2m} = O\left(B_{2T}^{-2m} T^{-m}\right),$$

see formula (1.9) expressing moments by cumulants, page 13. From now on following Brillinger [18] one gets the result. ■

Let us turn back to the proof of Theorem 73. Introduce

$$\Gamma(\omega_1, \omega_2, \vartheta) = S_2(\omega_1, \vartheta) S_2(\omega_2, \vartheta) S_2(\omega_1 + \omega_2, \vartheta),$$

now

$$\begin{aligned}
Q_T(\vartheta) - Q(\vartheta) &= \frac{pB_{1T}}{2} \sum_{b_{1k} \in \Lambda_1^{\square}} \left[\left(\frac{S_2(b_{1k}, \vartheta) - S_{2T}(b_{1k})}{S_2(b_{1k}, \vartheta)} \right)^2 \right. \\
&\left. - \left(\frac{S_2(b_{1k}, \vartheta) - S_2(b_{1k}, \vartheta_0)}{S_2(b_{1k}, \vartheta)} \right)^2 \right] \\
&+ \frac{pB_{1T}}{2} \sum_{b_{1k} \in \Lambda_1^{\square}} \left(\frac{S_2(b_{1k}, \vartheta) - S_2(b_{1k}, \vartheta_0)}{S_2(b_{1k}, \vartheta)} \right)^2 \\
&- p \int_{\Lambda_1} \left(\frac{S_2(\omega, \vartheta) - S_2(\omega, \vartheta_0)}{S_2(\omega, \vartheta)} \right)^2 d\omega \\
&+ \frac{qB_{2T}^2}{12} \sum_{(b_{2m}, b_{2l}) \in \Lambda_2^{\square}} \left[\frac{|S_3(b_{2m}, b_{2l}, \vartheta) - S_{3T}(b_{2m}, b_{2l})|^2}{\Gamma(b_{2m}, b_{2l}, \vartheta)} \right. \\
&\left. - \frac{|S_3(b_{2m}, b_{2l}, \vartheta) - S_3(b_{2m}, b_{2l}, \vartheta_0)|^2}{\Gamma(b_{2m}, b_{2l}, \vartheta)} \right] \\
&+ \frac{qB_{2T}^2}{12} \sum_{(b_{2m}, b_{2l}) \in \Lambda_2^{\square}} \frac{|S_3(b_{2m}, b_{2l}, \vartheta) - S_3(b_{2m}, b_{2l}, \vartheta_0)|^2}{\Gamma(b_{2m}, b_{2l}, \vartheta)} \\
&- q \iint_{\Lambda_2} \frac{|S_3(\omega_1, \omega_2, \vartheta) - S_3(\omega_1, \omega_2, \vartheta_0)|^2}{\Gamma(\omega_1, \omega_2, \vartheta)} d\omega_1 d\omega_2 \\
&= Q_T^{(1)} + Q_T^{(2)} + Q_T^{(3)} + Q_T^{(4)}.
\end{aligned}$$

By the previous lemmas the second term $Q_T^{(2)}$is $O(B_{1T})$ the fourth one $Q_T^{(4)}$ is $O\left(B_{2T}^2\right)$. Let us consider $Q_T^{(1)}$

$$\begin{aligned}
Q_T^{(1)} &= \frac{pB_{1T}}{2} \sum_{b_{1k} \in \Lambda_1^{\square}} \frac{(S_2(b_{1k}, \vartheta_0) - S_{2T}(b_{1k}))}{S_2^2(b_{1k}, \vartheta)} \\
&\quad \times \frac{(2S_2(b_{1k}, \vartheta) - S_{2T}(b_{1k}) - S_2(b_{1k}, \vartheta_0))}{S_2^2(b_{1k}, \vartheta)}.
\end{aligned}$$

By Lemma 98

$$\sup_k |S_2(b_{1k}, \vartheta_0) - S_{2T}(b_{1k})| = o_P\left(B_{1T}^{-\varepsilon} B_{1T}^{-1/2} T^{-1/2}\right),$$

for any $\varepsilon > 0$, and under the assumption of the theorem $Q_T^{(1)} \xrightarrow{P} 0$ as $T \to \infty$. The same argument shows that $Q_T^{(3)} \xrightarrow{P} 0$ as $T \to \infty$ indeed

$$Q_T^{(3)} = \frac{qB_{2T}^2}{12} \sum_{(b_{2m}, b_{2l}) \in \Lambda_2^{\square}} \frac{S_3(b_{2m}, b_{2l}, \vartheta_0) - S_{3T}(b_{2m}, b_{2l})}{\Gamma(b_{2m}, b_{2l}, \vartheta)} \times \left(S_3(b_{2m}, b_{2l}, \vartheta) - S_3(b_{2m}, b_{2l}, \vartheta_0) - S_{3T}(b_{2m}, b_{2l})\right).$$

By Lemma 98 we have

$$\sup_{m,l} |S_3(b_{2m}, b_{2l}, \vartheta_0) - S_{3T}(b_{2m}, b_{2l})| = o_P\left(B_{2T}^{-\varepsilon} B_{1T}^{-1} T^{-1/2}\right),$$

for any $\varepsilon > 0$, and under the assumption of the theorem $Q_T^{(3)} \xrightarrow{P} 0$ as $T \to \infty$. As the final step in the proof of Theorem 73 and Lemma 97 are used. All the assumptions required by the Lemma 97 are fulfilled due to the compactness of the set Θ and by setting $\vartheta' = \vartheta_0$ and $\hat{\vartheta} = \vartheta_T$. The convergence of $Q_T(\vartheta) \to Q(\vartheta)$ in probability has been proved above. This completes the proof of Theorem 73. ■

E.0.1 Proof for Section 4.4.

To prove the formula (4.21) on page 170, we show the technique that can be applied. Consider the integral

$$\iota = \iiint_{\Lambda_2^{\square} \times \Lambda_1^{\square}} S_5(\omega, -\omega, -\omega_1, -\omega_2, \vartheta_0) \times \frac{\frac{\partial}{\partial \vartheta} S_3(\omega_1, \omega_2, \vartheta_0) \frac{\partial}{\partial \vartheta} \log S_2(\omega, \vartheta_0)}{\prod S_2(\omega_k, \vartheta_0) S_2(\omega, \vartheta_0)} d\omega_1 d\omega_2 d\omega,$$

say. Recall that for a linear process the spectral density of pth order is given by

$$S_p(\omega_1, \omega_2, \ldots, \omega_p) = const \prod_1^p a\left(z_k^{-1}\right),$$

where $z_k^{-1} = \exp(-2\pi i \omega_k)$ and again $\sum_1^p \omega_k = 0$. Therefore

$$S_5(\omega, -\omega, -\omega_1, -\omega_2, \vartheta_0) = const \times S_2(\omega, \vartheta_0) S_3(-\omega_1, -\omega_2, \vartheta_0),$$

and so

$$\iota = const \times \iiint_{\Lambda_2^{\square} \times \Lambda_1^{\square}} \frac{\partial}{\partial \vartheta} \log S_3(\omega_1, \omega_2, \vartheta_0) \frac{\partial}{\partial \vartheta} \log S_2(\omega, \vartheta_0) d\omega_1 d\omega_2 d\omega$$
$$= 0.$$

Now take another term to calculate

$$\begin{aligned}
\iota &= \iint_{\Lambda_{21}^{\square}\times\Lambda_{1}^{\square}} \frac{S_3\left(-\omega_1,-\omega_2,\vartheta_0\right)\frac{\partial}{\partial\vartheta}S_3(\omega_1,\omega_2,\vartheta_0)}{\prod S_2\left(\omega_k,\vartheta_0\right)} \\
&\qquad \times \frac{\partial}{\partial\vartheta}\log S_2(\omega_1,\vartheta_0)d\omega_1 d\omega_2 \\
&= \varsigma^2 \iint_{\Lambda_{21}^{\square}\times\Lambda_{1}^{\square}} \frac{\frac{\partial}{\partial\vartheta}\prod_1^3 a\left(z_k^{-1}\right)}{\prod_1^3 a\left(z_k^{-1}\right)} \frac{\frac{\partial}{\partial\vartheta}\left|a\left(z_1^{-1}\right)\right|^2}{\left|a\left(z_1^{-1}\right)\right|^2} d\omega_1 d\omega_2 \\
&= \varsigma^2 \iint_{[0,1]^2} \sum_1^3 \frac{\frac{\partial}{\partial\vartheta}a\left(z_k^{-1}\right)}{a\left(z_k^{-1}\right)}\left(\frac{\frac{\partial}{\partial\vartheta}a\left(z_1^{-1}\right)}{a\left(z_1^{-1}\right)}+\frac{\frac{\partial}{\partial\vartheta}a\left(z_1\right)}{a\left(z_1\right)}\right) d\omega_1 d\omega_2 \\
&= \varsigma^2\eta.
\end{aligned}$$

The term

$$\begin{aligned}
&const \times \iiiint_{\Lambda_{2}^{\square}\times\Lambda_{2}^{\square}} S_6\left(\omega_1,\omega_2,-\lambda_1,-\lambda_2,\omega_3,\vartheta_0\right)B(\omega_1,\omega_2,\vartheta_0) \\
&\qquad \times B(-\lambda_1,-\lambda_2,\vartheta_0)d\omega_1 d\omega_2 d\lambda_1 d\lambda_2 \\
&= const \times \iiiint_{\Lambda_{2}^{\square}\times\Lambda_{2}^{\square}} \sum_{k=1}^{3} \frac{\frac{\partial}{\partial\vartheta}a\left(z_k^{-1}\right)}{a\left(z_k^{-1}\right)} \sum_{k=1}^{3} \frac{\frac{\partial}{\partial\vartheta}a\left(u_k\right)}{a\left(u_k\right)} d\omega_1 d\omega_2 d\lambda_1 d\lambda_2 \\
&= 0,
\end{aligned}$$

the next one is

$$\begin{aligned}
&const \times \iiint_{\Lambda_{21}^{\square}\times\Lambda_{2}^{\square}} S_3(\lambda_1,\lambda_2,\vartheta)B(\lambda_1,\lambda_2,\vartheta_0)\bar{S}_3(-\lambda_3,\omega_2,\vartheta) \\
&\qquad \times \bar{B}(-\lambda_3,\omega_2,\vartheta_0)d\lambda_1 d\lambda_2 d\omega_2 \\
&= const \times \iiiint_{\Lambda_{2}^{\square}\times\Lambda_{2}^{\square}} \sum_{k=1}^{3} \frac{\frac{\partial}{\partial\vartheta}a\left(z_k^{-1}\right)}{a\left(z_k^{-1}\right)} \\
&\qquad \times \left(\frac{\frac{\partial}{\partial\vartheta}a\left(u_2\right)}{a\left(u_2\right)}+\frac{\frac{\partial}{\partial\vartheta}a\left(z_3^{-1}\right)}{a\left(z_3^{-1}\right)}+\frac{\frac{\partial}{\partial\vartheta}a\left(u_2 z_3^{-1}\right)}{a\left(u_2 z_3^{-1}\right)}\right) d\lambda_1 d\lambda_2 d\omega_2 \\
&= 0,
\end{aligned}$$

and so

$$\begin{aligned}\iiint_{\Lambda_{21}^{\square}\times\Lambda_2^{\square}} & S_4(\lambda_1,\lambda_2,-\omega_2,\vartheta_0)\bar{B}(-\omega_2-\lambda_3,\,\omega_2,\,\vartheta_0)\\ &\times B(\lambda_1,\,\lambda_2,\vartheta_0)S_2(\lambda_3,\,\vartheta_0)d\lambda_1 d\lambda_2 d\omega_2\\ &= \kappa\varsigma\iiint_{\Lambda_{21}^{\square}\times\Lambda_2^{\square}}\sum_{k=1}^{3}\frac{\frac{\partial}{\partial\vartheta}a\left(z_k^{-1}\right)}{a\left(z_k^{-1}\right)}\\ &\times\left(\frac{\frac{\partial}{\partial\vartheta}a\left(u_2\right)}{a\left(u_2\right)}+\frac{\frac{\partial}{\partial\vartheta}a\left(z_3\right)}{a\left(z_3\right)}+\frac{\frac{\partial}{\partial\vartheta}a\left(u_2^{-1}z_3^{-1}\right)}{a\left(u_2^{-1}z_3^{-1}\right)}\right)d\lambda_1 d\lambda_2 d\omega_2\\ &= \kappa\varsigma^2\eta,\end{aligned}$$

and

$$\begin{aligned}\iint_{\Lambda_2^{\square}} & \frac{\left|\frac{\partial}{\partial\vartheta}S_3(\omega_1,\,\omega_2,\,\vartheta_0\right|^2}{S_2\left(\omega_1,\vartheta_0\right)S_2(\omega_2,\vartheta_0)S_2(\omega_3,\vartheta_0)}d\omega_1 d\omega_2\\ &= \varsigma^2\iint_{\Lambda_2^{\square}}\left|\sum_{k=1}^{3}\frac{\frac{\partial}{\partial\vartheta}a\left(z_k^{-1}\right)}{a\left(z_k^{-1}\right)}\right|^2 d\omega_1 d\omega_2 = 3\varsigma^2\eta.\end{aligned}$$

For the same reason

$$\iint_{\Lambda_{21}^{\square}\times\Lambda_1^{\square}} S_3\left(\omega_1,0,\vartheta_0\right)\bar{B}(\omega_1,0\,,\vartheta_0)\frac{\partial}{\partial\vartheta}\log S_2(\omega_2,\,\vartheta_0)d\omega_1 d\omega_2 = 0.$$

Appendix F

Proof for the Chapter Linearity Test

Proof of Theorem 78, page 181 . Let us use the following notation.

$$\begin{aligned} f_1(x,y) &= \frac{d}{d\lambda}B(\lambda,y)\mid_{\lambda=x}, \\ L(c,b) &= f_1(c,-b-c), \\ G(c) &= f_1(c,-c), \\ C(x) &= \int_0^x G(y)dy. \end{aligned}$$

On the basis of the properties of a bispectrum it is direct to prove the following

$$f_1(-x,-y) = -f_1^*(x,y) \tag{F.1}$$

$$f_1(-x,0) = -f_1(x,0) \tag{F.2}$$

$$f_1(x,-x-y) = f_1(x,y) - f_1(y,x). \tag{F.3}$$

For L we can write

$$\begin{aligned} \int_0^x L(c,x+y-c)dc &= \int_0^x f_1(c,-x-y)dc \\ &= B(x,-x-y) - B(0,-x-y) \\ &= B(x,y) - B(0,-x-y). \end{aligned} \tag{F.4}$$

If (F.1) holds, differentiating both sides of (5.7) with respect to a and setting a equal to zero and using (F.2), (F.3) we get

$$f_1(b,-b-c) = f_1(c,-c) - f_1(b,-b).$$

Using the above notation we have

$$L(c,b) = G(c) - G(b),$$

and hence

$$\begin{aligned}\int_0^x L(c, x+y-c)dc &= \int_0^x (G(c) - G(x+y-c))dc \\ &= C(x) + C(y) - C(x+y). \qquad \text{(F.5)}\end{aligned}$$

From (F.4) and (F.5) we obtain

$$B(x,y) = B(0,-x-y) + C(x) + C(y) - C(x+y). \qquad \text{(F.6)}$$

Setting $x = -y = z$ and then $z = -x - y$,

$$B(0,z) = B(z,-z) = B(0,0) + C(z) + C(-z),$$

$$B(0,-x-y) = B(0,0) + C(-x-y) + C(x+y). \qquad \text{(F.7)}$$

Substituting (F.7) in (F.6) we have the form

$$B(x,y) = B(0,0) + C(x) + C(y) + C(-x-y). \qquad \text{(F.8)}$$

This is almost the final form we are looking for but $C(\lambda)$ is not periodic. It follows from the definition that $G(-\lambda) = -G^*(\lambda)$ and $C(-\lambda) = C^*(\lambda)$ and because of the periodicity of both $B(\lambda, 0)$ and $f_1(\lambda, -\lambda)$, we have

$$\begin{aligned}\int_\alpha^{\alpha+2\pi} G(\lambda)d\lambda &= \int_{-\pi}^{\pi} G(\lambda)d\lambda \qquad \text{(F.9)} \\ &= 2i\int_0^{\pi} ImG(\lambda)d\lambda = i\theta,\end{aligned}$$

where θ is a real constant. As $C(\lambda + 2\pi) = C(\lambda) + i\theta$, we put

$$H(\lambda) = \frac{B(0,0)}{3} + C(\lambda) - \frac{i\theta\lambda}{2\pi},$$

and $H(\lambda)$ satisfies (5.8) and (5.9). If 2. holds, it is easy to show that 1. also holds. ■

References

[1] H. Akaike. Stochastic theory of minimal realization. *IEEE Trans.*, AC-19:667–673, 1974.

[2] S. I. Akamanam and M. B. Rao. On the ergodicity of bilinear time series models. *Journ. of Time Series Analysis*, 7:157–163, 1986.

[3] V. G. Alekseev. Some problems of the estimation of the bispectral density of a stationary random process. *Probl. Transmiss. Inform.*, 19(3):38–51, 1983. (in Russian).

[4] P. d. Allessandro, A. Isidori, and A. Ruberti. *Lectures on Bilinear System Theory.* Springer-Verlag, New York, 1973.

[5] M. Arató. *Linear Stochastic Systems with Constant Coefficients.* Lecure Notes on Control and Information Series. Springer-Verlag, New York, 1982.

[6] T. Baranyi and A. Ludmány. Role of the solar main magnetic dipole field in the solar-tropospheric relations. Part 1., Semianual fluctuations in europe. *Annales Geophysicae-Atmospheres Hydrospheres and Space Sciences*, 13(4):427–436, 1995.

[7] T. Baranyi and A. Ludmány. Role of the solar main magnetic dipole field in the solar-tropospheric relations. Part 2., Dependence on the types of solar sources. *Annales Geophysicae-Atmospheres Hydrospheres and Space Sciences*, 13(8):886–892, 1995.

[8] T. Baranyi, A. Ludmány, and G. Terdik. Semiannual fluctuation depending on the polarity of the solar main magnetic dipole fileld. *Journal of Geophysical Research*, 100:14801–14814, Aug. 1995.

[9] W. A. Barnett, A. R. Gallant, M. J. Hinich, J. A. Jungeilges, D. T. Kaplan, and M. J. Jensen. A single blind controlled competition among tests for nonlinearity and chaos. *Journal of Econometrics*, 77:297–302, 1997. Working paper at http://wueconb.wustl.edu/ barnett/.

[10] Y. Benghabrit and M. Hallin. Locally asymptotically optimal tests for ar(p) against diagonal bilinear dependence. *J. Statist. Plann. Inference*, 68(1):47–63, 1998.

[11] J. Beran. *Statistics for Long-Memory Processes*, volume 61 of *Monographs on Statistics and Applied Probability.* Chapman & Hall, 1994.

[12] T. Bollerslev. Generalized autoregressive conditional heteroskedasticity. *Journ. of Econometrics*, 31:307–327, 1986.

[13] G. Bonnet. Transformation des signaux aléatoires. *Ann. Télécom.*, 19(9-10):203–220, 1964.

[14] G. E. P. Box and G. M. Jenkins. *Time Series Analysis, Forecasting and Control.* Holden-Day, San Francisco, 1976.

[15] R. C. Bradley. On the spectral density and asymptotic normality of weakly dependent random fields. *Journ. of Theoretical Probability*, 5(2):355–373, 1992.

[16] D. R. Brillinger. An introduction to polyspectra. *Ann. Math. Statistics*, 36:1351–1374, 1965.

[17] D. R. Brillinger. The identification of polynomial systems by means of higher order spectra. *Journ. of Sound and Vibration*, 12:301–314, 1970.

[18] D. R. Brillinger. Cross-spectral analysis of processes with stationary increments including the stationary $G/G/\infty$ queue. *Annales of Probability*, 2(5):815–827, 1974.

[19] D. R. Brillinger. Statistical inference for stationary point processes, stochastic processes and related topics. In *Proceedings of the summer research Institute on statistical inference for stochastic processes*, pages 55–99, New York, 1975. Academic Press.

[20] D. R. Brillinger. *Time Series: Data Analysis and Theory.* Mc Graw Hill, New York, expanded edition, 1981.

[21] D. R. Brillinger. Fourier inference: Some methods for the analysis of array and nongaussian series data. *Water Resources Bull.*, 21:743–756, 1985.

[22] D. R. Brillinger. A study of second-and-third-order spectral procedures and maximum likelihood in the identification of a bilinear system. *IEEE Trans. Acoustics, Speech and Signal Processing*, 38:1238–1245, 1990.

[23] D. R. Brillinger. Some history of the study of higher-order moments and spectra. *Statistica Sinica*, 1:465–476, 1991.

[24] D. R. Brillinger. Some basic aspects and uses of higher-order spectra. *Signal Processing*, 36:239–249, 1994.

[25] D. R. Brillinger. Some uses of cumulants in wavelet analysis. *Nonparametric Statistic*, 6:93–114, 1996.

[26] D. R. Brillinger and R. Kaiser. Fourier and likelihood analysis in NMR spectroscopy. In R. D. Brillinger, P. Caines, J. Geweke, E. Parzen, M. Rosenblatt, and M. S. Taqqu, editors, *New Directions in Time Series*, pages 14–64. Springer-Verlag, 1992.

[27] D. R. Brillinger and M. Rosenblatt. Asymptotic theory of k-th order spectra. In B. Harris, editor, *Spectral Analysis of Time Series*, pages 153–188, New York, 1967. Wiley.

[28] P. L. Brockett, M. J. Hinich, and D. Patterson. Bispectral- based test for the detection of gaussianity and linearity in time series. *JASA*, 83(403):657–664, Sept. 1988.

[29] P. J. Brockwell and R. A. Davis. *Time Series: Theory and Methods.* Springer-Verlag, New York, 1987.

[30] K. C. Chanda. Stationarity and central limit theorem associated with bilinear time series models. *J. Time Ser. Anal.*, 12(4):301–313, 1991.

[31] J. M. Cohen and M. J. Cohen. *Dictionary of Quotations.* Penguin, Harmondsworth, 1960.

[32] U. Desai. Realization of bilinear stochastic systems. *IEEE Trans. Aut. Contr.*, AC-31:189–192, 1986.

[33] R. L. Dobrushin. Gaussian and their subordinated generalized fields. *Ann. of Probability*, 7(1):1–28, 1979.

[34] R. L. Dobrushin and P. Major. Noncentral limit theorems for non-linear functionals of Gaussian fields. *Z. Wahrscheinlichkeitstheorie verw. Gebiete*, 50:27–52, 1979.

[35] K. Dzhaparidze. *Parameter estimation and hypothesis testing in spectral analysis of stationary time series.* Springer-Verlag, New York, 1986. Translated from the Russian by Samuel Kotz.

[36] W. Feller. *An Introduction of Probability Theory and its Application*, volume II. John Wiley, New York, London, 1966.

[37] C. Francq and M. Menvielle. A model for the am (Km) planetary geomagnetic activity index and application to prediction. *Geophys. J. Int.*, 125:729–746, 1996.

[38] A. Frazho. A shift operator approach to bilinear system theory. *SIAM Journ. on Control*, 18:640–658, 1980.

[39] M. M. Gabr. Nonlinearity tests for bilinear systems. *Comput. Appl. Math.*, 40(3):313–322, 1992.

[40] M. M. Gabr and T. Subba Rao. On the identification of bilinear system from operating records. *Int. Journ. of Control*, 40:121–128, 1984.

[41] C. Gouriéroux. *ARCH Models and Financial Applications.* Springer Verlag, New York, 1997.

[42] C. Granger and A. P. Andersen. Nonlinear time series modeling. In D. Findely, editor, *Applied Time Series Analysis*, pages 25–38. Academic Press, New York, 1978.

[43] C. W. J. Granger and A. P. Andersen. *An Introduction to Bilinear Time Series Models.* Vandenhoek and Ruprecht, Gottingen, 1978.

[44] U. Grenander and M. Rosenblatt. *Statistical Analysis of Stationary Time Series.* Wiley, New York, 1957.

[45] D. Guégan and W. J. Ngatchou. Power of the lagrange multiplier test for certain subdiagonal bilinear models. *Statist. Probab. Lett.*, 29(3):201–212, 1996.

[46] D. Guégan and D. T. Pham. Correction note: "a note on the estimation of the parameters of the diagonal bilinear model by the method of least squares". *Scand. J. Statist.*, 16(4):363, 1989.

[47] D. Guégan and D. T. Pham. A note on the estimation of the parameters of the diagonal bilinear model by the method of least squares. *Scand. J. Statist.*, 16(2):129–136, 1989.

[48] P. R. Halmos. *Finite Dimensional Vector Spaces*, volume 7 of *Annals of Mathematics Studies.* Princeton University Press, Princeton, 1948.

[49] E. J. Hannan. *Multiple Time Series.* Springer-Verlag, New York, 1970.

[50] E. J. Hannan. The asymptotic theory of linear time series models. *J. Applied Proability*, 10:130–145, 1973.

[51] E. J. Hannan. A note on bilinear time series models. *Stochastic Proc. and their Appl.*, 12:221–224, 1982.

[52] E. J. Hannan. Remembrance of things past. In J. Gani, editor, *The Craft of Probabilistic Modeling*, Ser. of Appl. Probability, pages 190–212. Springer-Verlag, New York, 1986.

[53] G. H. Hardy and E. M. Wright. *An Introduction to the Theory of Number.* Oxford University Press, London, 1954.

[54] C. He and T. Terasvirta. Properties of the autocrrelation function of squared observations for second order GARCH processes under two sets of parameter constrains. *Journ. of Time Series Anal.*, 20(1):23–30, 1999.

[55] T. Hida. *Brownian Motion.* Springer-Verlag, New York, 1980.

[56] T. Hida and G. Kallianpur. The square of a Gaussian Markov process and nonlinear prediction. *Journal of Multivariate Analysis*, 5:451–461, 1975.

[57] T. H. Hildebrandt. *Introduction to the Theory of Integration.* Academic Press, New York and London, 1963.

[58] M. J. Hinich. Testing for Gaussianity and linearity of a stationary time series. *Journal of Time Series Analysis*, 3:169–176, 1982. JTSA.

[59] B. Holmquist. The d-variate vector Hermite polynomial of order. *Linear Alg. and its Appl.*, 237/238:155–190, 1996.

[60] A. Hong-Zhi and C. Bing. A Kolmogorov-Smirnov type statistic with application to test for nonlinearity in time series. *International Statistical Review*, 59(3):287–307, 1991.

[61] I. A. Ibragimov and Y. A. Rozanov. *Gaussian Stochastic Processes.* Nauka, Moskow, 1970. In Russian.

[62] E. Iglói and G. Terdik. Bilinear modelling of chandler wobble. *Theory of Probability and its Applications*, 44(2):398–400, 1997.

[63] A. Isidori. Direct construction of minimal bilinear realizations from nonlinear input output maps. *IEEE Trans. Aut. Contr.*, AC-18:626–631, 1973.

[64] M. P. Jia and R. Huang. On the stationarity and invertibility of a general bilinear time series model. *Gaoxiao Yingyong Shuxue Xuebao*, 7(3):381–390, 1992.

[65] G. W. Johnson and G. Kallianpur. Remarks on the existence of k-traces. In C. Houdré and V. Pérez-abreu, editors, *Chaos Expansions, Multiple Wiener-Ito Integrals and Their Applications*, pages 47–73. CRC Press, London, 1994.

[66] T. Kailath. *Linear Systems.* Englewood Cliffs, Prentice-Hall, N. J., 1980.

[67] G. Kallianpur. Some ramifications of Wiener's ideas on nonlinear prediction. In P. Masani, editor, *Norbert Wiener: Collected Works*, vol.III, pages 402–425. MIT Press, Cambridge, Mass., 1981.

[68] D. M. Keenan. A Tukey nonadditive-type test for time series nonlinearity. *Biometrika*, 72(1):39–44, 1985.

[69] W. K. Kim and L. Billard. Asymptotic properties for the first-order bilinear time series model. *Comm. Statist. Theory Methods*, 19(4):1171–1183, 1990.

[70] A. N. Kolmogorov. Interpolation and extrapolation of stationary sequences. *Izvestiya Akad. Nauk., Math. Ser.*, 72(5):3–14, 1941.

[71] J. Kormos and G. Pap. Nearly unstable multidimensional AR processes. *Computers Math. Applic.*, 34(10):1–7, 1997.

[72] N. Leonenko, A. Sikorskii, and G. Terdik. On spectral and bispectral estimator of the parameter of nonGaussian data. *Random Oper. and Stoch. Equ. (ROSE)*, 6(2):159–182, 1998. Correction: ROSE(1999), vol. 7, p. 107.

[73] V. P. Leonov and A. N. Shiryaev. On a method of calculation of semi-invariants. *Theor. Prob. Appl.*, 4:319–329, 1959.

[74] A. Lindquist and M. Pavon. On the structure of state space models for discrete-time stochastic vector processes. *IEEE Trans. on Aut. Contr.*, AC-29:418–431, 1984.

[75] J. Liu. On stationarity and asymptotic inference of bilinear time series models. *Statist. Sinica*, 2(2):479–494, 1992.

[76] J. Liu and P. J. Brockwell. On general bilinear time series model. *J. Appl. Prob.*, 25:553–564, 1988.

[77] S. I. Liu. Theory of bilinear time series models. *Comm. Statist.-Theory Meth.*, 14:2549–2561, 1985.

[78] P. Major. *Multiple Wiener–Itô integrals*, volume 849 of *Lecture Notes in Mathematics*. Springer–Verlag, New York, 1981.

[79] V. A. Malishev and R. A. Minlos. *Gibbs Random Fields*. Nauka, Moskow, 1985.

[80] R. N. Mantegna and H. E. Stanley. Scaling behaviour in the dynamics of an economic index. *Nature*, 376(6):46–49, July 1995.

[81] C. M. Martins. A note on the autocorrelations related to a bilinear model with non-independent shocks. *Statist. Probab. Lett.*, 36(3):245–250, 1997.

[82] P. Masani and N. Wiener. *Nonlinear prediction*, pages 190–212. The Harald Cramer Volume. Stockholm, 1959.

[83] J. Máth. Linear and quadratic predictability for homogeneous bilinear time series of hermite degree two. *Publ. Math. Debrecen*, 49(1-2):47–58, 1996.

[84] G. Michaletzky, J. Bokor, and P. Várlaki. *Representability of Stochastic Systems*. Akadémiai Kiadó, Budapest, 1998.

[85] C. Miller. A clt for the periodograms of "mixing random fields. *Stochastic Processes and Applications*, 45:169–182, 1989.

[86] G. Mitzel, S. Clancy, and W. J. Rugh. Transfer function representations for homogeneous nonlinear systems. *IEEE Trans.*, AC-24:242–249, 1979.

[87] R. R. Mohler. *Nonlinear Time Series and Signal Processing*, volume 849 of *Lecture Notes in Control*. Springer–Verlag, New York, 1988.

[88] R. R. Mohler and W. J. Kolodziej. An overview of bilinear system theory and applications. *IEEE Trans. Systems Man. Cybern.*, SMC-10:683–688, 1980.

[89] J. Z. Nelson and J. van Ness. Formulation of a nonlinear prediction. *Technometrics*, 15:1–17, 1973.

[90] J. W. v. Ness. Asymptotic normality of bispectral estimates. *Ann. Math. Statist*, 37:1257–1272, 1966.

[91] D. F. Nicholls and B. G. Quinn. *Random Coefficient Autoregressive Models: An Introduction*, volume 11 of *Lecture Notes in Statistics*. Springer–Verlag, New York, 1982.

[92] C. L. Nikias and J. M. Mendel. *Signal Processing with Higher-Order Spectra.* United Signal and Systems, (USS) Inc., 1990.

[93] C. L. Nikias and J. M. Mendel. Signal processing with higher-order spectra. *IEEE Signal Processing Magazine*, pages 10–37, jul 1993.

[94] C. L. Nikias and A. P. Petropulu. *Higher-Order Spectra Analysis, A Nonlinear Signal Processing Framework.* PTR Prentice Hall, Englewood Cliffs, New Jersey 07632, 1994.

[95] D. T. Pham. Bilinear markovian representation and bilinear models. *Stochastic Proc. and their Appl.*, 20:295–306, 1985.

[96] D. T. Pham. Bilinear time series models. In H. Tong, editor, *Dimension Estimation and Models*, pages 191–223. World Scientific, Singapore; London, 1993.

[97] G. Pólya and G. Szegő. *Problems and Theorems in Analysis*, volume I. Springer–Verlag, Berlin, Heidelberg, 1972.

[98] M. B. Priestley. Nonlinear models in time series analysis. *The Statistician*, 27:159–176, 1978.

[99] M. B. Priestley. State dependent models: A general approach to nonlinear time series analysis. *Journ. of Time Series Anal.*, 1:47–72, 1980.

[100] M. B. Priestley. *Spectral Analysis and Time Series.* Academic Press, New York, 1981.

[101] M. B. Priestley. *Non-Linear and Non-Stationary Time Series Analysis.* Academic Press, New York, 1988.

[102] B. G. Quinn. A note on the existence of strictly stationary solutions to bilinear equations. *Journ. of Time Series Analysis*, 3:249–252, 1982.

[103] B. G. Quinn. Stationarity and invertibility of simple bilinear models. *Stochastic Proc. and their Appl.*, 12:225–230, 1982.

[104] N. S. Raibman, V. V. Kapitonenko, F. A. Ovsepian, and P. Várlaki. *Dispersional Identification.* Nauka, Moskow, 1981. In Russian.

[105] M. B. Rao, T. Subba Rao, and A. M. Walker. On existence of some bilinear time series models. *Journ. of Time Series Analysis*, 4:95–110, 1983.

[106] J. Rice. On the estimation of the parameters of a power spectrum. *J. Multiv. Anal*, 9:378–392, 1979.

[107] M. Rosenblatt. *Stationary sequences and random fields.* Birkhauser, Boston, 1985.

[108] W. Rudin. *Functional analysis.* McGraw–Hill, New York, Düsseldorf, London, 1973.

[109] W. Rugh. *Nonlinear system theory.* J. Hopkins Univ. Press, Baltimore, 1981.

[110] F. Sakaguchi. A relation for 'linearity' of the bispectrum. *Journal of Time Series Analysis*, 12(3):53–83, 1991. JTSA.

[111] S. A. O. Sesay and T. Subba Rao. On the existence of some bilinear time series models. *Journ. of Time Series Analysis*, 4:95–116, 1983.

[112] S. A. O. Sesay and T. Subba Rao. Difference equations for higher-order moments and cumulants for the bilinear time series model bl$(p, 0, p, 1)$. *J. Time Ser. Anal.*, 12(2):159–177, 1991.

[113] S. A. O. Sesay and T. Subba Rao. Frequency-domain estimation of bilinear time series models. *Journ. of Time Series Analysis*, 13:521–545, 1992.

[114] A. N. Shiryaev. Some problems in spectral theory of higher order-moments i. *Theor. Prob. Appl.*, 5:293–313, 1960.

[115] Y. G. Sinai. Self-similar probability distributions. *Theor. Probability and Appl.*, 21:64–84, 1976.

[116] B. K. Stensholt and T. Subba Rao. On the theory of multivariate bilinear time series models. *J. Time Ser. Anal.*, pages 221–233, 1988.

[117] B. K. Stensholt and D. Tjostheim. Multiple bilinear time series. *J. Time Ser. Anal.*, pages 221–233, 1987.

[118] T. Subba Rao. On the theory of bilinear time series models. Technical Report 87, Department of Mathematics, UMIST, Manchester, UK, 1978.

[119] T. Subba Rao. On the theory of bilinear time series models. *J. R. Statist. Soc., B.*, 43:244–255, 1981.

[120] T. Subba Rao and M. E. A. Da Silva. Identification of bilinear time series models bl$(p, 0, p, 1)$. *Statist. Sinica*, 2(2):465–478, 1992.

[121] T. Subba Rao and M. M. Gabr. A test for linearity of stationary time series. *Journal of Time Series Analysis*, pages 145–158, 1980. JTSA.

[122] T. Subba Rao and M. M. Gabr. *An Introduction to Bispectral Analysis and BilinearTime Series*, volume 24 of *Lecture Notes in Statistics*. Springer–Verlag, New York, 1984.

[123] T. Subba Rao and W. K. Wong. Tests for Gaussianity and linearity of multivariate stationary time series. *Journ. of Statistical Planning and Inference*, 68:373–386, 1998.

[124] T. Subba Rao and W. K. Wong. Some contibutions to multivariate nonlinear time series and to bilinear models. In S. Gosh, editor, *Asymptotics, Nonparametrics and Time Series*, pages 1–42. Marcel Dekker Inc., New York, Febr. 1999.

[125] G. Szegö. *Orthogonal polynomials*, volume XXIII of *American Math. Soc., Colloquium Publ.* American Math. Soc., New York, 1936.

[126] G. Terdik. Transfer functions and conditions for stationarity of bilinear models with gaussian white noise. *Proc. R. Soc. London A*, 400:315–330, 1985.

[127] G. Terdik. Expectation of nonlinear functions of gaussian processes. *Publicationes Mathematicae, Debrecen*, 33:205–211, 1986.

[128] G. Terdik. Second order properties for multiple- bilinear models. *J. Multivar. Anal.*, 35:295–307, 1990.

[129] G. Terdik. Bilinear state space realization for polynomial systems. *Computer Math. Application*, 26(7):69–83, 1991.

[130] G. Terdik. Stationarity in fourth order and the marginal bispectrum for bilinear models with gaussian residuals. *Stochastic Processes and their Application*, 42:315–327, 1992.

[131] G. Terdik. On problem of identification for stochastic bilinear systems. *SAMS*, 17:85–102, 1995.

[132] G. Terdik. Linear and nonlinear modeling of the geomagnetic aa indices. In T. S. Rao, editor, *Applications of Time Series in Astronomy and Meterorology*, chapter 21, pages 329–339. Chapman & Hall, London, 1997.

[133] G. Terdik and M. Ispány. A note on stationarity of bilinear models. *Publicationes Mathematicae, Debrecen*, 38:165–173, 1991.

[134] G. Terdik and M. Ispány. Criteria for the existence of even order moments bilinear time series. *Stochastic Models*, 9(2):255–273, 1993. Communication in Statistics.

[135] G. Terdik and J. Máth. Bispectrum based checking of linear predictability for time series. In T. S. Rao, editor, *Developments in Time Series Analysis*, In Honour of Maurice B. Priestley, pages 274–282. Chapman & Hall, London, 1993.

[136] G. Terdik and J. Máth. Linear prediction for discrete stationary bilinear processes. In M. Arató and M. I. Yadrenko, editors, *Proceedings of second Ukrainian-Hungarian conference on new trends in probability theory and mathematical statistics*, pages 270–279. VSP/TBiMC, Kiev, 1993.

[137] G. Terdik and J. Máth. A new test of linearity for time series based on its bispectrum. *Journal of Time Series*, 19(6):737–749, 1998.

[138] G. Terdik and L. Meaux. The exact bispectra for bilinear realizable processes with hermite degree 2. *Adv. of Applied Probability*, 23:798–808, 1991.

[139] G. Terdik and T. Subba Rao. On Wiener-Itô representation and the best linear predictors for bilinear time series. *Journal of Applied Probability*, 26:274–286, 1989.

[140] H. Tong. *Non-Linear Time Series; A Dynamical System Approach.* Oxford University Press, Oxford, 1990.

[141] K. F. Turkman and M. A. Turkman. Extremes of bilinear time series models. *J. Time Ser. Anal.*, 18(3):305–319, 1997.

[142] A. M. Walker. Asymptotic properties of least squares estimates of parameters of the spectrum of stationary non-deterministic time series. *J. Austr. Math. Soc.*, 4:363–384, 1964.

[143] A. M. Walker. Some asymptotic results for the periodogram of a stationary time series. *J. Austr. Math. Soc.*, 5:107–128, 1965.

[144] P. Whittle. Estimation and information in stationary time series. *Ark. Math.*, 2:423–434, 1953.

[145] N. Wiener. *Extrapolation, Interpolation and Smoothing of Stationary Time Series with Engineering Applications.* MIT Press, Cambridge, Mass., 1949.

[146] N. Wiener. *Nonlinear problems in random theory.* Wiley and Sons, New York, 1958.

Index

Lecture Notes in Statistics

For information about Volumes 1 to 67, please contact Springer-Verlag

Vol. 68: M. Taniguchi, Higher Order Asymptotic Theory for Time Series Analysis. viii, 160 pages, 1991.

Vol. 69: N.J.D. Nagelkerke, Maximum Likelihood Estimation of Functional Relationships. V, 110 pages, 1992.

Vol. 70: K. Iida, Studies on the Optimal Search Plan. viii, 130 pages, 1992.

Vol. 71: E.M.R.A. Engel, A Road to Randomness in Physical Systems. ix, 155 pages, 1992.

Vol. 72: J.K. Lindsey, The Analysis of Stochastic Processes using GLIM. vi, 294 pages, 1992.

Vol. 73: B.C. Arnold, E. Castillo, J.-M. Sarabia, Conditionally Specified Distributions. xiii, 151 pages, 1992.

Vol. 74: P. Barone, A. Frigessi, M. Piccioni, Stochastic Models, Statistical Methods, and Algorithms in Image Analysis. vi, 258 pages, 1992.

Vol. 75: P.K. Goel, N.S. Iyengar (Eds.), Bayesian Analysis in Statistics and Econometrics. xi, 410 pages, 1992.

Vol. 76: L. Bondesson, Generalized Gamma Convolutions and Related Classes of Distributions and Densities. viii, 173 pages, 1992.

Vol. 77: E. Mammen, When Does Bootstrap Work? Asymptotic Results and Simulations. vi, 196 pages, 1992.

Vol. 78: L. Fahrmeir, B. Francis, R. Gilchrist, G. Tutz (Eds.), Advances in GLIM and Statistical Modelling: Proceedings of the GLIM92 Conference and the 7th International Workshop on Statistical Modelling, Munich, 13-17 July 1992. ix, 225 pages, 1992.

Vol. 79: N. Schmitz, Optimal Sequentially Planned Decision Procedures. xii, 209 pages, 1992.

Vol. 80: M. Fligner, J. Verducci (Eds.), Probability Models and Statistical Analyses for Ranking Data. xxii, 306 pages, 1992.

Vol. 81: P. Spirtes, C. Glymour, R. Scheines, Causation, Prediction, and Search. xxiii, 526 pages, 1993.

Vol. 82: A. Korostelev and A. Tsybakov, Minimax Theory of Image Reconstruction. xii, 268 pages, 1993.

Vol. 83: C. Gatsonis, J. Hodges, R. Kass, N. Singpurwalla (Editors), Case Studies in Bayesian Statistics. xii, 437 pages, 1993.

Vol. 84: S. Yamada, Pivotal Measures in Statistical Experiments and Sufficiency. vii, 129 pages, 1994.

Vol. 85: P. Doukhan, Mixing: Properties and Examples. xi, 142 pages, 1994.

Vol. 86: W. Vach, Logistic Regression with Missing Values in the Covariates. xi, 139 pages, 1994.

Vol. 87: J. Müller, Lectures on Random Voronoi Tessellations.vii, 134 pages, 1994.

Vol. 88: J. E. Kolassa, Series Approximation Methods in Statistics. Second Edition, ix, 183 pages, 1997.

Vol. 89: P. Cheeseman, R.W. Oldford (Editors), Selecting Models From Data: AI and Statistics IV. xii, 487 pages, 1994.

Vol. 90: A. Csenki, Dependability for Systems with a Partitioned State Space: Markov and Semi-Markov Theory and Computational Implementation. x, 241 pages, 1994.

Vol. 91: J.D. Malley, Statistical Applications of Jordan Algebras. viii, 101 pages, 1994.

Vol. 92: M. Eerola, Probabilistic Causality in Longitudinal Studies. vii, 133 pages, 1994.

Vol. 93: Bernard Van Cutsem (Editor), Classification and Dissimilarity Analysis. xiv, 238 pages, 1994.

Vol. 94: Jane F. Gentleman and G.A. Whitmore (Editors), Case Studies in Data Analysis. viii, 262 pages, 1994.

Vol. 95: Shelemyahu Zacks, Stochastic Visibility in Random Fields. x, 175 pages, 1994.

Vol. 96: Ibrahim Rahimov, Random Sums and Branching Stochastic Processes. viii, 195 pages, 1995.

Vol. 97: R. Szekli, Stochastic Ordering and Dependence in Applied Probability. viii, 194 pages, 1995.

Vol. 98: Philippe Barbe and Patrice Bertail, The Weighted Bootstrap. viii, 230 pages, 1995.

Vol. 99: C.C. Heyde (Editor), Branching Processes: Proceedings of the First World Congress. viii, 185 pages, 1995.

Vol. 100: Wlodzimierz Bryc, The Normal Distribution: Characterizations with Applications. viii, 139 pages, 1995.

Vol. 101: H.H. Andersen, M.Højbjerre, D. Sørensen, P.S.Eriksen, Linear and Graphical Models: for the Multivariate Complex Normal Distribution. x, 184 pages, 1995.

Vol. 102: A.M. Mathai, Serge B. Provost, Takesi Hayakawa, Bilinear Forms and Zonal Polynomials. x, 378 pages, 1995.

Vol. 103: Anestis Antoniadis and Georges Oppenheim (Editors), Wavelets and Statistics. vi, 411 pages, 1995.

Vol. 104: Gilg U.H. Seeber, Brian J. Francis, Reinhold Hatzinger, Gabriele Steckel-Berger (Editors), Statistical Modelling: 10th International Workshop, Innsbruck, July 10-14th 1995. x, 327 pages, 1995.

Vol. 105: Constantine Gatsonis, James S. Hodges, Robert E. Kass, Nozer D. Singpurwalla(Editors), Case Studies in Bayesian Statistics, Volume II. x, 354 pages, 1995.

Vol. 106: Harald Niederreiter, Peter Jau-Shyong Shiue (Editors), Monte Carlo and Quasi-Monte Carlo Methods in Scientific Computing. xiv, 372 pages, 1995.

Vol. 107: Masafumi Akahira, Kei Takeuchi, Non-Regular Statistical Estimation. vii, 183 pages, 1995.

Vol. 108: Wesley L. Schaible (Editor), Indirect Estimators in U.S. Federal Programs. viii, 195 pages, 1995.

Vol. 109: Helmut Rieder (Editor), Robust Statistics, Data Analysis, and Computer Intensive Methods. xiv, 427 pages, 1996.

Vol. 110: D. Bosq, Nonparametric Statistics for Stochastic Processes. xii, 169 pages, 1996.

Vol. 111: Leon Willenborg, Ton de Waal, Statistical Disclosure Control in Practice. xiv, 152 pages, 1996.

Vol. 112: Doug Fischer, Hans-J. Lenz (Editors), Learning from Data. xii, 450 pages, 1996.

Vol. 113: Rainer Schwabe, Optimum Designs for Multi-Factor Models. viii, 124 pages, 1996.

Vol. 114: C.C. Heyde, Yu. V. Prohorov, R. Pyke, and S. T. Rachev (Editors), Athens Conference on Applied Probability and Time Series Analysis Volume I: Applied Probability In Honor of J.M. Gani. viii, 424 pages, 1996.

Vol. 115: P.M. Robinson, M. Rosenblatt (Editors), Athens Conference on Applied Probability and Time Series Analysis Volume II: Time Series Analysis In Memory of E.J. Hannan. viii, 448 pages, 1996.

Vol. 116: Genshiro Kitagawa and Will Gersch, Smoothness Priors Analysis of Time Series. x, 261 pages, 1996.

Vol. 117: Paul Glasserman, Karl Sigman, David D. Yao (Editors), Stochastic Networks. xii, 298, 1996.

Vol. 118: Radford M. Neal, Bayesian Learning for Neural Networks. xv, 183, 1996.

Vol. 119: Masanao Aoki, Arthur M. Havenner, Applications of Computer Aided Time Series Modeling. ix, 329 pages, 1997.

Vol. 120: Maia Berkane, Latent Variable Modeling and Applications to Causality. vi, 288 pages, 1997.

Vol. 121: Constantine Gatsonis, James S. Hodges, Robert E. Kass, Robert McCulloch, Peter Rossi, Nozer D. Singpurwalla (Editors), Case Studies in Bayesian Statistics, Volume III. xvi, 487 pages, 1997.

Vol. 122: Timothy G. Gregoire, David R. Brillinger, Peter J. Diggle, Estelle Russek-Cohen, William G. Warren, Russell D. Wolfinger (Editors), Modeling Longitudinal and Spatially Correlated Data. x, 402 pages, 1997.

Vol. 123: D. Y. Lin and T. R. Fleming (Editors), Proceedings of the First Seattle Symposium in Biostatistics: Survival Analysis. xiii, 308 pages, 1997.

Vol. 124: Christine H. Müller, Robust Planning and Analysis of Experiments. x, 234 pages, 1997.

Vol. 125: Valerii V. Fedorov and Peter Hackl, Model-oriented Design of Experiments. viii, 117 pages, 1997.

Vol. 126: Geert Verbeke and Geert Molenberghs, Linear Mixed Models in Practice: A SAS-Oriented Approach. xiii, 306 pages, 1997.

Vol. 127: Harald Niederreiter, Peter Hellekalek, Gerhard Larcher, and Peter Zinterhof (Editors), Monte Carlo and Quasi-Monte Carlo Methods 1996, xii, 448 pages, 1997.

Vol. 128: L. Accardi and C.C. Heyde (Editors), Probability Towards 2000, x, 356 pages, 1998.

Vol. 129: Wolfgang Härdle, Gerard Kerkyacharian, Dominique Picard, and Alexander Tsybakov, Wavelets, Approximation, and Statistical Applications, xvi, 265 pages, 1998.

Vol. 130: Bo-Cheng Wei, Exponential Family Nonlinear Models, ix, 240 pages, 1998.

Vol. 131: Joel L. Horowitz, Semiparametric Methods in Econometrics, ix, 204 pages, 1998.

Vol. 132: Douglas Nychka, Walter W. Piegorsch, and Lawrence H. Cox (Editors), Case Studies in Environmental Statistics, viii, 200 pages, 1998.

Vol. 133: Dipak Dey, Peter Müller, and Debajyoti Sinha (Editors), Practical Nonparametric and Semiparametric Bayesian Statistics, xv, 408 pages, 1998.

Vol. 134: Yu. A. Kutoyants, Statistical Inference For Spatial Poisson Processes, vii, 284 pages, 1998.

Vol. 135: Christian P. Robert, Discretization and MCMC Convergence Assessment, x, 192 pages, 1998.

Vol. 136: Gregory C. Reinsel, Raja P. Velu, Multivariate Reduced-Rank Regression, xiii, 272 pages, 1998.

Vol. 137: V. Seshadri, The Inverse Gaussian Distribution: Statistical Theory and Applications, xi, 360 pages, 1998.

Vol. 138: Peter Hellekalek, Gerhard Larcher (Editors), Random and Quasi-Random Point Sets, xi, 352 pages, 1998.

Vol. 139: Roger B. Nelsen, An Introduction to Copulas, xi, 232 pages, 1999.

Vol. 140: Constantine Gatsonis, Robert E. Kass, Bradley Carlin, Alicia Carriquiry, Andrew Gelman, Isabella Verdinelli, Mike West (Editors), Case Studies in Bayesian Statistics, Volume IV, xvi, 456 pages, 1999.

Vol. 141: Peter Müller, Brani Vidakovic (Editors), Bayesian Inference in Wavelet Based Models, xi, 394 pages, 1999.

Vol. 142: György Terdik, Bilinear Stochastic Models and Related Problems of Nonlinear Time Series Analysis: A Frequency Domain Approach, xi, 258 pages, 1999.

GPSR Compliance

The European Union's (EU) General Product Safety Regulation (GPSR) is a set of rules that requires consumer products to be safe and our obligations to ensure this.

If you have any concerns about our products, you can contact us on ProductSafety@springernature.com

In case Publisher is established outside the EU, the EU authorized representative is:

Springer Nature Customer Service Center GmbH
Europaplatz 3
69115 Heidelberg, Germany

Batch number: 10704713

Printed by Printforce, the Netherlands